Battery Powered

The Social, Economical, and Environmental Impacts of the Lithium Ion Battery

Richard Skiba

AFTER MIDNIGHT PUBLISHING

Skiba, Richard (author)

Battery Powered: The Social, Economical, and Environmental Impacts of the Lithium Ion Battery

ISBN 978-0-9756446-5-2 (paperback) 978-0-9756446-6-9 (eBook)

Non-fiction

Contents

Preface

A s we stand on the cusp of a new era defined by technological advancements and sustainable energy solutions, the significance of lithium-ion batteries cannot be overstated. This book embarks on a comprehensive investigation into lithium-ion batteries, delving into their introduction, operational principles, applications, and overarching effects on the environment, economy, and society. Each chapter is structured to fulfill specific objectives, providing a systematic and in-depth exploration of this pivotal technology.

Chapter 1, "Introduction," offers an insightful overview of lithium-ion batteries, introducing key terms and delving into their historical evolution. This foundational chapter serves as a springboard for the subsequent exploration of the technology. Moving on to Chapter 2, "Working Principle," the operational mechanisms of lithium-ion batteries are meticulously examined, encompassing their function, chemistry, and materials, providing a comprehensive understanding of their inner workings.

Chapter 3, "Composition and Manufacturing," directs its focus towards the intricate composition and manufacturing processes of lithium-ion batteries. It meticulously details stages ranging from electrode manufacturing to cell assembly, including discussions on innovations and raw materials, offering a comprehensive insight into the production of these essential energy storage devices. Chapter 4, titled "Applications," delves into the diverse uses of lithium-ion batteries, spanning consumer electronics, electric vehicles, energy storage systems, and more, highlighting the versatility and impact of this technology across various sectors.

Chapter 5, "Challenges and Issues," confronts the various obstacles associated with lithium-ion batteries, addressing limitations, safety concerns, material dependencies, and factors impacting their lifespan, capacity, and cost. The chapter also explores opportunities and challenges specific to vehicle-based batteries, providing a comprehensive analysis of the hurdles and potential advancements in this field. In Chapter 6, "Environmental Impact," the book scrutinizes the environmental footprint of lithium-ion batteries throughout their lifecycle, drawing comparisons to fossil fuels and renewable energy sources, offering a comprehensive understanding of their environmental implications.

The economic facet of lithium-ion batteries is thoroughly examined in Chapter 7, titled "Economic Impact." This section delves into economic factors, quantifiable aspects, and the repercussions of transitioning to battery storage for renewable energy. A potential comparison with other energy sources, such as nuclear energy, may be explored, providing a comprehensive analysis of the economic implications of lithium-ion batteries. Finally, Chapter 8, "Social Impact," concludes the book by exploring positive and negative adaptations, social structuring, and interaction related to lithium-ion batteries. The technology is analysed through various social theories, underscoring a comprehensive examination of its societal implications, offering a multidimensional understanding of its societal impact.

This book offers a range of benefits to diverse audiences, making it a valuable resource for various purposes.

For students and researchers studying materials science, chemistry, physics, engineering, sustainability, environment, or related fields, this comprehensive book serves as an educational resource. It provides foundational knowledge encompassing the principles, materials, and technologies integral to lithium-ion batteries.

Industry professionals, particularly engineers and technologists engaged in battery design, manufacturing, and research and development, stand to gain from a book covering the latest advancements, manufacturing processes, and emerging technologies in lithium-ion batteries.

Entrepreneurs and startups seeking to enter the battery market or develop new technologies can find valuable insights in such a book. It offers an understanding of the historical context, market trends, and challenges associated with lithium-ion batteries, guiding strategic decision-making and innovation.

For policy makers and regulators shaping energy and environmental policies, this book provides insights into the role of lithium-ion batteries in energy storage, transportation, and renewable energy integration. This knowledge is fundamental for making informed decisions about regulations and incentives.

General readers, including consumers, benefit from a book tailored to a broader audience. It helps them comprehend the technology behind the batteries used in devices, electric vehicles, and renewable energy systems. This understanding empowers consumers to make informed choices and appreciate the significance of battery technologies.

Maintenance professionals and safety experts dealing with lithium-ion batteries in various applications can benefit from information about the safety aspects, including handling, storage, and disposal, provided in the book.

Inventors and innovators interested in developing new applications or exploring innovative uses for lithium-ion batteries can find inspiration and guidance in the book delving into the latest developments and potential future directions of this technology.

Lastly, investors and business analysts seeking to understand the lithium-ion battery market, including factors influencing its growth and potential disruptors, find valuable insights in the book. This information aids in making decisions related to investments, partnerships, and market positioning.

In summary, this comprehensive book about lithium-ion batteries caters to a broad audience, providing valuable insights, knowledge, and perspectives for students, professionals, entrepreneurs, policymakers, and general readers interested in various aspects of this critical technology.

Chapter One

Introduction

The Lithium-ion Battery - Overview

Lithium-ion batteries have been a significant technological advancement that has enabled numerous innovations and improvements in various industries. They are a type of rechargeable battery that has become widely popular for various electronic devices, ranging from smartphones and laptops to electric vehicles and renewable energy storage systems. Lithium-ion batteries are known for their high energy density, long cycle life, and relatively low self-discharge rate and have become the dominant power source in various applications, including portable electronics, electric vehicles, and grid-energy storage (Manthiram, 2017). The unique features of lithium-ion batteries, such as high energy density, high operating voltage, and long cycle life, have contributed to their widespread adoption (Li & Zhou, 2019; Santos et al., 2012). Additionally, retired lithium-ion batteries from electric vehicles offer a new option for battery energy storage systems (Meng, 2021).

The performance of lithium-ion batteries is indeed influenced by various factors, including temperature. Lower temperatures can decrease the activity of the active electrolyte and slow down the deintercalation rate of lithium ions during the discharge process (Lv et al., 2021). It is crucial for lithium-ion batteries to operate within an optimum temperature range to achieve maximum utilization, especially in electric vehicles (Ali et al., 2019; Madani et al., 2021a). Furthermore, safety and reliability are of utmost importance in the use of lithium-ion batteries, especially in demanding applications (Finegan, Darcy, Keyser, Tjaden, Heenan, Jervis, Bailey, Malik, et al., 2017).

Research efforts have also been directed towards upgrading existing lithium-ion batteries and developing alternate technologies, such as sodium-ion, metal-air, and lithium-sulphur batteries (Hibino et al., 2013). Additionally, nanomaterials are increasingly playing an active role in improving the efficiency and performance of lithium-ion batteries (Santos et al., 2012). The use of lithium-ion batteries in forklifts is also expanding, although they are sensitive to excessive current pulses (Özcan et al., 2023).

While lithium-ion batteries offer numerous advantages, there are challenges and areas for further research. For instance, the toxicity of gases released during lithium-ion battery combustion and the fire and explosion theory of lithium-ion batteries require further studies to ensure their safe use (Chen et al., 2017; Wang et al., 2005). Moreover, the dissimilarities between lithium and sodium characteristics may strongly affect the electrochemical processes and overall battery performance, especially when considering the adaptation of lithium-ion battery electrodes to sodium-ion technology (He et al., 2015).

The development of lithium-ion batteries has indeed revolutionized the portable electronics industry, particularly in the context of smartphones, laptops, tablets, and wearables. This revolutionary impact can be attributed to several key aspects that highlight the transformative nature of lithium-ion batteries. Firstly, lithium-ion batteries offer high energy density, enabling manufacturers to pack more power into smaller and lighter battery packs, thus enhancing the portability of devices (H. Kim et al., 2014). Additionally, these batteries provide extended battery life and support quick charging, addressing the need for longer-lasting power and reducing downtime associated with recharging (Cheng et al., 2021). Moreover, the form factor flexibility of lithium-ion batteries allows for the creation of sleek and compact designs, as shown in Figure 1, without compromising on battery performance, further enhancing the usability of portable electronics (Cheng et al., 2021). Furthermore, the development of high-capacity electrode materials for lithium-ion batteries has been significant for technological improvements in portable electronics and electric vehicles that rely on lithium-ion batteries as the power source (Dirican et al., 2014).

Figure 1: A Motorola HF5X Lithium Ion Polymer Battery, taken from a Motorola Photon 4G. Mbrickn, CC BY 4.0, via Wikimedia Commons.

The combination of these factors has made lithium-ion batteries the preferred power source for the portable electronics industry, significantly improving the user experience and facilitating the development of increasingly powerful and feature-rich devices in a compact and lightweight form. As a result, the impact of lithium-ion batteries on portable electronics and electric vehicles has been substantial, driving advancements in energy storage and contributing to the widespread adoption of these technologies (Cheng et al., 2021).

The integration of lithium-ion batteries into electric vehicles (EVs) has been a transformative development, significantly contributing to the growth and acceptance of electric transportation. The pivotal role of lithium-ion batteries in the EV industry is underscored by several key aspects. Firstly, lithium-ion batteries offer a higher energy density compared to other battery technologies, allowing them to store more energy per unit of weight or volume (J. M. Tarascon & M. Armand, 2001). This characteristic is crucial for electric vehicles, where maximizing the range on a single charge is a top priority. The high energy density of lithium-ion batteries allows EVs to cover more distance without requiring excessively large and heavy battery packs.

Secondly, lithium-ion batteries are relatively lightweight compared to traditional battery technologies, such as lead-acid batteries (Manthiram, 2017). This lightweight design is critical in the design of electric vehicles, as reducing weight contributes to overall energy efficiency and performance. Lighter vehicles typically require less energy to move, allowing for a more extended range and improved energy efficiency. Moreover, the ability of lithium-ion batteries to store a large amount of energy in a compact package directly translates to increased driving range for electric vehicles (Camargos et al., 2022).

Modern EVs equipped with lithium-ion battery packs can achieve competitive ranges, alleviating the range anxiety that has historically been a concern for potential electric vehicle adopters. Additionally, when properly managed, lithium-ion batteries can have a longer lifespan compared to some other battery technologies (S.-h. Lee et al., 2014). This longevity and durability are important factors for the automotive industry, where durability and longevity are paramount. The ability of lithium-ion batteries to withstand thousands of charge-discharge cycles contributes to the long-term viability of electric vehicles, making them a reliable and cost-effective option for consumers.

Furthermore, many electric vehicles utilize regenerative braking systems, which capture and store energy during braking (Wang et al., 2014). Lithium-ion batteries are well-suited for this application, as they can efficiently absorb and release energy during rapid changes in charging and discharging. This regenerative braking capability enhances the overall energy efficiency of electric vehicles and contributes to extended battery life. Ongoing research and development in lithium-ion battery technology continue to drive improvements, including innovations in materials, manufacturing processes, and battery management systems (Zhang et al., 2020). These advancements are crucial for making electric vehicles more accessible and appealing to a broader consumer base.

Figure 2: EnBW electric car charging station in Stuttgart with a 3rd generation Smart electric car of carsharing company Car2Go. Julian Herzog, CC BY 4.0, via Wikimedia Commons.

Finally, while there are environmental considerations in the production and disposal of lithium-ion batteries, the overall environmental impact of electric vehicles, powered by clean energy sources, is generally lower than traditional internal combustion engine vehicles (Mozaffarpour et al., 2022). The shift to electric mobility plays a crucial role in reducing greenhouse gas emissions and addressing environmental concerns associated with traditional transportation.

The high energy density, lightweight design, improved range, longevity, regenerative braking capabilities, technological advancements, and positive environmental impact make lithium-ion batteries indispensable for the development and widespread adoption of electric vehicles. As the automotive industry continues to invest in battery technology, the future holds the promise of even more efficient, cost-effective, and sustainable electric transportation solutions.

Lithium-ion batteries are pivotal in the realm of renewable energy storage, facilitating the integration and utilization of energy from intermittent renewable sources such as solar and wind power. These batteries play a crucial role in addressing several key aspects of renewable energy storage, including grid stabilization, time-shifting energy, smoothing power fluctuations, microgrid integration, frequency regulation, decentralized energy storage, promoting renewable energy integration, and capacity factor improvement (Etacheri et al., 2011).

Grid stabilization is achieved through the storage of excess energy generated during periods of high renewable energy production, which is then released during periods of low renewable energy production or high demand, contributing to a stable grid (Etacheri et al., 2011). Additionally, lithium-ion batteries enable the time-shifting of energy production and consumption, ensuring a consistent power supply and reducing dependence on fossil fuel-based backup systems during low renewable energy availability (Etacheri et al., 2011). Moreover, these batteries assist in smoothing out natural fluctuations in renewable energy production by storing excess energy during high production periods and discharging it during sudden drops in renewable energy output, thereby maintaining a stable power supply (Etacheri et al., 2011).

In the context of microgrid integration, lithium-ion batteries are employed to store energy locally, promoting energy independence and resilience during grid outages or disruptions (Etacheri et al., 2011). Furthermore, these batteries offer rapid response times, making them suitable for providing frequency regulation services to the electrical grid, ensuring a stable and reliable power supply as more renewable energy sources are integrated into the grid (Etacheri et al., 2011). The decentralized energy storage solutions supported by lithium-ion batteries help reduce transmission losses and improve overall grid efficiency, enhancing grid reliability and resilience (Etacheri et al., 2011).

Moreover, the use of lithium-ion batteries encourages the greater integration of renewable energy sources into the overall energy mix, facilitating the transition towards a cleaner and more sustainable energy infrastructure (Etacheri et al., 2011). Additionally, these batteries contribute to improving the capacity factor of intermittent renewable sources by storing and releasing excess energy when needed, thereby increasing the effective capacity factor of renewable sources (Etacheri et al., 2011). Lithium-ion batteries are critical for a sustainable and reliable energy future, efficiently addressing the intermittency of renewable sources and enabling the smoother integration of solar and wind power into the electrical grid while contributing to a more resilient and responsive energy infrastructure (Etacheri et al., 2011).

The integration of lithium-ion batteries into power tools has indeed revolutionized the landscape of cordless tools, offering users enhanced performance, convenience, and efficiency. This transformation is attributed to several key aspects that highlight the impact of lithium-ion batteries on the power tool industry. Firstly, lithium-ion batteries exhibit a remarkable power-to-weight ratio, crucial for power tools, enabling the creation of compact and lightweight battery packs that deliver substantial power output (Kang & Ceder, 2009). This high power-to-weight ratio contributes to the reduced tool weight, a noticeable benefit of lithium-ion batteries in power tools, enhancing the overall ergonomics and user-friendliness of the tools (Manthiram, 2017). Moreover, the lightweight nature of lithium-ion batteries enhances the portability of cordless power tools, allowing users to easily carry these tools to different job sites without being encumbered by the weight of the battery. This portability is particularly valuable for professionals who need to move between locations or work in tight spaces where corded tools might be impractical.

Additionally, lithium-ion batteries offer longer operating times between charges compared to traditional battery technologies, allowing users to work for extended periods before needing to recharge the battery. This extended runtime is especially beneficial for professionals in construction, woodworking, or other fields where uninterrupted work is essential for productivity. Furthermore, the fast charging capability of lithium-ion batteries minimizes downtime, contributing to increased overall efficiency on the job, as users can recharge their batteries quickly during breaks or between tasks. This feature ensures that the tools are ready for use when needed, enhancing productivity.

Consistent power output throughout the discharge cycle is another advantage of lithium-ion batteries, ensuring that power tools maintain a reliable level of performance until the battery is depleted, offering consistent torque and speed for various tasks. This reliability is crucial for professionals who rely on consistent tool performance for their work. Moreover, lithium-ion batteries generally have a longer lifespan and can withstand a higher number of charge-discharge cycles compared to older battery technologies, making them durable for frequent and demanding usage on construction sites or in workshops. Additionally, the virtually maintenance-free nature of lithium-ion batteries eliminates the need for frequent maintenance tasks, contributing to the overall ease of use and reducing the total cost of ownership for users.

The integration of lithium-ion batteries into medical devices has significantly advanced healthcare technology, enabling the development of more sophisticated and portable electronic medical equipment, as well as supporting critical implantable devices. Lithium-ion batteries contribute to the portability and mobility of various medical devices, such as ultrasound machines, infusion pumps, and portable monitors, benefiting from their high energy density and lightweight characteristics (Fu et al., 2016). Additionally, these batteries are well-suited for implantable medical devices, including pacemakers, defibrillators, and neurostimulators, due to their compact size and high energy density, ensuring long battery life and reliable operation (Fu et al., 2016). The long lifespan of lithium-ion batteries reduces the frequency of battery replacements, particularly crucial for implantable devices, minimizing the need for invasive procedures (Fu et al., 2016). Moreover, the high energy efficiency of lithium-ion batteries allows medical devices to operate effectively while consuming minimal power, essential for continuous glucose monitors or wearable health trackers, without frequent recharging or battery replacements (Fu et al., 2016).

Furthermore, the safety features of lithium-ion batteries, including built-in protection circuits, contribute to the overall reliability of medical devices, crucial in critical healthcare applications (Fu et al., 2016). The use of lithium-ion batteries in medical devices aligns well with the integration of smart technologies, enabling real-time data transmission to healthcare providers for remote patient monitoring (Fu et al., 2016). These batteries also contribute to the advancement of remote patient care by enabling the development of wearable and portable medical devices, facilitating continuous monitoring of patients' health metrics, allowing for early detection of medical issues and improved management of chronic conditions (Fu et al., 2016). Additionally, lithium-ion batteries are utilized in portable diagnostic imaging devices, such as handheld ultrasound machines and portable X-ray machines, benefiting from their lightweight and high energy density, making them more versatile and accessible for point-of-care diagnostics (Fu et al., 2016).

The widespread adoption of lithium-ion batteries in consumer electronics has significantly influenced the design, performance, and user experience of a variety of devices. Lithium-ion batteries offer a high energy density, allowing for compact and lightweight designs in devices like digital cameras and camcorders (H. Kim et al., 2014). They are rechargeable, providing a cost-effective and environmentally friendly power solution for consumer electronics, especially beneficial for devices that experience frequent use, such as digital cameras and handheld gaming devices. Additionally, lithium-ion batteries generally have a long cycle life, maintaining their performance through numerous charge and discharge cycles, ensuring that the devices can be used for an extended period without the need for frequent battery replacements. Furthermore, lithium-ion batteries support fast charging, reducing downtime and ensuring that devices are ready for use when needed, which is particularly relevant for devices with short usage cycles, such as digital cameras and gaming devices.

The compact size and slim profile of lithium-ion batteries allow manufacturers to design sleek and aesthetically pleasing consumer electronics, benefiting devices like digital cameras, camcorders, and handheld gaming devices. Moreover, lithium-ion batteries exhibit minimal memory effect compared to some other battery technologies, contributing to more consistent and reliable performance over time. The versatility of lithium-ion batteries enables the integration of advanced features in consumer electronics, enhancing the overall user experience by providing reliable and long-lasting power for devices like digital cameras and camcorders.

Lithium-ion batteries have become integral in the aerospace industry due to their unique characteristics, particularly their lightweight design and high energy density. These features make them well-suited for the demanding requirements of aerospace applications, such as satellite systems and electric aircraft. The lightweight design of lithium-ion batteries contributes significantly to weight reduction efforts in aerospace systems, directly impacting factors such as fuel efficiency, payload capacity, and overall mission success (Hashemi et al., 2020). Additionally, their high energy density allows them to store a large amount of energy in a relatively compact space, making them valuable in aerospace applications with significant space constraints (Tang et al., 2014).

In the aerospace industry, lithium-ion batteries play a crucial role in powering satellites during periods when they are not exposed to sunlight, ensuring uninterrupted satellite operations (Liu, Wang, et al., 2013). Moreover, they are essential for electric propulsion systems in aircraft, providing a cleaner and more sustainable alternative to traditional aviation fuels, thus contributing to improved efficiency, reliability, and sustainability in the aerospace industry (Qu et al., 2019). Furthermore, lithium-ion batteries are used to power various avionics and onboard systems in aircraft, delivering a reliable and consistent power supply crucial for safe and efficient operation (Hashemi et al., 2020).

Aerospace applications often require rapid charging and discharging capabilities, especially during critical mission phases. Lithium-ion batteries can meet these requirements, providing the necessary power on demand (Hashemi et al., 2020). Additionally, their long cycle life and durability ensure that the power systems in satellites and electric aircraft can endure the harsh conditions of space or extended flight durations (Liu, Wang, et al., 2013). Furthermore, the aerospace industry's increasing focus on sustainability and reducing its environmental footprint aligns with the use of lithium-ion batteries, which contribute to cleaner and more environmentally friendly operations in electric aircraft (Qu et al., 2019).

Large-scale lithium-ion battery systems have emerged as crucial components of modern energy infrastructure, offering a multitude of benefits for grid energy storage. These systems play a pivotal role in grid stabilization by swiftly managing fluctuations in energy supply and demand, thereby ensuring grid stability (Stroe et al., 2015). Additionally, they are effective in peak demand management, providing rapid discharge of stored energy during periods of high energy consumption, thus reducing strain on conventional power generation infrastructure (Sessa et al., 2018). Furthermore, large-scale lithium-ion battery systems facilitate the integration of renewable energy sources into the grid by storing excess energy and releasing it when demand is high, contributing to a more sustainable energy mix (Deng et al., 2016). Their rapid response capabilities make them well-suited for frequency regulation on the grid, ensuring the stability of the electrical system (Stroe et al., 2015). Moreover, these systems enhance grid resilience by providing a flexible and rapidly deployable source of energy during disruptions, such as natural disasters or equipment failures (Sessa et al., 2018).

The widespread adoption of lithium-ion batteries has had a transformative impact on various industries, improving the efficiency, reliability, and sustainability of a wide range of technologies and applications. However, it's worth noting that ongoing research and development in battery technology continue to explore alternative materials and designs to further enhance performance and address environmental concerns.

The ongoing demand for lithium-ion batteries is closely linked to the global emphasis on sustainability, the electrification of transportation, and advancements in technology across various sectors. Additionally, efforts to improve battery technology, increase energy density, and reduce costs are likely to contribute to the continued growth in demand for lithium-ion batteries. In 2022, the demand for automotive lithium-ion (Li-ion) batteries experienced a substantial increase of approximately 65%, reaching 550 GWh compared to around 330 GWh in 2021. This surge was predominantly driven by the growth in electric passenger car sales, which saw new registrations rise by 55% in 2022 compared to the previous year (International Energy Agency, 2023). In 2022, the market size of lithium-ion batteries worldwide reached $46.2 billion. The lithium-ion battery industry is anticipated to achieve a valuation of $189.4 billion by 2032, with a projected compound annual growth rate (CAGR) of 15.2% from 2023 to 2032 (Allied Market Research, 2023).

China witnessed a remarkable growth of over 70% in battery demand for vehicles, accompanied by an 80% increase in electric car sales in 2022 compared to 2021 (International Energy Agency, 2023). However, the rising share of plug-in hybrid electric vehicles

(PHEVs) slightly tempered the overall growth in battery demand. In the United States, battery demand for vehicles saw an approximately 80% increase, outpacing the growth in electric car sales, which rose by around 55% in 2022. Despite a modest 7% growth in the average battery size for battery electric cars in the United States, the average battery size remains approximately 40% higher than the global average (International Energy Agency, 2023). This disparity is attributed in part to the higher prevalence of SUVs in U.S. electric car sales compared to other major markets and manufacturers' strategies to provide extended all-electric driving ranges.

Globally, sales of battery electric vehicles (BEVs) and plug-in hybrid electric vehicles (PHEVs) continue to surpass those of hybrid electric vehicles (HEVs). As BEV and PHEV battery sizes tend to be larger, the demand for batteries experiences further escalation (International Energy Agency, 2023).

In 2022, lithium demand outpaced supply, despite a 180% increase in production since 2017. Notably, 60% of lithium, 30% of cobalt, and 10% of nickel demand were attributed to electric vehicle (EV) batteries in 2022, a significant rise from shares of 15%, 10%, and 2%, respectively, in 2017 (Fleischmann et al., 2023). As witnessed with lithium, the extraction and processing of these crucial minerals must escalate rapidly to support the energy transition, extending beyond EVs to meet the growing demand for clean energy technologies. To enhance supply chain sustainability, resilience, and security, reducing reliance on critical materials is imperative. Innovation, including advanced battery technologies requiring fewer critical minerals, and measures promoting optimized battery sizes and battery recycling, can contribute to this reduction (Fleischmann et al., 2023).

The global demand for batteries is on an upward trajectory, primarily fuelled by the imperative to combat climate change through the electrification of mobility and broader energy transition. Battery demand forecasts often underestimate market size, regularly necessitating upward corrections. A 2022 analysis by the McKinsey Battery Insights (Fleischmann et al., 2023) team suggests that the entire lithium-ion battery chain, encompassing mining through recycling, could grow by over 30% annually from 2022 to 2030. This growth is projected to result in a market value exceeding $400 billion and a market size of 4.7 TWh (Fleischmann et al., 2023).

Despite the environmental and social benefits of battery growth, challenges remain. To avoid shortages, battery manufacturers must secure a stable supply of raw materials and equipment, strategically invest, and commit to extensive decarbonization and true sustainability. Taking an offensive approach, prioritizing green initiatives, and adopting collaborative actions, standardized processes, and regulations, along with greater data transparency, can make the battery industry sustainable, circular, and resilient across the entire value chain. Emphasizing sustainability will differentiate leading battery players, generating value while safeguarding the environment (Fleischmann et al., 2023).

Looking ahead to 2030, global demand for lithium-ion batteries is projected to soar, from approximately 700 GWh in 2022 to around 4.7 TWh (Fleischmann et al., 2023). Mobility applications, especially electric vehicles (EVs), are expected to account for the bulk of demand. This surge is driven by regulatory shifts toward sustainability, increased consumer adoption of greener technologies, and commitments by major Original Equipment Manufacturers (OEMs) to ban internal combustion engine (ICE) vehicles. Battery energy storage systems (BESS) are also set to grow, with China anticipated to lead demand, followed by significant growth in the EU and the United States (Fleischmann et al., 2023). The entire lithium-ion battery value chain is projected to experience a five-fold increase in revenues, reaching over $400 billion in 2030 (Fleischmann et al., 2023). Recycling is identified as a viable option for sourcing battery materials, with the recycling segment expected to grow significantly in the following decade as more batteries reach their end-of-life.

Despite their popularity and utility, lithium-ion batteries come with inherent risks. All categories of batteries entail potential hazards and safety risks, but contemporary lithium-ion batteries, commonly found in the market, typically include a liquid electrolyte solution containing dissolved lithium salts like ethylene carbonate (Martin, 2023). This composition generates lithium ions, contributing to enhanced battery performance and the ability to store substantial energy in a compact space, making them widely used and practical.

However, the liquid electrolyte housing these lithium ions is highly flammable and volatile, posing a significant risk of fire or explosion, especially under elevated temperatures. Additionally, the power generation process of lithium-ion batteries produces heat as a by-product, further intensifying the risk of uncontrolled failures (Martin, 2023).

In instances of battery malfunction, the released energy and heat escalate the potential dangers, potentially fuelling a fire. The heat generated during lithium-ion battery failures can quickly reach temperatures of up to 400 degrees Celsius, with peak fire temperatures even surpassing this level (Martin, 2023). Unfortunately, containing lithium-ion battery fires proves challenging, as they are self-sustaining and considered more volatile compared to other battery types.

Factors contributing to lithium-ion battery failures include overheating, often caused by issues such as using faulty chargers, overcharging, or short circuits (Martin, 2023). Excessive heat can lead to internal damage in the battery cell, triggering failure. Physical damage to lithium-ion battery cells can also result in electrolyte leakage, introducing an additional hazard.

The danger of lithium-ion battery failures lies in the phenomenon of thermal runaway, where excessive heat reinforces the chemical reactions within the battery, intensifying the heat and chemical reactions in a destructive cycle (Martin, 2023). These failures are particularly challenging to manage, as lithium-ion battery fires from thermal runaway are difficult to extinguish. Water-based fire extinguishers may cool the battery but cannot completely extinguish the fire until the battery's energy is dissipated. Although specialized lithium-ion gel extinguishers exist, their availability for all lithium-ion battery applications is limited (Martin, 2023). Even after appearing extinguished, lithium-ion battery fires can reignite hours or days later. Furthermore, failed lithium-ion batteries can emit highly toxic gases, and in some cases, excessive heat can lead to explosions, adding another layer of risk to their failures (Martin, 2023).

Some Key Terms and Definitions

Lithium, represented by the symbol Li and atomic number 3, is a soft, silvery-white alkali metal. It possesses various notable characteristics and applications. Chemically, lithium belongs to the alkali metal group on the periodic table, displaying high reactivity. With a single valence electron, it readily donates this electron to form positive ions.

In terms of occurrence, lithium is relatively scarce in the Earth's crust, typically found in trace amounts within specific minerals like spodumene, lepidolite, and petalite. The primary sources of lithium are mineral deposits and lithium-containing brines, extracted through mining and processing these minerals to obtain lithium compounds.

Lithium and its compounds find diverse industrial applications. Notably, lithium-ion batteries, crucial for powering electronic devices, electric vehicles, and renewable energy storage systems, represent one of its most significant uses. Beyond industrial applications, lithium compounds, such as lithium carbonate, serve medicinal purposes as mood-stabilizing drugs for treating bipolar disorder. Lithium also plays a role in certain nuclear applications, being used in specific types of nuclear reactors and serving as a component in thermonuclear weapons.

Physically, lithium is the lightest metal and the least dense solid element under standard conditions. Due to its high reactivity with water, it necessitates storage in oil or an inert atmosphere to prevent corrosion.

Designated by the symbol Li and possessing an atomic number of 3, lithium's importance has surged in recent years. This is attributed to the escalating demand for lithium-ion batteries across various technologies, including smartphones, laptops, electric vehicles, and renewable energy storage. The distinctive properties of lithium render it a critical element in the advancement of modern energy storage and electronic devices.

A lithium-ion (Li-ion) battery operates as a rechargeable battery, facilitating the movement of lithium ions between the negative electrode (anode) and the positive electrode (cathode) during discharge and charging, respectively. This electrochemical process enables the storage and release of electrical energy.

Key components characterise lithium-ion batteries, as shown in Figure 3. The anode, typically constructed from graphite, serves as the negative electrode where lithium ions are released during discharge and absorbed during charging. On the other hand, the cathode, composed of materials like lithium cobalt oxide ($LiCoO_2$), lithium manganese oxide ($LiMn_2O_4$), or lithium iron phosphate ($LiFePO_4$), acts as the positive electrode, receiving lithium ions during charging and releasing them during discharge.

In electrochemical systems, such as batteries, fuel cells, and electrolysis cells, the terms "anode" and "cathode" refer to specific electrodes with distinct roles:

Anode:

- The anode is the electrode where oxidation occurs during an electrochemical reaction. In simple terms, it is where electrons are released or where a substance loses electrons. The anode is negatively charged during the operation of a cell. In a galvanic cell (like a battery), the anode is where the chemical reaction generates electrons that flow through an external circuit, providing electrical power.

- In a lithium-ion battery, for example, during discharge, lithium ions move from the anode to the cathode, and electrons are released at the anode. The anode is typically made of materials that can easily release electrons, such as graphite or lithium-based compounds.

Cathode:

- The cathode is the electrode where reduction occurs during an electrochemical reaction. It is where electrons are accepted or gained. The cathode is positively charged during the operation of a cell. In a galvanic cell, the cathode is where the electrons that travelled through the external circuit re-enter the cell to combine with ions or molecules, completing the circuit.

- In the context of a lithium-ion battery, during discharge, lithium ions move from the anode to the cathode, and electrons are accepted at the cathode. The cathode is typically made of materials that can easily accept electrons, such as lithium cobalt oxide ($LiCoO_2$) or other lithium-containing compounds.

The anode and cathode play complementary roles in an electrochemical cell, with the anode being the site of oxidation (electron release) and the cathode being the site of reduction (electron acceptance). The movement of electrons between the anode and cathode is what generates electrical energy in a cell.

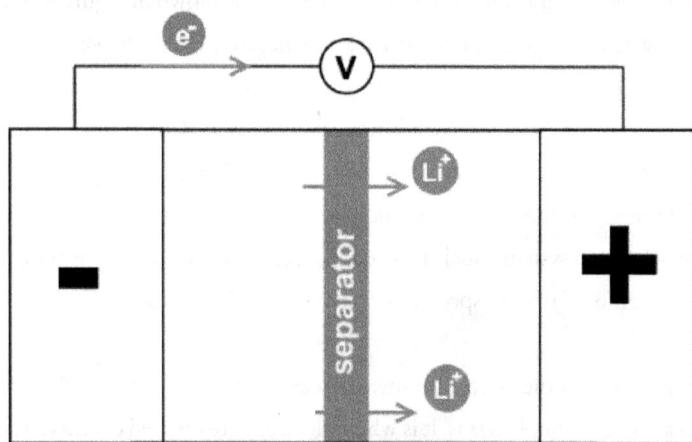

Figure 3: Structure of the Lithium-ion Battery. The electrons flow from the anode (-) to the cathode (+). A separator is situated between the anode and cathode. Menthy.denayer, CC BY-SA 4.0, via Wikimedia Commons.

Lithium cobalt oxide ($LiCoO_2$) is a compound extensively employed as a cathode material in lithium-ion batteries, playing a pivotal role in the electrochemical reactions occurring within the battery during charging and discharging cycles. This compound is composed of lithium (Li), cobalt (Co), and oxygen (O), with its chemical formula being $LiCoO_2$.

Functioning as the cathode material in lithium-ion batteries, lithium cobalt oxide operates at the positive electrode, receiving lithium ions during the charging process and releasing them during discharging. This mechanism is integral to the battery's overall performance.

An inherent characteristic of lithium cobalt oxide is its high energy density, allowing it to store a substantial amount of energy in a relatively compact and lightweight form. This property makes it particularly well-suited for applications requiring compact and lightweight energy storage solutions, such as in portable electronic devices like laptops, smartphones, and digital cameras.

Despite its widespread usage, concerns exist regarding lithium cobalt oxide cathodes. Notably, it is recognized for having limited thermal stability, and safety issues may arise under specific conditions, particularly in instances of high temperatures or overcharging. Ongoing research endeavours aim to address these concerns and explore alternative cathode materials with enhanced safety and performance attributes.

Lithium manganese oxide ($LiMn_2O_4$) is a compound frequently employed as a cathode material in lithium-ion batteries, playing a pivotal role in the electrochemical reactions integral to the battery's charging and discharging cycles.

This compound is characterized by a spinel structure, a crystal arrangement that involves a cubic close-packed configuration of oxygen ions with metal ions occupying tetrahedral and octahedral sites.

Comprising lithium (Li), manganese (Mn), and oxygen (O), lithium manganese oxide operates as the cathode material in lithium-ion batteries. In this capacity, it accepts and releases lithium ions during the electrochemical reactions at the positive electrode, or cathode.

Notably, lithium manganese oxide is often regarded as a safer alternative to certain other cathode materials, such as lithium cobalt oxide. Its propensity for being less susceptible to thermal runaway and related safety issues renders it suitable for applications where safety is of paramount concern.

Widely utilized in lithium-ion batteries for various purposes, including portable electronic devices, power tools, and electric vehicles, lithium manganese oxide contributes to the advancement of safer and more efficient energy storage systems.

Despite its safety advantages, lithium manganese oxide may entail slight trade-offs in terms of energy density and specific capacity compared to some alternative cathode materials. Researchers are actively engaged in ongoing efforts to optimize the performance of lithium manganese oxide and address any associated limitations.

Lithium iron phosphate (LiFePO₄) is a compound widely employed as a cathode material in lithium-ion batteries, specifically falling under the category of lithium iron phosphate-based cathodes. Its popularity in various applications is attributed to distinctive features that set it apart in the realm of battery technology.

The composition of lithium iron phosphate encompasses lithium (Li), iron (Fe), phosphorus (P), and oxygen (O), with the chemical formula LiFePO4 defining its molecular structure. In the context of lithium-ion batteries, it functions as the cathode material, actively participating in the electrochemical reactions integral to the battery's charging and discharging processes.

An inherent advantage of lithium iron phosphate lies in its structural stability. The compound exhibits a robust crystal structure, mitigating concerns related to electrode degradation and performance deterioration over successive charge and discharge cycles.

Safety considerations elevate lithium iron phosphate as a preferable alternative to certain other cathode materials, such as lithium cobalt oxide. Its reduced susceptibility to thermal runaway and associated safety issues makes it particularly suitable for applications prioritizing safety as a paramount concern.

Despite potentially having a lower energy density compared to some alternative cathode materials, lithium iron phosphate's safety and structural stability properties make it well-suited for specific applications. These applications span various fields, including electric vehicles, renewable energy storage systems, and portable electronic devices. The emphasis on safety positions it as the preferred choice in scenarios where safety and longevity take precedence over achieving the maximum energy density.

While acknowledging these trade-offs, the unique combination of safety, structural stability, and suitability for specific applications has propelled lithium iron phosphate into widespread use in the development of lithium-ion batteries. Ongoing research efforts are directed toward enhancing its performance and addressing any limitations that may exist.

A conductive solution known as the electrolyte facilitates the movement of lithium ions between the anode and cathode. Usually, the electrolyte consists of a lithium salt dissolved in a solvent. To prevent short circuits while enabling the flow of lithium ions, a permeable membrane, called the separator, keeps the anode and cathode apart. This describes a crucial aspect of the operation of a lithium-ion battery, specifically focusing on the role of the electrolyte and separator in facilitating the movement of lithium ions within the battery.

In a lithium-ion battery, the electrolyte serves as a conductive solution that allows for the transport of lithium ions between the anode and cathode. Typically, the electrolyte comprises a lithium salt dissolved in a solvent. The lithium ions move through the electrolyte during the battery's charging and discharging processes, participating in the electrochemical reactions that generate electrical energy.

To maintain the functionality and safety of the battery, it is essential to prevent direct contact between the anode and cathode, which could lead to short circuits. This is where the separator comes into play. The separator is a permeable membrane situated between the anode and cathode. Its primary function is to physically keep the anode and cathode apart, preventing a direct electrical connection while still allowing the flow of lithium ions. This separation is crucial for maintaining the integrity of the battery and preventing potential issues like short circuits that could compromise its performance or safety.

As such, the electrolyte facilitates the movement of lithium ions, serving as a conductive medium within the battery. Simultaneously, the separator ensures a physical barrier between the anode and cathode, preventing direct contact and potential short circuits while permitting the essential flow of lithium ions during the battery's operation.

Each individual unit of a lithium-ion battery is referred to as a cell, encompassing the anode, cathode, electrolyte, and separator. The voltage of a lithium-ion cell typically hovers around 3.7 volts, and connecting multiple cells in series forms a battery pack with a higher voltage.

A notable advantage of lithium-ion batteries is their rechargeability, allowing for numerous charge and discharge cycles without significant degradation. These batteries also boast a high energy density, enabling them to store a substantial amount of energy in a compact and lightweight package.

Lithium-ion batteries find widespread applications across various sectors, including portable electronic devices (e.g., smartphones, laptops, cameras), electric vehicles (EVs), and renewable energy storage systems. Despite being generally safe, improper handling can pose safety risks, such as thermal runaway and fires, due to issues like overcharging, overheating, or physical damage. Manufacturers implement safety features and guidelines to mitigate these risks.

History of the Lithium-ion Battery

Lithium, discovered in 1817 by Arfwedson and Berzelius through the analysis of petalite ore ($LiAlSi_4O_{10}$), was isolated in 1821 by Brande and Davy via the electrolysis of lithium oxide (Reddy, 2020; Takeda & Yamamoto, 2016). Despite its discovery, it took a century for Lewis to delve into its electrochemical properties (Takeda & Yamamoto, 2016). Due to lithium's advantageous physical characteristics—such as low density (0.534 g cm−3), high specific capacity (3860 mAh g−1), and low redox potential (−3.04 V vs. SHE)—it became evident that lithium could effectively function as a battery anode (Imanishi et al., 2012; Minami et al., 2021).

The statement emphasizes the advantageous physical properties of lithium that render it well-suited for use as a battery anode. Breaking down these key attributes, we first consider its low density of 0.534 g cm−3, signifying its lightweight nature. In the realm of batteries, a material with low density contributes to the overall construction of lightweight batteries, proving particularly beneficial in weight-sensitive applications like portable electronic devices and electric vehicles.

Moving on to the high specific capacity of 3860 mAh g−1, specific capacity measures the amount of electric charge a battery can store per unit mass of the active material. Lithium's high specific capacity indicates its ability to store a substantial amount of energy relative to its weight. This characteristic is pivotal for batteries, enabling the storage of more energy in a smaller and lighter package.

Considering the low redox potential of −3.04 V vs. SHE, redox potential gauges a material's tendency to gain or lose electrons during a chemical reaction. Lithium's low redox potential signifies its readiness to give up electrons. In the context of a battery, this property is vital for efficient electron flow during the electrochemical reactions that transpire when the battery undergoes discharging and charging.

The amalgamation of these characteristics positions lithium as an ideal candidate for a battery anode—the electrode where electrons are released during the discharging phase. The low density contributes to lightweight construction, the high specific capacity facilitates efficient energy storage, and the low redox potential promotes the smooth flow of electrons. This collective understanding has played a pivotal role in the development and widespread adoption of lithium-ion batteries across various applications.

In the early stages of 1958, Harris investigated the solubility of lithium in diverse non-aqueous (aprotic) electrolytes, including cyclic esters, molten salts, and inorganic lithium salt ($LiClO_4$), all dissolved in propylene carbonate (PC) (Takeda & Yamamoto,

2016). He observed the development of a passivation layer, capable of preventing a direct chemical reaction between lithium and the electrolyte while facilitating ionic transport (Takeda & Yamamoto, 2016). This discovery prompted research into the stability of lithium-ion batteries and fuelled interest in the commercialization of primary lithium-ion batteries (Takeda & Yamamoto, 2016).

Efforts to suppress lithium dendrite growth have been made, including the use of lithium alloys, modification of the interface between lithium metal and the solid electrolyte, and introduction of a lithium halide (Takahashi et al., 2021). Additionally, research on aqueous lithium-air rechargeable batteries has demonstrated high power density at room temperature under air atmosphere, further highlighting the advantageous physical characteristics of lithium (Imanishi et al., 2012; Minami et al., 2021). The discovery and properties of lithium, along with its advantageous physical characteristics, have paved the way for extensive research into its electrochemical properties, stability in batteries, and potential commercial applications (Reddy, 2020).

The history of the lithium-ion battery dates back to the 1970s when M. Stanley Whittingham, a British chemist, began researching energy storage devices. His work led to the development of the first functional lithium-ion battery in 1976 (Zubi, Dufo-López, Carvalho, & Paşaoğlu, 2018). This initial lithium battery was based on a titanium disulfide cathode and a lithium anode. However, due to the reactive nature of metallic lithium, safety concerns arose, leading to further research and development.

The first rechargeable batteries had solid materials in the electrodes, which broke down when they reacted chemically with the electrolyte. This destroyed the batteries. The advantage of Whittingham's lithium battery, as shown in Figure 1Figure 4: Whittingham's lithium battery. Source: © Johan Jarnestad/The Royal Swedish Academy of Sciences., was that lithium ions were stored in spaces in the titanium disulphide in the cathode. When the battery was used, lithium ions flowed from the lithium in the anode to the titanium disulphide in the cathode. When the battery was charged, the lithium ions flowed back again.

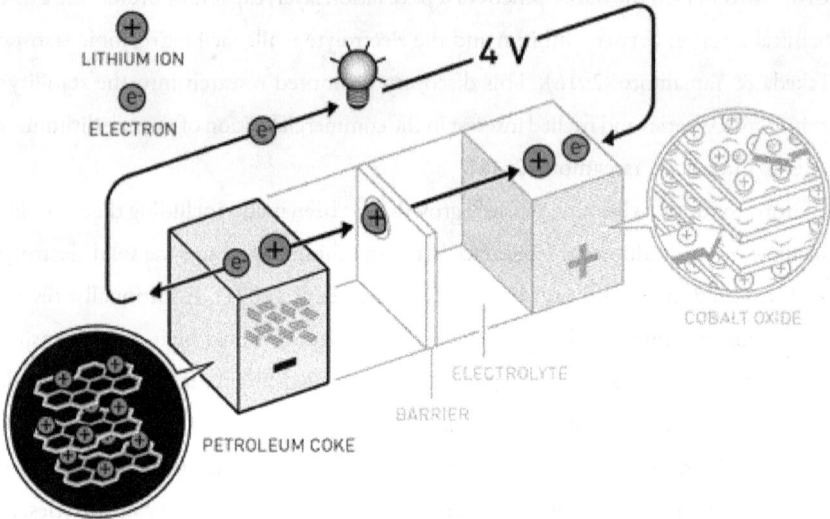

Figure 4: Whittingham's lithium battery. Source: © Johan Jarnestad/The Royal Swedish Academy of Sciences.

Stanley 's contributions were pivotal in establishing the foundation for lithium-ion battery (LIB) technology. In the 1970s, 's research on energy storage systems led to the exploration of intercalation, a process involving the insertion of ions into the crystal lattice of a host material during the battery's charging phase. This work laid the groundwork for the development of the first functional lithium battery in 1976, where utilized a lithium-titanium disulfide ($Li-TiS_2$) cathode and metallic lithium anode. The choice of the lithium-titanium disulfide cathode was based on its ability to undergo intercalation, allowing the reversible insertion and extraction of lithium ions during charging and discharging cycles (Whittingham, 2021).

Manthiram also highlighted the significance of intercalation compounds in the development of rechargeable batteries, forming the basis of essentially all lithium rechargeable batteries (Manthiram, 2020). Furthermore, Manthiram's early contributions not only led to the development of the first lithium battery but also laid the groundwork for subsequent advancements in LIB technology. His work has had a profound impact on the field, as evidenced by the 2019 Nobel Prize in Chemistry being jointly awarded to John B. Goodenough, M. Stanley , and Akira Yoshino for their significant contributions to the development of LIBs. This recognition underscores the enduring legacy of 's research, which has paved the way for the widespread use of lithium-ion batteries in various applications, including portable electronics, electric vehicles, and renewable energy storage. In conclusion, Stanley 's pioneering research on intercalation and the development of the first lithium battery has been instrumental in shaping the landscape of lithium-ion battery technology. His foundational work has not only advanced the fundamental understanding of battery chemistry but has also catalysed the evolution of rechargeable battery technologies, contributing to the sustainable and widespread adoption of lithium-ion batteries in modern society.

In the 1980s, John Goodenough significantly contributed to the advancement of lithium-ion batteries by developing a high-energy cathode material, lithium cobalt oxide (LiCoO2) (Goodenough & Kim, 2009). This innovation significantly improved the energy density and safety of lithium-ion batteries, making them more suitable for commercial applications.

Throughout the 1990s, significant efforts were made to further enhance the performance and safety of lithium-ion batteries. Sony commercialized the first lithium-ion battery in 1991, marking a major milestone in the history of energy storage technology. This development paved the way for the widespread use of lithium-ion batteries in portable electronic devices and set the stage for their future applications in electric vehicles and renewable energy storage systems.

In the early 2000s, researchers faced challenges related to the thermal management and safety of lithium-ion batteries. Studies by Tarascon and Armand (Armand & Tarascon, 2008; J.-M. Tarascon & M. Armand, 2001) highlighted the issues and challenges facing rechargeable lithium batteries, emphasizing the importance of addressing safety concerns and energy density. The 2010s witnessed a surge in research and development efforts aimed at improving the performance and sustainability of lithium-ion batteries. The environmental and human health impacts of rechargeable lithium batteries were extensively studied, emphasizing the need for proper management of end-of-life batteries to mitigate potential hazards. Furthermore, the development of lithium-ion batteries for environmental vehicles became a focal point, with ongoing research and development programs dedicated to enhancing the efficiency and environmental friendliness of lithium-ion batteries.

In recent years, the focus has expanded to include the economic analysis of centralized reused battery energy storage systems, indicating a shift towards sustainable and cost-effective energy storage solutions. Additionally, advancements in thermal management models and solutions for lithium-ion batteries have been a key area of research, reflecting the growing importance of efficient thermal control in maximizing battery performance and lifespan.

Overall, the history of the lithium-ion battery is characterised by continuous innovation and improvement, driven by the pursuit of higher energy density, enhanced safety, and environmental sustainability. From its early development in the 1970s to its widespread commercialization in the 1990s and the ongoing research efforts in the 21st century, the evolution of lithium-ion batteries has been shaped by the collaborative efforts of scientists, engineers, and industry stakeholders.

Chapter Two

Working Principle

Function

A battery is a complex system composed of several essential components, each playing a crucial role in its operation. The key components of a battery include the anode, cathode, separator, electrolyte, and positive and negative current collectors (Gandoman et al., 2021). The anode and cathode are responsible for storing lithium, while the electrolyte facilitates the movement of positively charged lithium ions between them through the separator (Etacheri et al., 2011). This movement generates free electrons in the anode, leading to the creation of a charge at the positive current collector. The electrical current then flows through the device being powered to the negative current collector, while the separator blocks electron flow inside the battery (Gandoman et al., 2021).

The current collectors, as highlighted in the literature, are vital components in both the cathode and anode of the battery and need to be well-designed to ensure efficient operation (Jin et al., 2018). The choice of current collector material is crucial as it can impact the capacity and flexibility of the battery (Y. Zhao et al., 2021). Additionally, the chemical and physical state of the current collector has been shown to impact the cycling ability of a battery (Zheng et al., 2014). Furthermore, the current collector strongly influences the electrochemical performance of the battery and needs to be properly selected for optimal performance (Ünal et al., 2023).

The electrolyte is another critical component of the battery, and its properties are essential for the overall performance and safety of the battery system (Wei et al., 2019).

The chemical stability of battery components, particularly in the presence of specific species, is crucial for the utilization of high energy density in batteries (Zheng et al., 2022). Moreover, the utilization of renewable but intermittent energy sources calls for high-performing storage batteries, emphasizing the importance of the electrolyte in advanced battery technologies (Wei et al., 2019).

In addition to the individual components, the overall design and integration of the battery components are also crucial. For instance, the fully integrated design of a stretchable solid-state lithium-ion battery, where all components are stretchable, demonstrates the importance of considering the mechanical properties of the components for flexible electronic devices and wearable systems (X. Chen et al., 2019). Furthermore, the engineering of solvation complex-membrane interaction has been shown to suppress cation crossover in batteries, highlighting the significance of the interaction between components for improved battery performance (H. Wang et al., 2020).

In the context of a lithium-ion battery, the operation involves charge/discharge cycles. During charging, lithium ions move from the cathode to the anode, and during discharging, they move from the anode back to the cathode. The cathode, typically made of lithium metal oxide compounds, releases lithium ions during charging, while the anode, usually made of graphite, receives and stores these ions. The conductive electrolyte, often a lithium salt dissolved in a solvent, facilitates ion movement, and a permeable separator prevents direct contact between the cathode and anode.

The operation of the rechargeable Li-ion battery depends critically on repeated and highly reversible formation/decomposition of lithium compounds at the cathode upon cycling (Peng et al., 2012). The consumption via reductive and oxidative decompositions of electrolyte additives at the anode and cathode in a lithium-ion battery was quantitatively analysed with charge-discharge tests, NMR spectroscopy, and gas chromatography (Haruna et al., 2016). It could be inferred that in the case of periodic charge and discharge pulses applied to the lithium-ion battery, important parameters including state of charge, current rates, initial cycling, and temperature have a significant influence on total generated heat (Madani et al., 2021b).

The unexpected discovery of a thermodynamically driven, yet kinetically controlled, surface modification in the widely explored lithium nickel manganese oxide cathode material may inhibit the battery charge/discharge rate (Gu et al., 2012). Acoustic emission detection and analysis during the discharge process can be used to characterize performance degradation and aging monitoring of lithium-ion batteries (K. Zhang et al., 2021).

Furthermore, lithium-ion batteries with large capacity are also promising power sources for hybrid electric vehicles, plug-in hybrid electric vehicles, pure electric vehicles, and storage of wind, solar, and tidal energy in smart grids (Zhu et al., 2013).

In order to ensure the safe, reliable, and economic operation of lithium-ion battery-powered devices and systems, it is important to predict life and other performances for lithium-ion batteries (Fang et al., 2015). During the initial charging cycles of the lithium-ion battery, a Solid Electrolyte Interphase (SEI) is generated on the graphite surface (Nie et al., 2015). The lithium-ion battery's heat loss throughout operating was approximated by accumulating two essential sources: reversible and irreversible heat (Madani et al., 2021a). The lithium-ion battery cells generate more heat during the high current charge and discharge process (Liu et al., 2019).

A long cycle life is critical for lithium-ion batteries when used in applications such as electric vehicles, which is different from portable devices that require fewer cycles (Nishijima et al., 2014). The conductive processes involving lithium ions are analysed in detail from a mechanistic perspective, and demonstrate that single ion polymeric electrolyte (SIPE) membranes can be used in lithium-ion batteries with a wide operating temperature range (25-80 °C) through systematic optimization of electrodes and electrode/electrolyte interfaces (Cai et al., 2014).

The search for new multifunctional cathode materials for lithium-ion batteries is extremely important to mitigate the drawbacks associated with the current electrode materials used in rechargeable lithium-ion batteries (Shreenivasa et al., 2020). The discharge product Li_2Se has been reported to be partially oxidized into polyselenides and dissolved in the ether-based electrolyte in the voltage range of 2-2.4 V, and the application of carbonate-based electrolytes has been proven to effectively inhibit the production of polyselenides, providing the opportunity for certain electrodes to match suitable electrolytes (W. Dong et al., 2021).

Rechargeable lithium-ion batteries have been widely used in consumer electronics and as power sources for sensors, portable terminals, electric vehicles, and various electronic equipment due to their high energy density and long cycle life (S.-h. Lee et al., 2014). Several characterizations were carried out to determine battery performance, including Electrochemical Impedance Spectrometry (EIS), Cyclic Voltammetry (CV), Charge/Discharge (CD), and Lithium Transference Number (LTN) (Muzadi et al., 2023). Unlike the lithium secondary batteries with lithium metal foil or lithium alloy as anodes, lithium-ion batteries with Li x C n anodes prohibit dendrite formation during charge-discharge processes (Xie et al., 2006).

In addition, batteries incorporating sodium or magnesium rather than lithium have also been investigated as alternative energy sources (Hibino et al., 2013). The emergence of a new flame-retardant material with the additive ethoxy (pentafluoro) cyclotriphosphazene can ameliorate the performance of lithium-ion batteries while ensuring their safety (Wu et al., 2021). Lithium-ion batteries exhibiting high energy and high power density have fast become the attractive power source for many portable electronic devices, as well as plug-in hybrid electric vehicles and electric vehicles (Yang et al., 2012).

The environment humidity, anode materials, and water content of the electrolyte must be strictly controlled for the preparation process of lithium-ion batteries to ensure their electrochemical performances (Li et al., 2016). The management of lithium-ion batteries in space applications will be significantly different than the familiar nickel-hydrogen and nickel-cadmium space battery systems (Lurie).

The movement of lithium ions from the anode to the cathode during the discharge process of a lithium-ion battery results in the release of electrons at the anode, generating an electric current that powers devices (Gu et al., 2012). This process is reversible, allowing the battery to undergo multiple charge and discharge cycles (Gu et al., 2012). The voltage of a lithium-ion cell typically hovers around 3.7 volts, and its high energy density enables the storage of a significant amount of energy in a compact package (Gu et al., 2012).

The cathode materials in lithium-ion batteries, particularly intercalation compounds, have been extensively studied due to their structural flexibility and adaptable interlayer space, which facilitate the smooth insertion and extraction of Li+ ions, making them suitable for practical applications (Shreenivasa et al., 2020). Additionally, the temperature has a significant influence on the capacity of lithium-ion batteries, as a decrease in temperature leads to a decrease in the activity of the active electrolyte and an increase in concentration, thereby slowing down the deintercalation rate of lithium ions during the discharge process (Lv et al., 2021).

Furthermore, research has shown that the micro-changes in the materials for the anode and cathode of lithium batteries after over-charge and over-discharge can be explored through methods such as x-ray diffraction and scanning electron microscopy, providing measures to prevent potential safety problems during the use of lithium batteries (Wang et al., 2021). Additionally, the use of coated Li metal as an anode in aqueous rechargeable lithium batteries has been investigated, demonstrating values similar to those in organic electrolytes (Wang et al., 2013).

Moreover, the charge-discharge rate significantly impacts the lithium reaction distributions in commercial lithium coin cells, as visualized by Compton scattering imaging, with inhomogeneous reactions inducing an overvoltage during the charge/discharge cycle (Suzuki et al., 2018). Strategies such as the regulation of cathode material components, the construction of lithium ion and electron transport pathways within the composite cathode, and the interfacial modification of cathode materials have been shown to have significant effects on lithium-rich cathode materials in all-solid-state lithium batteries (Yang et al., 2023).

In addition, the compound Li ion creates a long-range stress field around itself by expanding the interlayer spacing of graphite, enhancing the thermal diffusion of Li in graphite by alternating vertical electric fields (Ohba et al., 2012). The reversible lithium intercalation in a lithium-rich layered rocksalt $Li2RuO_3$ cathode through a $Li3PO_4$ solid electrolyte has been demonstrated, leading to the high theoretical capacity of these materials (Zheng et al., 2015). Furthermore, the entropic heat coefficient is a crucial factor affecting the magnitude of reversible heat in lithium titanate oxide batteries (Madani et al., 2019).

The internal resistances of battery cells can be measured using the hybrid pulse power characteristic (HPPC), providing insights into the electrical properties of power lithium-ion batteries (Y. Zhang et al., 2023). Additionally, the characteristics of vanadium-doped and bamboo-activated carbon-coated $LiFePO_4$ and its performance for lithium-ion battery cathodes have been studied, revealing that the particle agglomeration decreases with increasing levels of vanadium concentrations (Sofyan et al., 2018). Moreover, the heat loss measurement of lithium titanate oxide batteries under fast charging conditions can be employed for battery thermal modelling and the design of thermal management systems (Madani et al., 2019).

The development of sodium-ion rechargeable batteries using sodium cobalt phosphate cathodes has been explored, with intercalation cathode materials such as metal chalcogenides, transition metal oxides, and polyanion compounds being among the materials studied (Wijesinghe et al., 2019). Furthermore, room temperature ionic liquids (RTILs) have been investigated as promising electrolytes for $Li/LiFePO_4$ batteries, offering non-volatility, non-flammability, and high thermal stability (Zhang et al., 2011).

Encapsulation of $LiNi_{0.5}Co0.2Mn_{0.3}O_2$ with a thin inorganic electrolyte film has been proposed to reduce gas evolution in the application of lithium-ion batteries (Kim, 2013).

$LiNi_{0.5}Co0.2Mn_{0.3}O_2$, often denoted as NMC532, is a type of cathode material used in lithium-ion batteries. It is a ternary oxide compound composed of lithium (Li), nickel (Ni), cobalt (Co), manganese (Mn), and oxygen (O). The numerical subscript values (0.5, 0.2, 0.3) represent the relative proportions of each metal element in the compound.

Here's a breakdown of the composition:

- Li: Lithium

- Ni: Nickel

- Co: Cobalt

- Mn: Manganese

- O_2: Oxygen

The specific composition of $LiNi0.5Co0.2Mn0.3O2$ is designed to optimize certain properties of the cathode material in lithium-ion batteries. NMC cathodes are known for offering a balanced combination of high energy density, good cycling stability, and reasonable cost compared to other cathode materials.

The numbers in the formula represent the atomic ratios of each metal element, indicating the stoichiometry of the compound. In this case, the atomic ratio is 5:2:3 for Ni:Co:Mn, respectively. The overall composition aims to achieve a suitable trade-off between the capacity, voltage, and stability of the cathode material during charge and discharge cycles in lithium-ion batteries. Different variations of NMC cathodes exist, such as NMC111, NMC622, NMC811, each with different metal ratios, offering varying performance characteristics for specific applications.

Modelling and state of charge (SOC) estimation of lithium iron phosphate batteries considering capacity loss have been studied, with the charge and discharge processes being affected by polarization, temperature, and other factors (J. Li et al., 2018). Additionally, transient temperature distributions on lithium-ion polymer SLI batteries have been investigated, revealing that the cells generate more heat during high current charge and discharge processes (Liu et al., 2019). Improper charge and discharge processes have been shown to decrease battery performance (Aji & Afianti, 2021).

The synthesis of nanostructured LiFePO$_4$/C cathode materials for lithium-ion batteries has been explored, demonstrating excellent electrochemical properties with various rates (Yang et al., 2012). Furthermore, the modelling of lithium-ion batteries for charging/discharging characteristics based on a circuit model has been proposed, with the active material lithium ion being provided by lithium-containing transition metal oxides in the positive electrode (Haizhou, 2017). The research on lithium-ion batteries encompasses various aspects, including the materials used in the cathode and anode, the influence of temperature, charge-discharge processes, and the development of new electrolytes and encapsulation techniques. These studies provide valuable insights into the fundamental processes and practical considerations for the design and optimization of lithium-ion batteries.

A lithium-ion battery functions by shuttling lithium ions between the cathode and anode during charge and discharge cycles, facilitated by the electrolyte and separator. This process allows the flow of electrons, generating electrical energy for various applications.

Chemistry

The chemistry underlying lithium-ion batteries centres on redox (oxidation-reduction) reactions occurring at the cathode and anode during the charging and discharging cycles. Let's delve into the crucial processes:

Cathode (Reduction): The cathode in a typical lithium-ion battery consists of a lithium metal oxide compound, often lithium cobalt oxide (LiCoO2). The reduction reaction during discharging can be represented as:

$$CoO_2 + Li^+ + e^- \rightarrow LiCoO_2$$

This half-reaction elucidates the reduction of cobalt oxide (CoO$_2$) at the cathode. Lithium ions (Li$^+$) from the electrolyte combine with electrons (e$^-$) at the cathode, forming lithium cobalt oxide (LiCoO2). This process occurs when the battery discharges, providing electrical energy to a device.

Anode (Oxidation): The anode, typically composed of graphite, undergoes oxidation during discharging, forming a graphite intercalation compound. The oxidation reaction is represented as:

$$LiC_6 \rightarrow C_6 + Li^+ + e^-$$

This half-reaction illustrates the oxidation of the graphite intercalation compound (LiC6) at the anode. Lithium ions (Li$^+$) are released along with electrons (e$^-$) during this process. It occurs when the battery discharges, supplying electrical energy.

Full Reaction (Discharging and Charging): The overall reaction during both discharging and charging cycles is encapsulated by:

$$LiC_6 + CoO_2 \rightleftharpoons C_6 + LiCoO_2$$

This equation captures the entire process, signifying the movement of lithium ions between the anode and cathode. During discharging (left to right), lithium ions move from the anode to the cathode. Conversely, during charging (right to left), lithium ions move from the cathode back to the anode.

In summary, during discharging, lithium ions are released from the anode, traverse the electrolyte to the cathode, and partake in redox reactions, generating electrical energy. During charging, the process reverses, with lithium ions moving from the cathode back to the anode. This redox chemistry forms the foundation of the reversible reactions essential for the operation of lithium-ion batteries.

Materials

Lithium-ion batteries are widely used in various applications, including electric vehicles (EVs), due to their high energy density (Nie et al., 2015), noting that these batteries have different standards in various regions, such as NMC/NMCA in Europe and North America and LFP in China. NMC/NMCA batteries have a higher energy density, while LFP batteries have a lower cost (Ying-hao et al., 2014). The mineral composition of a lithium-ion battery includes nickel, manganese, lithium, cobalt, copper, aluminium, graphite, and other materials (Gaikwad et al., 2014). The percentage of lithium found in a battery is expressed as the percentage of lithium carbonate equivalent (LCE) the battery contains, which is equal to 1g of lithium metal for every 5.17g of LCE (Fell et al., 2015).

Lithium-ion batteries work by collecting current and feeding it into the battery during charging. A graphite anode attracts lithium ions and holds them as a charge. Recent research shows that battery energy density can nearly double when replacing graphite with a thin layer of pure lithium (Feng et al., 2014). When discharging, the cathode attracts the stored lithium ions and funnels them to another current collector. The circuit can react as both the anode and cathode are prevented from touching and are suspended in a medium that allows the ions to flow easily (Gaikwad et al., 2014).

The use of different materials and configurations in lithium-ion batteries has been a subject of extensive research. For example, lithium titanate (LTO) anodes are promising for their higher rate capabilities, safety, and long cycle-life, owing to their zero volumetric growth during lithiation (Yongchao Liu et al., 2021). Additionally, the electrochemical performance of materials such as MnO_2 and copper-based current collectors has been investigated for their potential use in lithium-ion batteries (A. Gao et al., 2022; Kühnel et al., 2016). Furthermore, the use of "water-in-salt" electrolytes has been explored to enable the use of cost-effective aluminium current collectors for aqueous high-voltage batteries (Chen et al., 2017).

Cylindrical battery cathode materials are broadly classified into six categories: carbon cathode materials, alloy cathode materials, tin-based cathode materials, lithium transition metal nitride cathode materials, nanometer materials, and nanometer cathode materials.

Carbon Nanometer Material Cathode Material: Cathode materials actively utilized in lithium-ion batteries predominantly comprise carbon materials. This category includes artificial graphite, natural graphite, mesophase carbon microspheres, petroleum coke, carbon fibre, pyrolysis resin carbon, and other variants.

Alloy Cathode Materials: This category encompasses tin-based alloy, silicon-based alloy, germanium-based alloy, aluminium-based alloy, antimony-based alloy, magnesium-based alloy, and other alloy variations. Notably, there are currently no commercial products within this category.

Tin-Based Cathode Material: Tin-based cathode materials are further divided into tin oxide and tin-based composite oxide. Despite various valence tin metals constituting the oxide, no commercial products within this category have been introduced.

Lithium Transition Metal Nitride Cathode Materials: As of now, no commercial products featuring lithium transition metal nitride cathode materials have been introduced to the market.

Nanometer Materials: This category encompasses carbon nanotubes and nano-alloy materials.

Nanometer Cathode Material: This category involves nanometer oxide material.

This classification offers a comprehensive overview of the diverse cathode materials utilized in cylindrical batteries. The focus is on delineating their current usage and highlighting the absence of commercial products in certain categories.

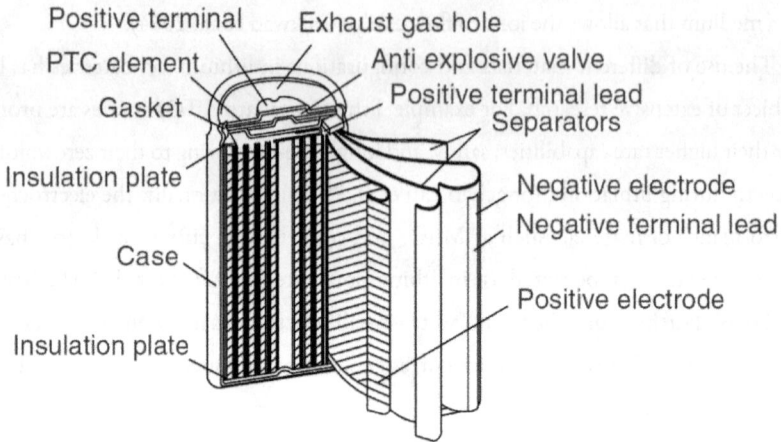

Figure 5: Schematic diagram of the structure of a lithium-ion battery (Large, 2019).

Cylindrical lithium-ion batteries have gained popularity among Japanese, Korean, and Chinese enterprises. The first cylindrical lithium-ion battery was pioneered by the Japanese company SONY in 1992. Recognized brands in this category include SONY, Panasonic, Sanyo, Samsung, LG, Wanxiang A123, Bic, Lishen, and others.

These batteries are classified based on their dimensions, represented by five digits. The first and second digits denote the battery diameter, the third and fourth digits indicate the battery height, and the fifth digit signifies the circular shape (Large, 2019). Various types of cylindrical lithium-ion batteries include 10400, 14500, 16340, 18650, 21700, 26650, and 32650.

- 10440 Battery: The 10440 battery has a diameter of 10mm and a height of 44mm, equivalent to the size of an AAA battery. Typically, it has a small capacity, often a few hundred mAh, and finds applications in mini electronic products like flashlights, mini stereos, and amplifiers.

- 14500 Battery: With a diameter of 14mm and a height of 50mm, the 14500 battery operates at 3.7V or 3.2V. Its nominal capacity is slightly larger than the 10440 battery, typically around 1600mAh. Known for excellent discharge performance, it is commonly used in consumer electronics such as wireless audio, electric toys, and digital cameras.

- 16340 Battery: The 16340 battery features a diameter of 16mm and a height of

34mm. Due to its compact size and substantial capacity, it is often utilized in glare flashlights, LED flashlights, headlamps, laser lamps, and illuminators.

- 18650 Battery: An 18650 battery boasts an 18mm diameter and 65mm height, offering high energy density of nearly 170 watts/kg. Widely employed in various applications due to its mature technology and good quality, it is suitable for approximately 10 KWH battery capacity scenarios, including mobile phones, laptops, and small appliances.

- 21700 Battery: Featuring a diameter of 21mm and a height of 70mm, the 21700 battery provides increased space utilization and energy density compared to the 18650 battery. Widely used in digital devices, electric cars, balance cars, solar lithium battery street lamps, LED lights, and electric tools.

- 26650 Battery: The 26650 battery, with a diameter of 26mm and a height of 65mm, possesses a nominal voltage of 3.2V and a nominal capacity of 3200mAh. Recognized for its excellent capacity and consistency, it is becoming a viable alternative to the 18650 battery, especially in the power battery market.

- 32650 Battery: The 32650 battery, with a diameter of 32mm and a height of 65mm, exhibits strong continuous discharge capacity. Suited for electric toys, backup power supply, UPS batteries, wind power generation systems, and wind and photovoltaic complementary power generation.

The safety and thermal characteristics of lithium-ion batteries have also been a focus of research. Studies have been conducted on the thermal runaway behaviour of lithium-ion batteries under various external environments, and early monitoring and warning systems have been proposed to mitigate potential thermal runaway incidents. Additionally, the toxicity of gases released during the combustion of lithium-ion batteries has been studied to understand the potential environmental and health impacts.

In the context of standardization, the differences between domestic and overseas standards for lithium-ion traction batteries for electric vehicles have been compared, and suggestions for the development of China's lithium-ion traction battery standards have been proposed (Ying-hao et al., 2014). This highlights the importance of standardization in ensuring the quality and safety of lithium-ion batteries used in electric vehicles.

Currently, the prevalent commercial cylindrical battery anode materials encompass lithium cobalt oxide (LiCoO2), lithium manganese (LiMn2O4), ternary (NMC), lithium iron phosphate (LiFePO4), among others. Distinct material systems exhibit varying characteristics, as outlined in Table 1 (Large, 2019).

Table 1: Batteries with different material systems and their different characteristics.

Item	$LiCoO_2$	$LiNiCoMnO_2$	$LiMn_2O_4$	$LiFePO_4$
Tap Density (g/cm3)	2.8~3.0	2.0~2.3	2.2~2.4	1.0~1.4
Specific Surface Area (m2/g)	0.4~0.6	0.2~0.4	0.4~0.8	12~20
Gram Volume (mAh/g)	135~140	140~180	90~100	130~140
Voltage Platform (V)	3.7	3.5	3.8	3.2
Cycle Performance	≥500 times	≥500 times	≥300 times	≥2000 times
Transition Metal	Little	Little	A lot	Abundant
Material Cost	Very expensive	Expensive	Cheap	Cheap
Environmental Protection	Contain Co	Contain Ni and Co	Non-toxic	Non-toxic
Safety Performance	Bad	Good	Better	Best
Scope of Application	Small and medium-size battery	Small batteries/small power battery	Power battery, low-cost battery	Power battery/power supply with super capacity
Advantages	Stable charge and discharge, simple production process	Stable electrochemical performance and good cycling performance	Abundant manganese resource, relatively cheap, good safety performance	High safety, environmental protection, long life
Disadvantages	Cobalt is expensive and has low cycle life	Cobalt is expensive	Low energy density, poor electrolyte compatibility	Poor low-temperature performance, low discharge voltage

This comparative analysis provides insights into the distinctive features and trade-offs associated with each cylindrical battery anode material, aiding in the selection process based on specific application requirements.

Chapter Three

Composition and Manufacturing

Lithium-ion Battery Production

The production process of lithium-ion battery cells involves three primary stages: electrode manufacturing, cell assembly, and cell finishing, each encompassing various sub-processes (Ludwig et al., 2017). In the material preparation stage, the cathode, typically comprising metal oxide powder, is combined with a binder and conductive additives to form a paste, while the anode, often composed of graphite, undergoes a similar process (Li et al., 2020). Additionally, the electrolyte, a conductive solution containing a lithium salt dissolved in a solvent, is prepared (J. Li et al., 2020). The widely produced types of cells are prismatic, cylindrical, and pouch, with manufacturing processes generally similar across these designs (Ludwig et al., 2017). Major manufacturers and equipment suppliers are located in China, Japan, and South Korea, with manufacturing constituting approximately 25 percent of the Li-ion battery cost (Ludwig et al., 2017).

The Li-ion cells consist of four main components: two electrodes (anode and cathode), a separator to prevent contact and shorting, and an electrolyte medium facilitating the movement of lithium ions between the electrodes (Ludwig et al., 2017). The anode is typically constructed from graphite, while the cathode can be an alloy of multiple metals (nickel, cobalt, lithium, among others) (J. Li et al., 2020). All components are enclosed in a casing, with exposed tabs for positive and negative terminals, and subsequently arranged and connected to form a battery pack (Ludwig et al., 2017).

The manufacturing of lithium-ion battery electrodes involves a new additive manufacturing process based on dry powders (Ludwig et al., 2017). Conventional LIB electrodes, especially cathodes, are manufactured through a slurry processing method where N-methyl-2-pyrrolidone (NMP) is used as a solvent (J. Li et al., 2020). Additionally, a green solvent, triethyl phosphate, has been used to recover invaluable cobalt-containing cathodes, such as NMC622, by dissolving the polymeric binder of poly(vinylidene fluoride) (Bai et al., 2021). Furthermore, aerosol jet printed electrolytes composed of polyethylene oxide (PEO), lithium difluoro(oxalato)borate (LiDFOB), and alumina nanoparticles have been developed (Deiner et al., 2019). This process, as outlined by (Yangtao Liu et al., 2021), is summarised in Figure 6.

Reference: iScience 24, 102332, April 23, 2021

Figure 6: Schematic of LIB (Lithium-Ion Battery) manufacturing processes. Adapted from Liu et al. (2021).

Stage 1 - Electrode Manufacturing

Process Overview: The initial stage involves mixing electrode materials with a conductive binder to create a consistent slurry with a solvent (Gupta, 2021). To prevent cross-contamination between the anode (Carbon) and cathode (Lithium metal oxide) materials, separate processing rooms are utilized. The slurry is then applied continuously or intermittently to both sides of the current collector (Al foil for cathode and Cu foil for the anode) using application tools such as slot die, doctor blade, or anilox roller (Gupta, 2021). The coating thickness is controlled during this process.

The coated foil undergoes drying in a lengthy oven to eliminate the solvent, with recovered solvent from the cathode coating repurposed for thermal recycling. The calendaring process follows, employing rotating rollers to compress the coated foils and adjust physical properties such as bonding, conductivity, density, and porosity (Gupta, 2021).

Post-calendaring, the finished electrodes are cleaned, cut into narrow strips using slitting machines, and recoiled. Subsequently, the coils are sent to a vacuum oven to eliminate residual moisture and solvent. Components such as active materials, conductive additives, solvents, binders, aluminium foil, and copper foil are typically procured by the cell manufacturer.

Equipment Used in the Process: Machinery used in this initial cell manufacturing stage includes mixers for slurry creation, coating and drying machines, calendaring or roll pressing machines, electrode cutting or slitting machines, and vacuum drying ovens.

In the coating process, cathode and anode pastes are separately applied to thin metal foils, which act as the battery electrodes. Calendering follows to compress the coated electrodes, enhancing density and adhesion. During assembly, a separator is introduced between the cathode and anode to prevent direct contact, and the components are configured based on the battery design (Gupta, 2021).

Stage 2 - Cell Assembly

Process Overview: Upon completing the electrode preparation, they proceed to the dry room for sub-assembly. Here, the separator is placed between the anode and cathode to establish the internal structure of the cell. The choice between stacking for pouch cells and winding for prismatic and cylindrical cells depends on the cell case type. Highly automated equipment typically handles cell assembly (Gupta, 2021).

Following this, the assembled cell structure is linked to terminals or cell tabs, along with any safety devices, utilizing ultrasonic or laser welding. The sub-assembly is then inserted into the cell housing (pouch or metal case based on the cell design), which undergoes sealing through laser welding or heating, with an opening for electrolyte injection.

Electrolyte filling and sealing transpire in a dry room to prevent moisture-induced electrolyte decomposition and toxic gas emission. A high-precision dosing needle is used for electrolyte filling, activating the capillary effect in the cell through a pressure profile (inert gas supply and/or vacuum generation in alternating operation). The finalised cell is labelled with a batch or serial number on the case.

Components such as packaging materials, cell housing, insulation materials, and electrolyte are typically procured by the cell manufacturer.

Equipment Used in the Process: Machinery involved in the second stage of cell manufacturing encompasses a die-cutting machine, stacking machine (for pouch cells), winding machine (for cylindrical and prismatic cells), sealing and tab welding machine, and electrolyte filling machine.

In the cell formation step, assembled components are placed in a cell case, and electrolyte is injected to enable lithium ion movement (Gupta, 2021). Sealing is pivotal to prevent electrolyte leakage and maintain a controlled environment. The battery undergoes a formation process involving multiple charge and discharge cycles to stabilize performance and capacity. Rigorous testing follows, covering aspects such as capacity, voltage, and quality control measures. Subsequently, batteries are packaged into packs, often incorporating safety features, thermal management systems, and a battery management system (BMS) for individual cell monitoring and control (Gupta, 2021).

Stage 3 - Cell Finishing

Process Overview: The formation process initiates the initial charging and discharging cycles of the battery cell post-electrolyte injection. Cells are positioned in information racks and engaged by spring-loaded contact pins, undergoing precise charging or discharging as per defined current and voltage curves. During this phase, lithium ions integrate into the graphite's crystal structure on the anode, establishing the Solid Electrolyte Interface (SEI), a protective layer influencing the low self-discharge of Li-ion batteries and impacting battery performance and lifespan (Gupta, 2021).

In larger pouch cells, initial charging can produce significant gas evolution, directed into a gas bag during degassing. This gas bag, pierced into a vacuum chamber, allows the release of gases before final vacuum sealing, and the gas bag is then treated as hazardous waste.

Subsequent to formation, aging follows for quality assurance. This involves monitoring cell characteristics and performance by regularly measuring open-circuit voltage (OCV) for up to three weeks, distinguishing between high-temperature (HT) and normal temperature (NT) aging. Cells typically undergo HT aging before NT aging and are stored in aging shelves or cabinets. A fully functional cell exhibits no significant changes over the entire aging period.

After aging, cells undergo testing in an end-of-line (EOL) test rig. Discharged to the shipping state of charge, cells undergo capacity measurements, pulse tests, internal resistance measurements (DC), optical inspections, OCV tests, and leakage tests.

Upon successful completion of tests, cells can be assembled into battery packs based on specific requirements and end-use. The formation and aging process constitutes 32 percent of the total manufacturing process (Gupta, 2021).

Equipment Used in the Process: Machinery in this final stage includes battery formation testers/equipment, aging cabinets, grading machines, and battery testing machines. Generally, coater, winder, and grading & testing equipment contribute to 70 percent of the total cost of Li-ion cell production equipment, subject to variations based on automation levels.

Consistent quality control and safety measures are maintained throughout the manufacturing process to ensure reliability. Safety features, including pressure relief vents and thermal protection, are integrated to enhance overall battery safety. It's essential to acknowledge variations in manufacturing processes based on lithium-ion battery types, intended applications, and design specifications, with the outlined process offering a general overview applicable to numerous Li-ion batteries (Gupta, 2021).

Innovation in Product and Processes

Advancements in product design and manufacturing processes have led to significant improvements in the Li-ion cell industry, resulting in reduced production costs for raw materials and the final battery cell. These advancements have also contributed to enhanced overall performance and quality of the end product (Malik, 2017). Several noteworthy developments in technology and materials are influencing Li-ion cell manufacturing, including mixing technologies, simultaneous coating processes, dry coating application, infrared heating support, and lamination technology (Ludwig et al., 2017).

Mixing technologies such as intensive mixers, planetary mixers, and dispersers offer enhanced flexibility and precision in the mixing process, leading to improved battery performance and quality (Ludwig et al., 2017). Innovative coating processes that allow simultaneous coating of both the top and bottom sides of the foil have streamlined the manufacturing process, increasing efficiency and reducing production costs (Ludwig et al., 2017). The adoption of dry coating application for the active material on the carrier foil, in powder form without the use of solvents, has significantly reduced thermal activation time, contributing to the economic viability of large-scale Li-ion battery production (Ludwig et al., 2017). Additionally, the supplementation of traditional dryers with infrared heating has enhanced their efficiency and effectiveness in the production process, further reducing manufacturing costs (Ludwig et al., 2017). Lamination technology has emerged as a pivotal advancement in Lithium-ion battery production, improving production line speed and ensuring a more streamlined and efficient manufacturing workflow (Ludwig et al., 2017).

These innovations collectively signify a positive shift in Li-ion cell manufacturing, promising advancements in efficiency, cost-effectiveness, and overall product quality (Malik, 2017). The continuous innovation in product design and manufacturing processes not only enhances cost efficiency but also contributes to improved overall performance and quality of the end product (Malik, 2017).

Raw Materials

The raw materials for lithium-ion batteries are sourced from various locations around the world. Lithium, a crucial component in lithium-ion batteries, is primarily produced in Chile, known for its lithium-rich brine deposits, Australia, which produces lithium from both hard rock and brine deposits, and China, which extracts lithium from mineral deposits (Ruismäki et al., 2020). Cobalt, commonly used in the cathodes of lithium-ion batteries, is majorly produced in the Democratic Republic of Congo (DRC) as a byproduct of copper mining, and in Russia (Ruismäki et al., 2020). Nickel, another key component in the cathodes of lithium-ion batteries, is sourced from major producers such as Indonesia and the Philippines (Ruismäki et al., 2020). Graphite, used in the anodes of lithium-ion batteries, is primarily produced in China and Brazil (Ruismäki et al., 2020). Manganese, used in some lithium-ion battery cathodes, is sourced from major producers such as South Africa and Australia (Ruismäki et al., 2020). Aluminium, used in the construction of battery casings, is primarily produced in China and Russia (Ruismäki et al., 2020).

Figure 7: The open pit of the Greenbushes mine, Western Australia, seen from the public mine lookout. Calistemon, CC BY-SA 4.0, via Wikimedia Commons.

The sourcing of these raw materials has raised environmental and ethical concerns, particularly regarding issues such as mining practices, child labour, and environmental impact in some production areas (Ruismäki et al., 2020). Efforts are being made to address these concerns and promote sustainable and responsible sourcing practices in the battery supply chain (Ruismäki et al., 2020).

The demand for batteries, particularly for passenger electric vehicles (EVs) utilizing lithium-ion batteries, has been on the rise, accounting for 60% of the total demand for batteries (Diekmann et al., 2016). This increasing demand has led to a high requirement for raw materials for the production of lithium-ion batteries (Diekmann et al., 2016).

Figure 8: The Kwinana Lithium Plant, Western Australia, owned by Tianqi Lithium. The view is from Rockingham Road, just south of the Anketell Road turn off. Calistemon, CC BY-SA 4.0, via Wikimedia Commons.

The recycling and recovery of lithium-ion battery materials have also been a subject of research. Studies have focused on processes such as battery scrap recycling, recovery of lithium carbonate from cathode scrap, and sustainable processes for the recovery of anode and cathode materials derived from spent lithium-ion batteries (Gao et al., 2017; G. Zhang et al., 2019).

In addition to the primary raw materials, research has also been conducted on the development of advanced materials for lithium-ion batteries. This includes the study of porous $LiMn_2O_4$ as a cathode material, the use of bagasse-based porous carbon material, and the utilization of hierarchical porous carbon derived from biomass for advanced supercapacitors and lithium-ion batteries ((Fan et al., 2019; Lin et al., 2023; Qu et al., 2011). Furthermore, the recovery of specific elements from lithium-ion battery materials has been a focus of research. Studies have investigated the recovery of lithium, manganese, and iron from mixed waste lithium-ion battery cathode materials, as well as the recovery of lithium from Li-enriched slag through thermodynamic and kinetic studies (Klimko et al., 2020; Y. H. Wang et al., 2023).

The thermal management of lithium-ion batteries has also been a subject of research, with studies focusing on the use of phase change materials and metal foam for cooling lithium-ion battery packs (Madani et al., 2020b; Mohammadian et al., 2015).

The production and sourcing of raw materials for lithium-ion batteries have significant environmental and ethical implications. Efforts are being made to address these concerns and promote sustainable and responsible sourcing practices. Research is also focused on the development of advanced materials, recycling and recovery processes, and thermal management techniques to improve the sustainability and performance of lithium-ion batteries.

Lithium

Lithium, a crucial component in lithium-ion batteries and various other applications, is primarily produced from three main sources: brine deposits, hard rock minerals, and lithium-enriched clays, see Figure 9. Each source involves distinct extraction and processing methods. Brine deposits involve the extraction of lithium-rich brine from underground reservoirs, followed by evaporation to concentrate the lithium content and harvesting the lithium as lithium carbonate or lithium chloride through precipitation (P. Xu et al., 2020). Hard rock minerals, such as spodumene, are mined and then concentrated through processes like froth flotation, magnetic separation, or dense media separation to increase lithium content. The concentrated spodumene is further processed to produce lithium carbonate or lithium hydroxide, the forms commonly used in battery production (Meshram et al., 2014). Lithium-enriched clays, particularly in regions like Nevada, USA, are treated with sulfuric acid to extract lithium, which is then processed to create lithium carbonate or lithium hydroxide (Meshram et al., 2014).

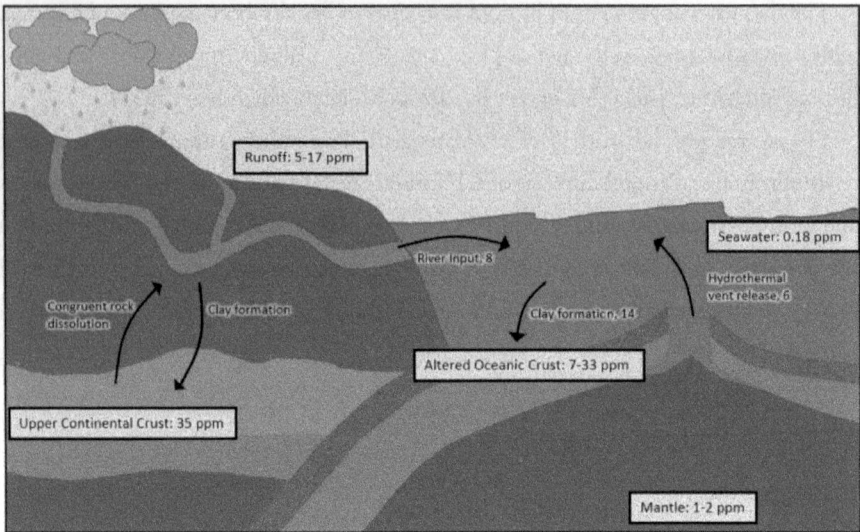

Figure 9: Lithium Cycle Diagram. Green desert scrub, CC BY-SA 4.0, via Wikimedia Commons.

In addition to these traditional methods, emerging technologies such as direct lithium extraction (DLE) are being explored to extract lithium more efficiently from brines without the need for extensive evaporation ponds. Methods like ion exchange, solvent extraction, and selective sorbents are being investigated (Aljarrah et al., 2022). The choice of lithium production method depends on factors such as the lithium concentration in the source, economic feasibility, and environmental considerations (P. Xu et al., 2020).

Furthermore, the lithium produced from these sources undergoes further refining and processing to meet the specifications required for use in lithium-ion batteries and other applications. This refining and processing are critical to ensure the quality and purity of the lithium for its intended applications (P. Xu et al., 2020).

The annual global production of lithium is a topic of significant interest due to the growing demand for lithium in various industries, particularly in the context of the increasing use of lithium-ion batteries in electric vehicles and energy storage systems. Several studies provide insights into the annual production of lithium and its implications for various sectors. Vikström et al. (2013) discuss the future production outlook for lithium and highlight the current annual global production of approximately 85 kilotons (kt) of lithium. This reference provides a comprehensive overview of lithium availability and production, making it relevant and appropriate for understanding the annual production of lithium. Furthermore, Chagnes and Pośpiech (2013) estimate that the annual demand for lithium carbonate could reach 420,000 tons if 60 million cars worldwide were to be replaced with plug-in hybrids. This figure is nearly five times the current annual lithium carbonate production, indicating a significant gap between demand and current production levels.

Moreover, Liang et al. (2021) present statistics from the United States Geological Survey (USGS), indicating that global lithium production decreased by 19% while global consumption increased by 18% in 2019. This data underscores the dynamic nature of lithium production and consumption on an annual basis. Additionally, (Yang & Rogach, 2020) mention that the annual global lithium production surpassed 0.085 million tons in 2018, further emphasizing the scale of lithium production on a yearly basis. In summary, the annual global production of lithium is estimated to be approximately 85 kilotons, with projections indicating a potential demand of 420,000 tons for lithium carbonate. These figures highlight the significant gap between current production levels and projected demand, underscoring the need for further research and investment in lithium production.

To calculate the energy required to produce 85,000 to 100,000 metric tons of lithium, we need to consider various factors such as the production process, recycling, and the energy consumption associated with lithium extraction and battery production. Several studies provide valuable insights into the energy requirements for lithium production and its environmental impacts. (Gaines & Singh, 1995) discuss the energy requirements for the production of different types of batteries, highlighting variations in energy consumption based on the battery chemistry. This information is relevant as it emphasizes the variability in energy requirements based on the type of battery being produced, which is crucial for understanding the energy needs for lithium production.

Mashtalir et al. (2018) provide specific energy requirements for lithium production, stating that the energy-intensive process requires 35–40 kWh per kilogram of lithium. This data is essential for estimating the overall energy consumption for producing 85,000 to 100,000 metric tons of lithium. A kilowatt-hour (kWh) is a unit of energy commonly used to measure electricity consumption. It represents the amount of energy consumed by a 1-kilowatt (kW) electrical appliance running for one hour. In other words, it is a measure of the total energy expended over the course of one hour at a rate of 1 kilowatt.

The formula to calculate energy consumption (in kilowatt-hours) is:

- Energy (kWh)=Power (kW)×Time (hours)

- Energy (kWh)=Power (kW)×Time (hours)

For example, if a 100-watt light bulb is used for 10 hours, the energy consumption would be:

- Energy (kWh)=0.1 kW×10 hours=1 kWh

- Energy (kWh)=0.1kW×10hours=1kWh

Kilowatt-hours are commonly used on electricity bills to measure and charge for the amount of energy consumed by households, businesses, and other entities.

Furthermore, Gangaja et al. (2021) estimate the generation of spent lithium-ion batteries, emphasizing the importance of recycling and regeneration. This information is relevant as it underscores the significance of recycling in reducing the overall energy demand for lithium production. Additionally, (Liao-ji et al., 2022) provide insights into the accumulation of lithium in new energy vehicles, indicating a specific quantity of lithium accumulated. This information is valuable for understanding the demand for lithium in the context of new energy vehicle production. Moreover, (Gallagher et al., 2014) discuss the promise of lithium–air batteries for electric vehicles, highlighting the high theoretical specific energy and energy density. While this reference does not directly provide energy requirements for lithium production, it offers insights into the potential energy benefits associated with lithium-based batteries. In summary, to calculate the energy required to produce 85,000 to 100,000 metric tons of lithium, it is crucial to consider the specific energy requirements for lithium production, the demand from new energy vehicle industries, and the potential energy benefits associated with lithium-based batteries. However using a basis of 35 kWh per kilogram of lithium, as noted by (Mashtalir et al., 2018), 2,975,000,000 kWh, or 2,975,000 MWh.

The amount of energy in 2,975,000 megawatt-hours (MWh) is a significant quantity that can be comprehended through various examples and contexts. To put this amount of energy into perspective, it is essential to consider its implications in different sectors such as electricity generation, industrial usage, and environmental impact. Firstly, to understand the magnitude of 2,975,000 MWh, it is crucial to compare it to the energy consumption of a specific region or country. For instance, according to the U.S. Energy Information Administration (EIA), the total annual electricity consumption in the United States in 2020 was approximately 3,801,000 MWh. Therefore, 2,975,000 MWh is equivalent to a substantial portion of the annual electricity consumption of the entire United States. This comparison highlights the immense scale of 2,975,000 MWh and its significance in meeting the energy needs of a large population.

Furthermore, in the context of industrial usage, 2,975,000 MWh of energy can be exemplified by its application in the manufacturing sector. For instance, the production of steel is an energy-intensive process, and the World Steel Association reports that the global steel industry consumed approximately 2,300 terawatt-hours (TWh) of energy in 2019. By converting TWh to MWh (1 TWh = 1,000,000 MWh), it can be inferred that 2,975,000 MWh is equivalent to the energy required for the production of a significant portion of the world's steel output. This comparison underscores the substantial energy demand of industrial processes and the role of 2,975,000 MWh in supporting industrial production on a global scale.

Moreover, in the context of environmental impact, 2,975,000 MWh of energy can be related to the generation of electricity from renewable sources. For example, according to the International Energy Agency (IEA), the average annual electricity generation from solar photovoltaic (PV) power plants in 2020 was approximately 3,000 TWh globally. By converting TWh to MWh, it can be deduced that 2,975,000 MWh represents a significant contribution to the annual electricity generation from solar PV. This correlation emphasizes the potential of 2,975,000 MWh to mitigate greenhouse gas emissions and reduce reliance on fossil fuels through the expansion of renewable energy capacity.

In addition, the context of transportation can also provide insight into the magnitude of 2,975,000 MWh. For instance, the U.S. Environmental Protection Agency (EPA) estimates that the average energy consumption of an electric car is 0.3 MWh per 100 miles (160 km). Therefore, 2,975,000 MWh would be adequate to power the average electric cars for a substantial distance, highlighting the potential of this energy quantity to support the transition towards electrified transportation and reduce reliance on conventional internal combustion engine vehicles.

Furthermore, in the context of global energy production, 2,975,000 MWh can be compared to the annual electricity generation of specific countries. For example, according to the International Energy Agency (IEA), the total annual electricity generation in India in 2020 was approximately 1,400,000 MWh. Therefore, 2,975,000 MWh exceeds the annual electricity generation of India, signifying its substantial magnitude in the global energy landscape.

The amount of energy in 2,975,000 MWh is a significant quantity that can be comprehended through various examples and contexts. By comparing it to the electricity consumption of a country, industrial energy usage, renewable electricity generation, transportation, and global energy production, the magnitude and implications of 2,975,000

MWh become evident. This comprehensive analysis underscores the substantial impact of 2,975,000 MWh in meeting energy demands, supporting industrial processes, mitigating environmental impact, and contributing to the global energy landscape.

For further context, to calculate the amount of coal required to produce 2,975,000 MWh of energy, we can use the material intensity metrics provided by (Demirel, 2018). According to their findings, the integrated clean-coal technology power plant with methanol production requires 0.52 MT of coal to generate 1 MWh of electricity. Therefore, to produce 2,975,000 MWh of energy, the amount of coal required can be calculated as: 2,975,000 MWh * 0.52 MT/MWh = 1,546,000 MT of coal

To comprehend the significance of 1,546,000 metric tons (MT) of coal, it is essential to consider various contexts that illustrate the scale of this amount. The energy content of coal, averaging approximately 24 million British thermal units (MMBtu) per ton, implies that this quantity of coal holds a total energy of roughly 37,104,000 MMBtu (Cao et al., 2005). This substantial energy reserve could be utilized for electricity generation in coal-fired power plants. In terms of residential heating, with 1,546,000 MT of coal, there is potential to theoretically heat tens of thousands of homes for a year (Bond et al., 2002). Furthermore, coal is a vital component in steel production, and with this amount of coal, a significant contribution to steel manufacturing could be made (Lin et al., 2022). However, it is crucial to note that the specific impact and applications of this amount of coal depend on factors such as the type of coal, its energy content, and the efficiency of the processes in which it is utilized (Klein et al., 2018).

Burning coal releases carbon dioxide (CO_2) into the atmosphere, with an average emission of approximately 2.4 metric tons of CO_2 per ton of coal. Consequently, 1,546,000 MT of coal would correspond to a substantial amount of CO_2 emissions (Huang et al., 2014). On a global scale, 1,546,000 MT of coal represents a substantial quantity in the context of international coal trade, where coal is a major commodity (Y. Li et al., 2022). The impact of this quantity of coal extends to carbon emissions, residential heating, steel production, and global coal trade, highlighting its significance in various sectors.

The annual cost of lithium production varies depending on the source and extraction method. Vikström et al. (2013) highlight the energy-intensive nature of lithium production, with theoretical calculations suggesting that producing a volume of lithium comparable to present world production would require a significant amount of electrical energy. On the other hand, (Warren, 2021) reports an estimated lithium production cost of $3,845/mt LCE using ion-imprinted polymer sorbents and synthetic Salton Sea brine. Additionally, Wagh et al. (2020) mention that the production cost of lithium from salt brine is less than half of the cost from ores or spodumene. These references provide insights into the energy and cost considerations associated with lithium production. Furthermore, Bardi (2010) discusses the potential of extracting lithium from seawater, which could provide a significant amount of the metal for economic use. This alternative source could impact the annual cost of lithium production if successfully implemented. Additionally, Ahmed et al. (2016) emphasize the importance of manufacturing process optimization to limit the cost of the final lithium-ion battery product.

Cobalt

Cobalt is primarily produced as a byproduct of copper and nickel extraction (Tisserant & Pauliuk, 2016). The extraction and production of cobalt from ores are closely related to the mining and processing of other metals, such as copper and nickel (Harper et al., 2011). The majority of the world's cobalt is mined in the Congo, with 60% of its production being processed in China (Boisvert et al., 2020). Cobalt mine production mainly occurs as a byproduct of the mining and processing of other metals, and it may be found in the residues of metallurgical processing, such as tailings and slag (Harper et al., 2011). The hard-rock mining at the Cobalt camps produced a substantial amount of waste rock and tailings that were rich in arsenic, cobalt, nickel, and mercury (Sivarajah et al., 2021).

The production of cobalt involves various processes such as leaching, solvent extraction, and precipitation. For instance, the recovery of cobalt from the Boleo deposit involves leaching, solvent extraction, and electrowinning processes (Dreisinger et al., 2012). Additionally, the selective acid leach of nickel from mixed nickel-cobalt hydroxide yields a concentrated nickel sulphate solution suitable for direct recovery of a final nickel product and a cobalt concentrate which can be processed further (Byrne et al., 2017).

Cobalt mining in the Democratic Republic of Congo (DRC) involves both industrial and artisanal mining practices, see Figure 10. The DRC is a major global producer of cobalt, and the mining operations are often concentrated in the southern part of the country, particularly in the Katanga region.

Figure 10: Artisan Mining in Kailo Congo. Julien Harneis, CC BY-SA 2.0, via Wikimedia Commons.

Many cobalt mines in Congo use open-pit mining techniques. This involves the extraction of ore from the surface, and it is commonly employed when the cobalt deposits are relatively shallow. The extracted ore is then transported to processing facilities. Underground Mining: Some cobalt deposits are deeper, requiring underground mining. This involves creating tunnels and shafts to access the ore. Underground mining is more complex and expensive than open-pit mining but may be necessary for deeper deposits.

In addition to large-scale industrial mining, there is a significant presence of artisanal miners in Congo. Artisanal miners often work in smaller operations, using manual techniques to extract cobalt-rich ore. This can involve digging shallow pits or tunnels. Hand Sorting: Artisanal miners manually sort and process the extracted material, separating cobalt ore from other rocks and minerals. This process is less mechanized compared to industrial mining.

Once the cobalt-rich ore is extracted, it undergoes a series of processing steps to isolate and refine the cobalt. This typically involves crushing, milling, and flotation processes to concentrate the cobalt minerals. The final stage of processing involves smelting, where the concentrated cobalt ore is heated to high temperatures to extract pure cobalt metal.

Cobalt mining in Congo has faced challenges related to environmental impact, safety concerns, and issues surrounding labour practices. Additionally, there have been ethical concerns regarding the use of child labour in some artisanal mining operations. Efforts have been made to address these challenges, including increased regulations, responsible sourcing initiatives, and transparency measures within the supply chain.

It's important to note that the cobalt supply chain has been a subject of scrutiny due to social and environmental concerns, prompting industry initiatives and increased attention to responsible sourcing practices.

Furthermore, the environmental impact of cobalt exploitation from slag in mines has been assessed, considering scenarios where the converter slag goes through a sulfuric acid roasting process with leaching for cobalt recovery (Merlo et al., 2021). The environmental legacy of the Cobalt mining camp represents an area that has been extremely impacted by arsenic-rich mine waste (Sprague, 2017).

The human cost of mining cobalt in Congo is a multifaceted issue that encompasses various aspects such as health, livelihoods, and social dynamics. The mining of cobalt in the Democratic Republic of the Congo (DRC) has led to widespread environmental contamination, raising concerns about its association with adverse health effects, including birth defects (Brusselen et al., 2020). Additionally, artisanal and small-scale mining (ASM) in the DRC provides a source of livelihood for a significant portion of the population, including millions of miners, and their families (Spira et al., 2017). This indicates that the mining industry plays a crucial role in the socio-economic landscape of the region.

Furthermore, the mining activities have resulted in high dust ingestion estimates for the general population in the mining districts of the DRC, particularly in the cobalt mining area of Lubumbashi (Smolders et al., 2019). This suggests that the local population is at risk of exposure to hazardous substances associated with mining activities. Moreover, the impacts of trace metal pollution from mining activities on population health have been highlighted, emphasizing the need for stringent monitoring of mining exploitation to protect the environment and ensure population well-being (Muimba-Kankolongo et al., 2022).

The involvement of children in the artisanal and small-scale mining sector has also been noted, indicating the complex dynamics of family involvement and the potential exploitation of children in mining activities (André & Godin, 2013). This raises concerns about the well-being and safety of children engaged in such labour.

In addition to health and social implications, the environmental impact of cobalt mining is evident, as seen in the contamination of lakes and surrounding areas with mine tailings and associated pollutants (Little et al., 2020; Nkongolo et al., 2023). This environmental contamination can have far-reaching consequences for the ecological balance and the well-being of communities relying on these natural resources. The scale of cobalt mining in the DRC is substantial, with the region producing a significant portion of the global cobalt supply (Furberg et al., 2018; Landrigan et al., 2022; Secrist & Fehring, 2022). This underscores the magnitude of the impact that the mining industry can have on the local population and the environment.

To determine the annual cobalt requirement for lithium-ion battery production, it is essential to consider the cobalt content in these batteries. Larcher and Jm (2014) highlighted that in today's lithium-ion batteries for portable electronics, cobalt content ranges between about 5 and 15 wt%. The term "wt%" refers to weight percent, which is a measure of the concentration of a substance in a mixture. Weight percent is calculated as the weight of a component divided by the total weight of the mixture, multiplied by 100 to express the result as a percentage.

According to Alonso et al. (2022), the total cobalt requirement in the United States for lithium battery production was approximately 28,700 metric tons, which is significantly larger than the reported raw material consumption value. This indicates a substantial demand for cobalt in the production of lithium batteries. Furthermore, Nkulu et al. (2018) highlight that over 50% of the world's current cobalt production is utilized in rechargeable batteries for smartphones, laptop computers, and electric vehicles. This emphasizes the significant role of cobalt in the battery industry and the high demand for this material in lithium battery production.

In addition, Muralidharan et al. (2022) discuss the goal of reducing the cobalt content in battery cathodes to less than 50 mg Wh−1 at the cell level, indicating efforts to minimize the usage of cobalt in battery manufacturing. In simpler terms, the statement is expressing an objective to minimize the amount of cobalt used in the cathodes of lithium-ion battery cells. The goal is to achieve a cobalt content of less than 50 milligrams per watt-hour of energy storage at the individual cell level. This reduction in cobalt content is often pursued for various reasons, including cost reduction, environmental concerns, and addressing ethical issues associated with cobalt mining. It also aligns with efforts to improve the energy density and sustainability of battery technologies. This suggests a trend towards reducing the reliance on cobalt in lithium battery production.

Moreover, point out the supply risks associated with high geopolitical concentrations of cobalt and the social and environmental impacts linked to cobalt mining, highlighting the challenges and considerations related to cobalt procurement for battery production (C. Xu et al., 2020). Considering these references, it is evident that cobalt plays a crucial role in lithium battery production, with substantial demand and supply chain considerations. Efforts are being made to reduce the reliance on cobalt in battery manufacturing, but its significance in the industry remains prominent.

To calculate the energy required to mine 28,700 metric tons of cobalt for the US market requirements alone, we need to consider various factors such as the energy consumption in mining operations, the recovery process, and the overall environmental impact. Cobalt is a critical component in renewable energy technologies, and its demand is expected to increase substantially by 2050 (Meide et al., 2022). The production of cobalt is hindered by high characterization costs (Crawford et al., 2022), and it is mainly obtained as a by-product in nickel and copper mining activities (Heidrich et al., 2022). The Boleo project, for example, is expected to produce approximately 1708 tonnes per annum of cobalt metal (Dreisinger et al., 2012).

The energy consumption in mining operations is a significant factor. For example, in the energy balance of a coal mine, up to 70-80% falls to electrical energy (Voronin et al., 2021). Additionally, the fuel consumption of mining dump trucks accounts for about 30% of total energy use in surface mines (Sahoo et al., 2014). Furthermore, the recovery of cobalt from the Boleo deposit involves leach, solvent extraction, and electrowinning processes (Dreisinger et al., 2012). These processes require significant energy inputs.

Moreover, the environmental impacts of key metals' supply and low-carbon technologies are likely to decrease in the future due to technological innovation, which may lead to increased energy efficiencies (Harpprecht et al., 2021). However, the recovery of cobalt from electronic waste and natural resources usually involves applying pyrometallurgical or hydrometallurgical processes, which also consume substantial energy (Patrício et al., 2011).

In terms of the specific energy consumption for mining, ALROSA and DeBeers use about 96 kWh/ct and 150 kWh/ct, respectively, including all energy required to mine (Zhdanov et al., 2021). This data provides insight into the energy intensity of mining activities for different companies.

To calculate the energy required to produce cobalt, you can use the given energy requirement per cubic tonne and multiply it by the total quantity of cobalt to be produced.

Given: Energy required per cubic tonne = 96 kWh Quantity of cobalt to be produced = 28,700 metric tons

First, convert metric tons to cubic tonnes. Assuming the density of cobalt is approximately 8.9 g/cm³ (which is the density of cobalt metal), you can use the conversion factor 1 metric ton = 1,000 kg to find the volume in cubic meters (m³).

$$\text{Volume} = \frac{\text{Mass}}{\text{Density}}$$
$$\text{Volume} = \frac{28,700 \text{ metric tons} \times 1,000 \text{ kg/metric ton}}{8.9 \text{ g/cm}^3 \times 1,000 \text{ g/kg}}$$

Now, you have the volume in cubic meters. Multiply this by the energy requirement per cubic tonne to get the total energy:

Energy=Volume×Energy requirement per cubic tonne

Energy=(Volume×Energy requirement per cubic tonne)×1,000

Now you can plug in the values and calculate:

$$\text{Energy} = \left(\frac{28,700 \text{ metric tons} \times 1,000 \text{ kg/metric ton}}{8.9 \text{ g/cm}^3 \times 1,000 \text{ g/kg}} \times 96 \text{ kWh/cubic tonne} \right) \times 1,000$$

Energy≈3.29×10⁹kWh

Therefore, it would take approximately 3.29 billion kilowatt-hours (or 3,290,000 mWh) of energy to produce 28,700 metric tons of cobalt with an energy requirement of 96 kWh per cubic tonne.

Nickel

Nickel is an important component in the cathodes of lithium-ion batteries (LIBs), serving as one of the two electrodes that influence the battery's overall performance. The cathode material plays a significant role in determining the characteristics of the battery, and various types of cathode materials incorporate nickel, contributing to diverse battery performances.

One prevalent cathode chemistry is the Nickel-Cobalt-Manganese (NCM) system, where nickel, cobalt, and manganese are combined in different proportions. For instance, NCM 111 denotes a cathode material with equal parts of nickel, cobalt, and manganese in a 1:1:1 ratio.

Another widely used cathode chemistry is Nickel-Cobalt-Aluminium (NCA), where nickel and cobalt are the primary metals, and aluminium serves as a stabilizing element. This particular chemistry finds common application in electric vehicle batteries, including certain Tesla models.

Nickel-Manganese-Cobalt (NMC) cathodes, which blend nickel, manganese, and cobalt, strike a balance between energy density, power capability, and cost. Different formulations, such as NMC 111, NMC 532, and NMC 622, signify variations in the ratio of nickel, manganese, and cobalt.

A noticeable trend involves developing cathodes with higher nickel content, represented by formulations like NMC 811 and beyond. This shift aims to enhance energy density while reducing cobalt content, considering cost considerations and ethical considerations associated with cobalt mining.

Some advanced battery designs explore the use of pure nickel cathodes, known as Nickel/Nickel-Lithium (Ni/Ni-Li) batteries. These designs prioritize maximizing energy density and reducing dependency on cobalt.

Nickel is favoured for its high energy density and effective storage and release of electrical energy. The choice of cathode chemistry, especially regarding nickel content, directly impacts factors such as energy density, power density, cycle life, and cost. It is essential to acknowledge that selecting cathode materials involves trade-offs between these factors, and ongoing research is focused on optimizing LIBs for diverse applications.

Nickel is mined extensively across the globe, with major production hubs situated in various countries on different continents. Here are some key contributors to the world's nickel production:

- Indonesia: Indonesia stands out as the leading global producer of nickel, boasting extensive nickel reserves. The country predominantly extracts nickel from laterite ores, solidifying its pivotal role in the nickel market.

- Philippines: The Philippines is a major player in the nickel production landscape, particularly renowned for its expertise in extracting nickel laterite ores, contributing significantly to the global supply.

- Russia: Russia possesses substantial nickel reserves and serves as a major producer, focusing primarily on the extraction of nickel from sulphide ores, thereby playing a key role in the global nickel market.

- Canada: Canada holds a noteworthy position in nickel production, with abundant deposits in provinces such as Ontario and Quebec. Canadian nickel extraction encompasses both sulphide and laterite ores.

- Australia: Australia, with considerable nickel resources, conducts mining operations across various states, including Western Australia, further contributing to the global nickel supply.

- New Caledonia: This French territory located in the Pacific Ocean is recognized for its sizable nickel deposits. Nickel mining, especially from laterite ores, forms a vital component of New Caledonia's economic activities.

- Brazil: Brazil plays a significant role in the global nickel market, primarily deriving nickel from lateritic deposits. The country's substantial nickel reserves contribute to its prominent position in the industry.

- Cuba: Cuba is a notable nickel producer within the Caribbean region, extracting nickel from both sulphide and laterite ores, thereby making a valuable contribution to the global nickel supply.

- China: China emerges as a major player in global nickel production, possessing significant deposits of both sulphide and laterite ores, further solidifying its impact on the global nickel market.

- Norway: Norway boasts active nickel mines and is recognized for its production of nickel from sulphide ores, making a distinctive mark in the global nickel industry.

These countries collectively constitute a substantial portion of the global nickel supply, and the distribution of nickel production is influenced by the geological characteristics of nickel deposits in various regions. The mining techniques employed vary based on the specific type and location of nickel ore.

Figure 11: Haul Truck used to transport mining material in the nickel mining of PT. Vale Indonesia in Sorowako, East Luwu, South Sulawesi Indonesia. Mirwanto Muda, CC BY-SA 4.0, via Wikimedia Commons.

Nickel is typically produced through two main processes: pyrometallurgical (smelting) and hydrometallurgical processes. The choice between these processes depends on the type of ore, its composition, and the desired nickel content. The pyrometallurgical process is commonly used for nickel sulphide ore, where the ore is concentrated, roasted to convert nickel sulphides into an oxide, and then smelted with reducing agents to produce nickel metal (X. Wang et al., 2017). On the other hand, the hydrometallurgical process is often preferred for lateritic ores, which are rich in iron and nickel but low in sulphur. This process involves leaching the ore with sulfuric acid or ammonia under high-pressure conditions, separating the leach solution from the solid residues, and selectively separating nickel and cobalt from the solution using techniques such as precipitation and solvent extraction (Kaya et al., 2017).

Regardless of the initial process used, the nickel obtained is usually not in its pure form and requires further refining steps to achieve high purity. Techniques such as solvent extraction, precipitation, and electrorefining are employed to purify the nickel to meet the required specifications (Han et al., 2019). Nickel production involves various refining and purification steps to obtain high-purity nickel suitable for different applications. The choice of process depends on the characteristics of the ore and the specific requirements of the final nickel product (Rizky et al., 2023).

Advancements and variations in production methods may occur over time as technology evolves. For instance, recent studies have focused on the mechanical properties of carbon fibre reinforced nanocrystalline nickel composite electroforming deposits, showing significant increases in microhardness and tensile strength when the grains of the composite were refined (Qian et al., 2019). Additionally, the effect of sulfuric acid concentration on nickel recovery from laterite ore by using atmospheric acid leaching method has been investigated, concluding that precipitation reduces nickel recovery (Muntaqin et al., 2022).

To determine the annual nickel requirement for lithium-ion battery production, we can draw from several reputable sources. Golroudbary et al. (2023) projected a 32% compound annual growth rate in demand for nickel from electric vehicle (EV) batteries, driving up nickel consumption in rechargeable batteries to 24% annually to 1.27 million tonnes by 2030. Mentus (2021)estimated that 400,000-800,000 tons of nickel would be required annually, representing 20-40% of the metal currently used. These figures provide a comprehensive understanding of the substantial and increasing demand for nickel in lithium-ion battery production.

Furthermore, highlighted that lithium-ion batteries typically consist of 5-10% nickel, indicating the significant reliance on nickel for their production (Nurqomariah & Fajaryanto, 2018b). This emphasizes the substantial annual requirement for nickel in the manufacturing of lithium-ion batteries. In addition, Wessells et al. (2011) discussed the potential of nickel hexacyanoferrate nanoparticle electrodes for use in aqueous sodium and potassium ion batteries, demonstrating the versatility of nickel in various battery technologies. This further underscores the importance of nickel in battery production, including lithium-ion batteries.

To estimate the energy required to mine 800,000 tons of nickel, it is essential to consider the energy consumption associated with nickel mining and processing. According to Mudd and Jowitt (2014), the energy use for nickel mining and processing ranges from about 30 to 600 GJ/t nickel (Mervine et al., 2023). This wide range indicates the variability in energy requirements based on different mining and processing methods. Additionally, highlighted that the mining and smelting of primary nickel raise concerns due to energy consumption, carbon dioxide emissions, and waste generation (Reck et al., 2008). These factors contribute significantly to the overall energy demand for nickel production.

Furthermore, the conventional approach of nickel mining commonly requires ores with a minimum grade level of about 30,000 mg/kg to be economically feasible (Akinbile et al., 2021). This indicates that the energy required for mining and processing nickel is influenced by the grade of the ore. Additionally, the production of waste rock from nickel mining contributes to the overall energy consumption and environmental impact. 's research focused on estimating the carbon sequestration potential, with a specific focus on waste rock production from nickel mining (Renforth, 2019). This highlights the interconnectedness of mining activities and their environmental implications, including energy usage.

Moreover, the energy consumption of mining dump trucks accounts for about 30% of the total energy use in surface mines (Sahoo et al., 2014). This emphasizes the significance of transportation-related energy consumption in the overall energy requirements for mining operations. Additionally, the global transition away from fossil energy will result in increased production of tailings (i.e., wastes) from the mining of nickel and platinum group metals (PGMs) (Woodall et al., 2021). This indicates that the energy demands associated with mining activities are subject to broader global energy transitions and their impacts on mining waste production.

To calculate the energy required to mine 800,000 tonnes of nickel at 30 GJ/t, you can use the following formula:

Energy=Mining Rate × Energy Intensity

Energy=Mining Rate × Energy Intensity

Where:

Mining Rate is the quantity of nickel mined (800,000 tonnes),

Energy Intensity is the energy intensity per ton of nickel (30 GJ/t).

The calculation:

Energy=800,000 tonnes×30 GJ/t

Energy=24,000,000 GJ

So, the energy required to mine 800,000 tonnes of nickel at 30 GJ/t is 24,000,000 GJ.

Kilowatt-hours (kWh) and gigajoules (GJ) are units of energy, but they belong to different measurement systems. The relationship between them involves conversion factors based on the energy content of different fuels.

1 kilowatt-hour (kWh) is equal to 3.6 gigajoules (GJ). This conversion is based on the fact that 1 kilowatt-hour is equivalent to 3.6 million joules (1 kWh = 3.6 x 10^6 J), and 1 joule is equal to 1/1,000,000,000 gigajoules (1 J = 1 x 10^-9 GJ). Therefore:

1kWh=3.6GJ

This conversion factor is useful when comparing energy quantities or when dealing with energy consumption and production on different scales. It's important to note that the specific energy content can vary depending on the source or type of energy (e.g., coal, natural gas, electricity), so the conversion factor may vary accordingly.

To convert gigajoules (GJ) to kilowatt-hours (kWh), you can use the conversion factor that 1 gigajoule is equivalent to 277.77778 kilowatt-hours. Therefore:

Energy in kWh=Energy in GJ×277.77778Energy in kWh=Energy in GJ×277.77778

Calculating it:

Energy in kWh=24,000,000 GJ×277.77778 kWh/GJ

Energy in kWh≈6,666,666,720 kWh

So, 24,000,000 gigajoules is approximately equivalent to 6,666,666,720 kilowatt-hours.

To determine the daily energy requirement from the annual energy requirement, you can divide the annual requirement by the number of days in a year. Assuming a standard year with 365 days, the daily energy requirement (D) can be calculated using the formula:

$$D = \frac{\text{Annual Energy Requirement}}{\text{Number of Days}}$$

Substitute in the values:

$$D = \frac{6,666,666,720 \text{ kWh}}{365 \text{ days}}$$

D≈18,273,287.07 kWh/day

Therefore, approximately 18,273,287.07 kilowatt-hours are required daily. Keep in mind that this is a simplified calculation, and actual daily energy consumption patterns may vary.

For context, to generate the equivalent required energy, to mine 800,000 tonnes of nickel, the number of required solar panels will be considered.

The number of solar panels required to produce a certain amount of energy depends on various factors, including the efficiency of the solar panels, the amount of sunlight in the location, and the time over which the energy is generated.

Assuming an average efficiency of 20% for solar panels, you can use the following formula to estimate the number of solar panels needed:

$$\text{Number of Panels} = \frac{\text{Total Energy Requirement}}{\text{Panel Efficiency} \times \text{Sunlight Hours}}$$

Let's assume an average of 4 sunlight hours per day (this can vary based on location and other factors):

$$\text{Number of Panels} = \frac{18,273,287.07 \, \text{kWh}}{0.20 \times 4 \, \text{hours/day}}$$

Number of Panels≈4,568,321.77

Therefore, approximately 4.57 million solar panels (rounded up) would be needed to generate 18,273,287.07 kilowatt-hours per day, assuming an average efficiency of 20% and 4 hours of sunlight per day. Again, keeping in mind that this is a simplified calculation, and actual requirements may vary based on local conditions and specific solar panel characteristics.

Further, and again for context and perception of scale, the number of homes that can be powered by a given quantity of solar panels depends on several factors, including the capacity of the solar panels, the average electricity consumption per home, and the local sunlight conditions. Additionally, the efficiency of the solar panels and the geographic location are significant considerations.

Assuming an average residential solar panel capacity of around 300 watts per panel, we can make a rough estimate. Keep in mind that this is a simplified calculation and actual results may vary.

Let's assume an average daily sunlight exposure of around 4 hours (a common estimate).

Total daily energy production per solar panel: 300 W/panel×4 hours=1200 Wh/panel=1.2 kWh/panel

Total daily energy production for 4,568,321.774,568,321.77 solar panels: 1.2 kWh/panel×4,568,321.77 panels=5,481,986.12 kWh

Assuming an average annual home electricity consumption of around 10,000 kWh, the number of homes powered by 5,481,986.125,481,986.12 kWh would be:

$$\frac{5,481,986.12 \, \text{kWh}}{10,000 \, \text{kWh/home}} = 548 \, \text{homes}$$

So, roughly 4,568,321.774,568,321.77 solar panels could power approximately 548 homes, given these assumptions. Actual results may vary based on local conditions, panel efficiency, and other factors.

Graphite

Graphite plays a significant role in lithium-ion batteries (LIBs) as the primary material used in the anode, one of the battery's electrodes. This anode material significantly influences the overall performance and characteristics of the battery. In LIBs, graphite is employed in various ways to optimize battery functionality.

Graphite is mined in various locations around the world, and several countries play a significant role in graphite production.

China is the largest producer of graphite globally, contributing substantially to the world's total graphite production. India is another major graphite producer, boasting notable graphite mines that contribute to the global supply. In Brazil, significant graphite deposits are present, and the country actively contributes to the global graphite market. Canada is recognized for its graphite production, with mines situated in provinces like Quebec and Ontario. Mozambique is emerging as a notable graphite producer, with new mining projects enhancing its role in the global graphite supply chain. Madagascar holds substantial graphite reserves and actively participates in graphite mining operations. Norway is known for its graphite mining activities, particularly in the southern part of the country. Russia also plays a role in the global graphite market, with graphite mines contributing to the overall supply.

It's crucial to note that the distribution of graphite mines depends on geological factors, and the emergence of new mining projects is a possibility over time. Additionally, the quality and characteristics of graphite can vary among different sources, influencing its applications and desirability in various industries.

Graphite, with its high theoretical specific reversible charge, is an attractive candidate for the negative electrodes of lithium-ion batteries (Buqa et al., 2005). The intercalation of lithium into graphite is a fundamental process in the operation of lithium-ion batteries, making graphite one of the most studied and commercially important anode materials (Qian et al., 2016). Additionally, natural graphite is considered a high-quality raw material for the negative electrode of lithium-ion batteries (S. Yang et al., 2022).

The anode of LIBs commonly features graphite as its primary component. This graphite is often derived from natural sources or produced synthetically through processes like the graphitization of carbon materials. Natural graphite is typically mined, while synthetic graphite is manufactured to meet specific requirements.

The structure of the anode involves applying a thin layer of graphite particles onto a copper foil. This graphite structure facilitates the reversible intercalation of lithium ions during the charging and discharging cycles of the battery. The ability of graphite to efficiently host lithium ions is fundamental to the proper functioning of lithium-ion batteries.

During the charging process, lithium ions are inserted or intercalated into the graphite structure of the anode. Conversely, during discharging, these lithium ions are released from the graphite structure. This dynamic interaction between graphite and lithium ions is at the core of the energy storage mechanism in lithium-ion batteries.

Cycling performance, specifically the stability of the graphite structure during repeated charge and discharge cycles, is a critical factor. Graphite's capability to maintain structural integrity contributes to the overall cycling stability of the battery, ensuring its longevity and reliable performance over time.

Graphite anodes are recognized for their high coulombic efficiency. This term refers to the capacity of the anode to efficiently store and release a significant percentage of lithium ions during each charging and discharging cycle. The high coulombic efficiency of graphite anodes enhances the overall efficiency and performance of lithium-ion batteries.

Graphite anodes are preferred for their capacity to provide a stable and reliable platform for lithium-ion intercalation. Researchers continue to explore and develop advanced anode materials, including graphite, to further optimize battery performance and contribute to ongoing advancements in battery technology.

Graphite is obtained through various methods, depending on the geological characteristics of the graphite deposit. The two main types of graphite deposits are natural graphite and synthetic graphite. Natural graphite is typically mined using open-pit mining methods, which involve removing overburden to expose the graphite-bearing ore, followed by extraction through drilling, blasting, and transportation to the processing plant. In cases where the graphite deposit is located at deeper levels, underground mining methods may be employed, involving the creation of tunnels and shafts to access the graphite ore. After extraction, the graphite ore undergoes a milling process to break it down into a fine powder, followed by a flotation process to separate graphite particles. The concentrate obtained from flotation may undergo additional processes to further concentrate the graphite content, and then it is dried and purified before sizing and packaging for distribution to end-users (Lee et al., 2015).

Synthetic graphite, on the other hand, is produced from carbon precursors through processes like graphitization, which does not involve traditional mining methods. Instead, it is manufactured using high-temperature processes to transform carbonaceous materials into crystalline graphite structures. The choice between natural and synthetic graphite depends on the specific requirements of the applications and the properties desired in the final product (Kulkarni et al., 2022).

The mining and processing of graphite have significant environmental and economic implications. A life cycle assessment for graphite production provides valuable insights into the environmental impact of both natural and synthetic graphite manufacturing processes. It highlights the energy-intensive nature of synthetic graphite production, which involves high temperatures and prolonged periods of time, contributing to its environmental footprint. Additionally, the assessment emphasizes the traditional methods of producing battery-grade graphite, which include processing naturally mined graphite or manufacturing synthetic graphite via the Acheson process, involving high temperatures for prolonged periods of time (Surovtseva et al., 2022).

Furthermore, the geographical distribution of graphite mines is an important aspect to consider. For instance, Sri Lanka is known for its highly crystalline vein graphite mined in large areas in the Central Highlands, with specific mines such as Kahatagaha, Kolongaha, and Bogala being the main sources of graphite in the region. Studies have been conducted to explore and characterize these vein graphite deposits, providing valuable insights into their geochemical, structural, and morphological characteristics (Hewathilake et al., 2018; Wickramasinghe et al., 2018).

In addition to traditional mining and processing methods, research has focused on innovative approaches to synthesize and utilize graphite-based materials. For example, studies have explored the synthesis of graphene from natural microcrystalline graphite minerals and its potential applications in energy storage. The scalable synthesis of graphene microsheets from natural microcrystalline graphite minerals has shown unusual energy storage capabilities, highlighting the potential for advanced energy materials derived from natural graphite sources (Wang et al., 2015).

By 2030, the demand for graphite in batteries is expected to surpass more than 5 times the amount mined in 2021 (Lasley, 2023). While lithium and nickel shortages for electric vehicle batteries have received significant attention, the demand for graphite, a crucial component in the anode of lithium-ion batteries, has been overlooked. According to experts, a megafactory producing 30 gigawatt-hours of battery storage annually requires about 33,000 metric tons of graphite (Lasley, 2023). With over 300 gigafactories in the pipeline, this would translate to a potential demand of up to 9.9 million metric tons of graphite annually (Lasley, 2023). Even considering a more realistic scenario, where 70% of these battery plants operate at an average of 70% design capacity, the global lithium battery sector would still require about 4.9 million metric tons of graphite per year. This contrasts with the approximately 1 million metric tons mined globally in 2021 to meet the demands of all industrial sectors. Experts predict a significant challenge for global miners in ramping up graphite production by 500 to 600% over the next decade, with concerns that supply may struggle to keep up with the increasing demand for graphite.

To calculate the energy required to mine 4.9 million metric tons of graphite, we need to consider the energy consumption involved in the production of graphite. Manufacturing 1 metric ton of graphite consumes over 4.0×10^4 MJ of energy, with over 10 metric tons of CO_2-eq emissions (S. Wang et al., 2023). This indicates that the energy consumption for mining 4.9 million metric tons of graphite would be substantial. Additionally, traditional artificial graphite production is known to have high energy consumption and high cost, making it a research hotspot to reduce the energy consumption of artificial graphite (J. Yang et al., 2022).

To convert megajoules (MJ) to megawatt-hours (mWh), you need to divide the energy value in megajoules by 3.6, as there are 3.6 megajoules in one kilowatt-hour (kWh), and there are 1,000 kilowatt-hours in one megawatt-hour.

$$4.0 \times 10^4 \, \text{MJ} = \frac{4.0 \times 10^4}{3.6 \times 1,000} \, \text{mWh}$$

Performing the calculation:

$$\frac{4.0 \times 10^4}{3.6 \times 1,000} = \frac{4.0 \times 10^4}{3,600} \, \text{mWh}$$

$$\frac{4.0 \times 10^4}{3,600} \approx 11.111 \, \text{mWh}$$

Therefore, $4.0 \times 104 4.0 \times 10^4$ megajoules is approximately equal to 11.111 megawatt-hours.

To calculate the requirements of production of 4.9 million metric tonnes of graphite at 11.111 megawatt-hours required per metric tonne, you can multiply these two values:

Energy in mWh=4.9 million×11.111 megawatt-hours

Energy in mWh =54,389,000 megawatt-hours

Therefore, the energy requirements to extract 4.9 million metric tonnes of graphite at 11.111 megawatt-hours per metric tonne is 54,389,000 megawatt-hours.

Graphite is widely used as an anode material in lithium-ion batteries due to its properties, but it faces a bottleneck due to its low theoretical capacity, which limits the energy output delivered for lithium-ion batteries (X. Wang et al., 2022). The low theoretical capacity of graphite as an anode material is a significant factor to consider when evaluating the energy requirements for mining graphite. Furthermore, the production of graphene nanoplatelets from graphite remains an energy- and time-intensive process, indicating the high energy demands associated with graphite processing (Awada et al., 2022).

In the context of energy applications, metal nanoparticles supported on graphite play central roles in promoting the utilization of green and renewable energy for a sustainable future (Pandey et al., 2019). This highlights the potential for graphite to contribute to energy storage and conversion technologies, further emphasizing the importance of understanding its energy consumption.

Moreover, the energy consumption of mining operations is a critical aspect to consider. For instance, up to 70-80% of the energy balance of a coal mine falls to electrical energy, indicating the significant energy demands of mining activities (Voronin et al., 2021). Fuel consumption of mining dump trucks accounts for about 30% of total energy use in surface mines, further underlining the substantial energy requirements of mining operations (Sahoo et al., 2014).

Manganese

Manganese is commonly used in cathodes to enhance the overall performance of lithium-ion batteries. It is often combined with lithium and other metals to form cathode materials, specifically lithium manganese oxide ($LiMn_2O_4$), which serves as an active material in the cathode and contributes to the reversible intercalation and deintercalation of lithium ions during the charging and discharging processes (Bak et al., 2014). This compound operates at a higher voltage compared to some other cathode materials, contributing to the overall voltage of the battery, which is advantageous for achieving higher energy density in lithium-ion batteries (Bak et al., 2014). Additionally, manganese-based cathodes offer good stability during cycling, contributing to the long-term performance and cycle life of the battery (Bak et al., 2014; Radin et al., 2019).

Manganese is often preferred for its cost-effectiveness compared to other transition metals used in cathodes, such as cobalt, making lithium-ion batteries more cost-competitive. Furthermore, manganese-based cathodes strike a balance between energy density and safety, offering reasonable energy density while exhibiting improved safety characteristics, reducing the risk of thermal runaway or safety incidents during battery operation.

Variations of cathode chemistries, such as NMC (Nickel Manganese Cobalt) cathodes, where manganese is combined with nickel and cobalt, aim to optimize specific aspects of battery performance, including energy density, power capability, and cost. Ongoing research and development in battery technology continue to explore ways to enhance the role of manganese and other materials for improved lithium-ion battery performance (T. Wang et al., 2020).

Manganese is mined in various countries around the world, with significant manganese-producing regions and mines located on different continents. South Africa stands out as a major producer of manganese, boasting some of the largest manganese mines globally. In China, another significant contributor to the global supply, manganese is extracted to meet the demand. Australia also plays a crucial role in the manganese market, possessing substantial manganese resources with mining operations scattered across various states.

Figure 12: An open pit operation of manganese at Nsuta, a town located in the Tark-wa-Nsuaem Municipal District of the Western Region in Ghana, West Africa. Asamoah Daniel Kwame Oware, CC BY-SA 4.0, via Wikimedia Commons.

Gabon, situated in Central Africa, is recognized for its manganese production and holds significant manganese reserves. Brazil is actively involved in manganese mining, operating several mines that contribute to the global manganese market. Ghana is gradually emerging as a manganese producer, with new mining projects underway to meet demand. India, with its manganese mines, actively participates in the global manganese supply chain. Ukraine is known for its manganese production and hosts manganese mines, further diversifying the global sources of this essential metal. The distribution of manganese mines is influenced by geological factors, and ongoing mining projects may continue to reshape the landscape of manganese production.

To determine the annual requirement of manganese for lithium-ion battery production, it is essential to consider the usage of manganese in the cathode materials of these batteries. Lithium-and manganese-rich nickel manganese cobalt oxides (LMR-NMCs) have been extensively studied as promising positive electrode (cathode) materials for rechargeable lithium-ion batteries due to their high initial specific capacities exceeding 280 mAh/g (Hendrickx et al., 2022). Additionally, lithium ion batteries utilizing manganese-based cathodes have received considerable interest in recent years for their lower cost and more favourable environmental friendliness relative to their cobalt counterparts (Vissers et al., 2016). Furthermore, the inclusion of manganese within the solid electrolyte interphase (SEI) layer of the graphite negative electrodes has been correlated with a substantial loss of 'cycleable' lithium (Gowda et al., 2014). These references highlight the significant role of manganese in lithium-ion battery technology.

The global demand for manganese production has been steadily increasing, driven primarily by the growth of the world steelmaking industry and the expanding applications of manganese in various sectors (Cao et al., 2017). This surge in demand is attributed to the concentration of global manganese production, with a few key mining companies dominating the market (S. Li et al., 2019). However, the demand for high-grade manganese ore has not been met, leading to a global shortage of manganese ferroalloys (Yi et al., 2017). China has emerged as the largest consumer of manganese resources, further contributing to the global demand for manganese (Sun et al., 2021).

The increasing demand for manganese has prompted the exploration and development of technologies to improve manganese beneficiation and production processes (Tastanova et al., 2021). Additionally, there is a growing focus on sustainable production strategies to ensure the efficient utilization of manganese resources and lower emissions (Digernes et al., 2018). The potential for sustainable manganese production has been highlighted, emphasizing the need for efficient material cycle systems to meet the demand for manganese (Nakajima et al., 2008).

In the transformed post-COVID-19 business environment, the global manganese market, which was estimated at 23.9 million metric tons in 2022, is anticipated to achieve a revised size of 39.3 million metric tons by 2030, with a compounded annual growth rate (CAGR) of 6.4% during the period from 2022 to 2030 (Reportlinker, 2023). Within the report, the Silicomanganese (SiMn) segment, upon analysis, is expected to exhibit a CAGR of 6.7% and attain a volume of 26.2 million metric tons by the conclusion of the analytical period. Considering the ongoing recovery post the pandemic, the growth projection for the Ferromanganese (FeMn) segment is adjusted to a revised CAGR of 5.3% for the succeeding 8-year period.

Talens Peiró and Villalba Méndez (2013) observed that the energy demand for beneficiation is contingent upon the specific mineral ore type. For instance, the production of 1 tonne of Rare Earth Metals (REM) from bastnäsite in Bayan Obo necessitates a minimal energy input of 12.06 GJ, while the same quantity of REM from Mountain

Pass requires 1.73 GJ. The most energy-intensive phases in REM production are the solvent extraction-based separation of each Rare Earth (RE) (15.60–22.70 GJ/tonne of REM) and the reduction of each RE, with energy consumption ranging from 38 to 48 GJ. For example, the electrolysis process for 1 tonne of REM from Mountain Pass, predominantly comprising cerium, lanthanum, and neodymium, as outlined in the supporting information, demands 47.34 GJ of energy. Overall, the mineral processing and reduction of 1 tonne of a specific REM exhibit an average energy intensity of 58.51 GJ, comparable to that of manganese (58 GJ/tonne) but higher than base metals such as iron (28 GJ/tonne) and lead (31 GJ/tonne).

Using an energy requirement basis of 58 GJ, to calculate the total energy required, you can use the following formula:

Total energy = Energy per tonne * Total production

Given the production rate of 58 GJ/tonne and the total production of 23.9 million metric tons:

Total energy = 58 GJ/tonne * 23,900,000 tonnes

Total energy = 1,384,200,000 GJ

This is the total energy required to produce 23.9 million metric tons of manganese at the given production rate.

To convert gigajoules (GJ) to megawatt-hours (MWh), you can use the conversion factor: 1 GJ = 277.778 MWh.

Total energy in MWh = Total energy in GJ * 277.778

Total energy in MWh = 1,384,200,000 GJ * 277.778 MWh/GJ

Total energy in MWh ≈ 384,435,600,000 MWh

Therefore, at a production rate of 58 GJ/tonne of manganese, approximately 384,435,600,000 megawatt-hours of energy would be required to produce 23.9 million metric tons of manganese.

According to the US Geological Survey (USGS), the predominant use of manganese ore, accounting for approximately 85-90%, is in the production of ferromanganese alloys. However, in the past decade, there has been a notable increase in its consumption for silicomanganese alloys compared to ferromanganese alloys (Moore Stephens, 2023). Around 4-5% of manganese is utilized in the production of hot metal, specifically in the steel production process through the blast furnace route. Another 5-10% finds application in various industries such as the manufacturing of dry cells (batteries) and chemicals. On this basis, to determine the energy used for battery production, we can calculate 5% of the total energy consumption for manganese production.

Total energy consumption for manganese production = 384,435,600,000 MWh Percentage used for battery production = 5%

Energy used for battery production = (5/100) * 384,435,600,000 MWh

Calculating this:

$$\text{Energy used for battery production} = \tfrac{5}{100} \times 384,435,600,000 \,\text{MWh}$$

$$\text{Energy used for battery production} = 0.05 \times 384,435,600,000 \,\text{MWh}$$

$$\text{Energy used for battery production} = 19,221,780,000 \,\text{MWh}$$

Therefore, approximately 19,221,780,000 megawatt-hours of energy are used for battery production, assuming 5% of the total energy consumption for manganese production is dedicated to battery development.

Aluminium

Aluminium is not typically used as a component in the electrodes (cathode or anode) of lithium-ion batteries. Instead, aluminium is primarily used in other components and aspects of battery technology. Aluminium foils serve as common current collectors in the cathode, providing a substrate for the cathode material and facilitating electron flow during electrochemical reactions.

The metal finds application in the construction of battery casings and enclosures, offering structural support and protecting internal components. Additionally, aluminium powder or flakes, while not a primary component, may act as conductive additives in electrode materials to enhance electrical conductivity.

Given its excellent thermal conductivity, aluminium is incorporated into some battery designs to aid in heat dissipation and thermal management, crucial for maintaining battery safety and performance. Furthermore, aluminium surfaces can undergo anodization, a process forming a protective oxide layer that enhances corrosion resistance. Anodized aluminium is utilized in various battery components, contributing to improved durability.

Although aluminium does not serve as a major active material in lithium-ion battery electrodes, its roles in current collectors, casings, conductive additives, thermal management, and anodization are crucial for the overall construction, performance, and safety of batteries.

Aluminium is not mined in the traditional sense like many other metals; instead, it is extracted from bauxite ore through a process called the Bayer process. The key steps involved in the extraction of aluminium include bauxite mining, where the primary source of aluminium is obtained from tropical and subtropical regions. During this phase, the topsoil is removed, and the ore is extracted from beneath the surface.

The mined bauxite is then subjected to crushing and grinding to create a fine powder, increasing the surface area for subsequent chemical processing. In the digestion phase, the crushed bauxite is mixed with a hot, concentrated solution of sodium hydroxide (caustic soda) in large pressure vessels. This process dissolves the aluminium compounds in the bauxite, leaving behind impurities.

Following digestion, the resulting solution undergoes clarification to remove solid impurities, resulting in a clear liquid containing dissolved aluminium compounds. The solution is then cooled, leading to the precipitation of aluminium hydroxide, which is subsequently separated from the remaining solution. The precipitate undergoes calcination, a heating process that converts it into alumina (aluminium oxide).

The alumina is then smelted in a high-temperature electrolytic cell using the Hall-Héroult process. This involves dissolving alumina in molten cryolite, and an electric current is passed through the solution. The electrolysis separates aluminium metal, which collects at the cathode. The final step involves casting the liquid aluminium obtained from the smelting process into ingots or other desired shapes.

It's crucial to note that both the Bayer process and the Hall-Héroult process are energy-intensive, requiring a significant amount of electricity for the extraction of aluminium from bauxite. Therefore, regions with abundant and affordable energy sources are often preferred for aluminium smelting operations.

In order to determine the energy required to extract one tonne of aluminium from bauxite ore, we can draw upon several relevant references. According to , the overall energy consumption for recycling aluminium is approximately 5% of the energy required for primary production, amounting to 8-10 GJ ton^{-1} of aluminium (Utgikar et al., 2006). This indicates that primary production of aluminium from bauxite ore requires significantly more energy. further support this, stating that a typical aluminium production requires almost 200 GJ/t, which is equivalent to 55,600 kWh/t (X. Y. Chen et al., 2022). Additionally, estimate that the energy consumption for aluminium melting in the metal casting industry is in the order of 6,000-17,000 MJ per tonne when using crucible and natural gas (Jolly & Dai, 2011). This demonstrates the substantial energy demand involved in the extraction and processing of aluminium from bauxite ore.

Furthermore, Haraldsson and Johansson (2018) emphasize that the refining of bauxite to alumina and the subsequent reduction of alumina to metallic aluminium are the two most energy- and CO_2-intensive processes in aluminium production. This highlights the energy-intensive nature of the entire process, from bauxite extraction to aluminium production. Additionally, Rivera et al. (2019) mention that the energy consumption of the process is about 3.5 GJ tonne^{-1}. This further underscores the high energy requirements for aluminium extraction from bauxite ore.

The extraction of one tonne of aluminium from bauxite ore is an energy-intensive process, with estimates ranging from 3.5 GJ to 200 GJ per tonne. The primary production of aluminium from bauxite ore demands a substantial amount of energy, making it an energy-intensive process.

To determine the percentage of aluminium used for battery production annually, we can refer to the assumption made in GREET 2018 (Greenhouse gases, Regulated Emissions, and Energy use in Transportation), which states that a recycled content of 11% is assumed for aluminium used in battery production (Dai et al., 2019). This indicates that a portion of the aluminium used in battery production is recycled, and this figure can be used to estimate the percentage of aluminium used for battery production annually.

In 2022, the global production volume of primary aluminium amounted to around 68.5 million metric tons, representing a two percent increase compared to the previous year (Statista, 2023c). As such, 7,535,000 metric tonnes was utilised in battery production in 2022.

To convert gigajoules (GJ) to megawatt-hours (mWh), you can use the conversion factor:

1 GJ=0.27778 MWh1 GJ=0.27778 MWh

Now, you can calculate the energy required to produce 7,535,000 metric tons of aluminium:

Energy (MWh)=Energy use per tonne (GJ/tonne)×Total production (tonnes)×Conversion factor

Energy (MWh)=3.5GJ/tonne×7,535,000tonnes×0.27778MWh/GJ

Now, calculate the result:

Energy (MWh)≈7,580,208MWh

Therefore, approximately 7,580,208 megawatt-hours (MWh) of energy is required to produce 7,535,000 metric tons of aluminium using 3.5 GJ per tonne.

Production Energy Demands

The production of batteries, including materials and recycling, has been estimated to require a significant amount of energy, with more than 400 kWh needed to make a 1 kWh Li-ion battery (Larcher & Jm, 2014). This highlights the substantial energy input associated with battery manufacturing, which indirectly impacts the usage of aluminium in battery production.

Based on the discussion above and the energy required to obtain the raw materials for lithium ion battery production annually, as summarised in Table 2, 19,296,680,874 mWh are required annually.

Table 2: Energy requirements to obtain raw materials for lithium ion battery production annually based on 2022 demands.

Raw Material	Production Energy Requirements Annually in megawatt-hours (mWh)
Lithium	2,975,000
Cobalt	3,290,000
Nickel	6,666,666
Graphite	54,389,000
Manganese	19,221,780,000
Aluminium	7,580,208
Total Energy (mWh)	19,296,680,874

To put the energy value of 19,296,680,874 megawatt-hours (mWh) into context, here are some examples:

Household Electricity Usage: The average annual electricity consumption for a U.S. household is around 10,766 kilowatt-hours (kWh). The provided energy value is equivalent to powering approximately 1,788,780 average U.S. households for one year.

Electric Vehicles: An electric car typically consumes around 0.3 kWh per mile. The energy value could power an electric vehicle to travel approximately 64.3 trillion miles, assuming an average efficiency.

Data Centres: Large data centres can consume substantial amounts of electricity. For context, a medium-sized data centre might use around 100,000 kWh per month. The provided energy value is equivalent to running a medium-sized data centre for approximately 1,602,228 months.

Renewable Energy Generation: A modern wind turbine can produce around 7.5 million kWh of electricity annually. The provided energy value is equivalent to the annual electricity generation of approximately 2,573 wind turbines.

Global Energy Consumption: According to some estimates, the total global energy consumption is around 170,000 terawatt-hours (TWh) per year. The provided energy value represents about 0.01% of the total global energy consumption for a year.

These examples help illustrate the scale and magnitude of the given energy value in different contexts. Keep in mind that the actual impact depends on the specific application and efficiency of energy use.

From the perspective of renewable energy, to estimate the number of wind turbines required to produce 19,296,680,874 megawatt-hours (MWh) annually, we need to consider the average annual electricity generation of a modern wind turbine.

Let's use the assumption that a modern wind turbine can produce around 7.5 million kWh of electricity annually. Now, we convert the total desired energy production to kilowatt-hours:

$$19{,}296{,}680{,}874 \text{ MWh} \times 1{,}000 \text{ kWh/MWh} = 19{,}296{,}680{,}874{,}000 \text{ kWh}$$

Next, we divide this by the annual electricity generation per wind turbine:

$$\frac{19{,}296{,}680{,}874{,}000 \text{ kWh}}{7{,}500{,}000 \text{ kWh/turbine}} \approx 2{,}573{,}557 \text{ turbines}$$

Therefore, it would take approximately 2,573,557 modern wind turbines to produce 19,296,680,874 megawatt-hours annually, based on the given assumptions. Keep in mind that this is a simplified calculation, and actual turbine performance can vary based on factors like location, wind speed, and turbine efficiency. Over two and a half million wind turbines would be required to generate the energy required to meet current demands for lithium ion batteries.

Similarly, in terms of solar panels, to estimate the number of solar panels required to produce 19,296,680,874 megawatt-hours (MWh) annually, we need to consider the average annual electricity generation of a solar panel.

Let's use the assumption that a solar panel can produce around 1 megawatt-hour per year, which is a simplified average value for illustrative purposes.

Now, we convert the total desired energy production to megawatt-hours:

19,296,680,874 MWh

This means you would need the same number of solar panels (given the assumption of 1 MWh per panel) to produce this amount of energy annually. Therefore, you would need approximately 19,296,680,87419,296,680,874 solar panels (over 19 billion).

It's important to note that the actual energy production of a solar panel can vary based on factors such as location, sunlight exposure, panel efficiency, and maintenance. This calculation is a simplified estimate for illustration.

For further context and comparison, to comprehensively assess overall energy consumption, it is essential to amalgamate data from diverse sources, including electricity, oil, natural gas, and coal. The top energy-consuming countries in 2020 were China, the United States, India, Russia, Japan, Canada, Germany, Iran, Brazil, and South Korea (Karlsson & Byström, 2005). China and the United States stand out in overall energy consumption, with China leading in electricity usage and the U.S. dominating oil consumption. Population emerges as a primary influencer, evident in the top three countries—China, the U.S., and India—being among the most populous (Şoavă et al., 2018). Per capita, Iceland emerges as the highest energy consumer, surpassing even the top 10 countries. Despite lower rankings in electricity and oil consumption, Iceland's per capita energy use in 2020 exceeded 167,000 kilowatt-hours per person annually. In contrast, China, despite leading in overall energy consumption, exhibits comparatively lower per capita energy use due to its large population (Bilgili & Öztürk, 2015).

In 2020, China emerged as the leading global energy consumer, recording a substantial consumption of 145.46 billion kWh (World Population Review, 2023). Following closely, the United States secured the second position with an energy consumption of 87.79 billion kWh. India, Russia, Japan, Canada, Germany, Iran, Brazil, and South Korea rounded out the top 10, with energy consumption figures ranging from 31.98 billion kWh to 11.79 billion kWh. These countries represent significant contributors to global energy demand, with China and the United States standing out as the foremost consumers on this list (World Population Review, 2023).

In 2019, China led the world in electricity consumption, registering a substantial figure of 6,875 billion kWh (World Population Review, 2023). The United States followed as the second-largest consumer with 3,989 billion kWh. India, Russia, Japan, Canada, South Korea, Brazil, Germany, and France comprised the remaining top 10, with electricity consumption ranging from 1,229 billion kWh to 449 billion kWh. These countries played key roles in shaping global electricity demand, reflecting their industrial and technological activities, with China and the United States prominently standing out as the primary contributors on this list.

In 2019, the United States emerged as the largest consumer of oil globally, with a substantial consumption rate of 20,543 million barrels per day (World Population Review, 2023). China followed as the second-largest consumer, registering 14,008 million barrels per day, while India, Japan, Russia, Saudi Arabia, Brazil, South Korea, Germany, and Canada comprised the remaining top 10, with oil consumption ranging from 4,920 million barrels per day to 2,303 million barrels per day. These figures underscored the significant role these countries played in shaping the demand for oil, reflecting their industrial activities, transportation needs, and overall energy reliance on fossil fuels (World Population Review, 2023).

Chapter Four

Applications

L ithium-ion batteries find widespread applications across various industries and devices due to their high energy density, lightweight design, and rechargeable nature. Lithium-ion batteries (LIBs) have become the power source of choice for a wide range of applications. They are extensively used in portable consumer electronics, such as mobile phones, digital cameras, and laptops (L. Zhang et al., 2017). Moreover, LIBs are dominating the electric vehicle market and are on the verge of entering the utility market for grid-energy storage (Manthiram, 2017). The versatility of LIBs is further demonstrated by their use in renewable energy storage systems (Saxena et al., 2021). Additionally, LIBs are increasingly utilized in onboard energy storage systems for the next generation of automotive vehicles, presenting new economic, technical, and environmental challenges (Filomeno & Feraco, 2020).

The application of LIBs extends beyond consumer electronics and electric vehicles. They are also used in stationary applications, such as electrical energy storage for peak shaving in industrial environments and home use of solar power (Giegerich et al., 2016). Furthermore, LIBs have been employed in railcars, as evidenced by the running test of a railcar using a lithium-ion battery (Matsuo et al., 2009). Research and development efforts have been dedicated to tailoring LIBs for environmental vehicles, highlighting their adaptability to various vehicle types (Abe et al., 2007). The potential of LIBs is not limited to their current applications. Significant research efforts are being devoted to upgrading existing batteries, including lithium-ion types, and developing alternate technologies, such as sodium-ion, metal-air, and lithium-sulphur batteries (Hibino et al., 2013). This underscores the continuous exploration of new frontiers for LIBs and their potential for further diversification in energy storage applications.

Lithium-ion batteries have become the power source of choice for portable electronics, electric vehicles, and are on the verge of entering the utility market for grid-energy storage (Manthiram, 2017; Shang & Cheng, 2022). The applications that depend on electrical energy storage include portable electronics, electric vehicles, and devices for renewable energy storage from solar and wind (Manthiram et al., 2012). The electrochemical properties and lithium morphologies of lithium batteries have been significantly improved, leading to their widespread adoption (R. Zhang et al., 2017).

Consumer Electronics

Lithium-ion (Li-ion) batteries have revolutionized the consumer electronics industry due to their exceptional characteristics, including high energy density, light weight, and rechargeable nature. These batteries have become the standard power source for a wide range of devices, transforming the way people interact with technology. Smartphones, laptops, tablets, cameras, wearable devices, portable audio devices, and gaming devices all benefit from the versatility and performance of Li-ion batteries. The high energy density of Li-ion batteries allows for the design of slim and lightweight devices, providing users with longer battery life and faster charging capabilities (Bruce et al., 2008). Additionally, the rechargeable nature of these batteries ensures convenience and portability, as users can charge their devices multiple times without the need for frequent battery replacements.

Figure 13: Smartphone use. Océanos y dados, CC0, via Wikimedia Commons.

In the realm of consumer electronics, Li-ion batteries power portable computing devices, offering users the flexibility to work or entertain themselves on the go. The ability to
recharge these batteries extends the usability of laptops and tablets without being tethered
to a power source. Furthermore, Li-ion batteries have largely replaced traditional alkaline
and nickel-based batteries in digital cameras, allowing photographers to capture more
photos on a single charge and eliminating the need for frequent battery replacements. The
rise of wearable technology, such as smartwatches and fitness trackers, is closely linked to
the use of Li-ion batteries, enabling the design of compact and lightweight wearables that
can perform various functions, including health monitoring and activity tracking.

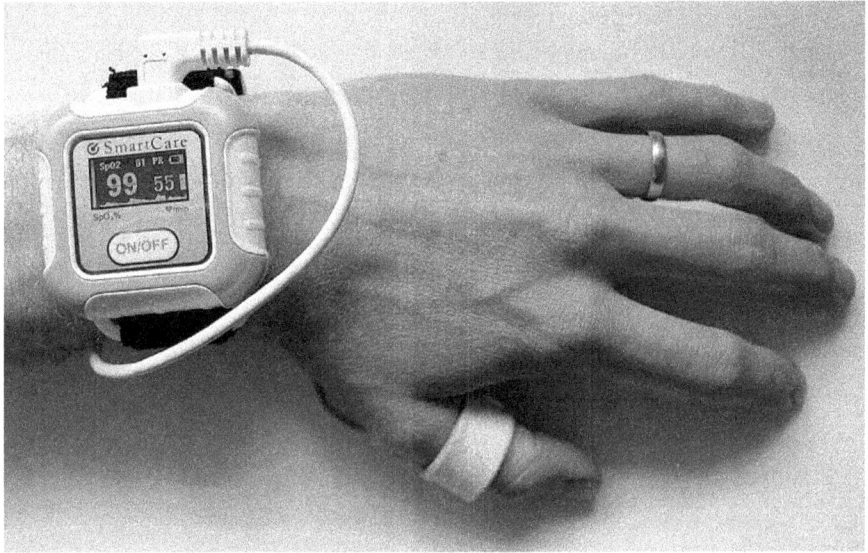

Figure 14: SmartCare wrist-worn pulse oximeter (SmartCare Analytics Ltd., London, UK). Peter H Charlton, CC BY 4.0, via Wikimedia Commons.

Moreover, Li-ion batteries provide a reliable power source for extended playback periods in portable audio devices, aligning with the convenience expected by users of these on-the-go devices. Handheld gaming consoles often rely on Li-ion batteries to support power-intensive demands and allow users to play for extended periods without the need for frequent recharging. The versatility and performance characteristics of Li-ion batteries have made them indispensable in the realm of consumer electronics, providing a seamless and mobile experience for users.

The adoption of lithium battery technology has significantly improved the performance and efficiency of smartphones. Stable power supply enables smartphones to handle demanding tasks like high-resolution video streaming and resource-intensive gaming. Longer battery life and quick charging capabilities have transformed the user experience, enhancing productivity and convenience. Lithium batteries have made smartphones more portable and convenient, allowing for greater mobility and flexibility in daily activities.

Ongoing advancements in battery technology aim to further enhance the efficiency, safety, and overall performance of Li-ion batteries in consumer electronics. Researchers are exploring new nanomaterials with unique properties for use as electrodes and electrolytes in lithium batteries, aiming to improve their performance and safety. Additionally, advancements in structure engineering aim to harness the volume expansion of anode materials to achieve high performance beyond lithium-based rechargeable batteries, addressing the limited resources of lithium in the Earth's crust.

Furthermore, research focuses on improving battery safety by developing battery management systems with excellent reliability and efficiency, aiming to prevent thermal runaway and enhance overall safety during battery operation. Lithium–air batteries are also being explored for their high theoretical energy density, which is about 10 times larger than that of lithium-ion batteries, with the potential to become the key battery system of the next generation. The development of new state-of-charge estimation methods and predictive models based on artificial intelligence and transformer models aims to improve battery management and prolong battery life.

Lithium battery technology has extended the usage time of laptops and tablets, enabling users to work, study, or entertain themselves for extended periods without constantly searching for a power source. These batteries can deliver the necessary power to support high-performance processors and graphics cards, improving computing capabilities. The lightweight and compact nature of lithium batteries have made laptops and tablets more portable, enabling productivity anytime, anywhere.

Lithium battery technology has profoundly impacted consumer electronics, making devices more efficient, portable, and convenient. With advancements like longer battery life, quick charging, and high energy density, lithium batteries have become indispensable tools in modern lives. As technology evolves, these advancements will play a pivotal role in shaping the future of consumer electronics.

The advent of lithium-ion batteries has had a profound impact on mobility and connectivity, extended productivity, convenience and portability, entertainment on the go, remote work and learning, health and fitness tracking, efficient communication, access to information, the emergence of the Internet of Things (IoT), and reduced environmental impact.

The mobility and connectivity aspect has been greatly enhanced by the portability of devices such as smartphones and laptops, made possible by lithium-ion batteries (Miles, 2001). This has allowed individuals to stay connected, work, and access information from virtually anywhere, thereby improving overall mobility and connectivity. Furthermore, the extended battery life provided by lithium-ion batteries has increased productivity, especially in work and educational settings, as users can engage with their devices for extended periods without the need for frequent recharging (Miles, 2001).

The portability and extended battery life of lithium-ion-powered devices have played a crucial role in facilitating remote work and online learning (Miles, 2001). Individuals can participate in virtual meetings, attend classes, and complete tasks without being tethered to a power source. Moreover, wearable devices powered by lithium-ion batteries, such as smartwatches and fitness trackers, have transformed personal health management, allowing users to monitor their physical activity, track health metrics, and receive real-time feedback, thereby promoting a healthier lifestyle (Miles, 2001).

Smartphones, driven by lithium-ion batteries, have revolutionized communication, making instant messaging, video calls, and social media platforms integral parts of daily communication, connecting people across the globe (Miles, 2001). Furthermore, lithium-ion batteries have powered the proliferation of devices like tablets and e-readers, providing users with easy access to a vast amount of information, thereby transforming how people consume news, books, and educational content (Miles, 2001). The emergence of the Internet of Things (IoT) has been significantly influenced by lithium-ion batteries, enabling seamless connectivity between smart devices and enhancing convenience and efficiency in various applications (Miles, 2001). While there are environmental considerations, the longer lifespan and rechargeability of lithium-ion batteries contribute to a reduction in electronic waste compared to disposable batteries, with efforts being made to improve recycling processes (Miles, 2001).

The number of smartphones produced globally is a topic of interest due to the widespread use and impact of these devices. While specific data on the total number of smartphones produced may not be readily available, it is possible to estimate the production based on various indicators. For instance, the smartphone penetration rate in Hong Kong was reported to have reached 85.8% in 2016, with an expected user base of over 6.1 million by 2022 (Cheung et al., 2020). Additionally, it was noted that smartphone ownership is approaching 50% globally and 80% in advanced economies, indicating a significant level of smartphone adoption (Cooke et al., 2021). Furthermore, the success of specific smartphone models, such as the iPhone, with over 42 million units sold, highlights the substantial scale of smartphone production and consumption (Laugesen & Yuan, 2010).

The increasing number of smartphones is also attributed to the advancement of individual communication needs and the shift of sales to online outlets, leading to the development of more mobile applications in recent years (Pop et al., 2022). This trend aligns with the widespread use of smartphones as essential tools in modern society (S.-K. Kim et al., 2016). Moreover, smartphones are particularly popular among the younger generation, with 81% of young people owning their own devices (Asbar et al., 2021). The emphasis on industrial mobile phones towards innovation in smartphone production further underscores the increasing production of smartphones in various types of mobile devices, setting new standards for companies (Winarti et al., 2021).

The global smartphone market is also influenced by the strategies and market positions of major players such as Apple and Samsung. Unique patterns in the iPhone market, including the thriving second-hand market, contribute to the overall volume of smartphones in circulation (Hawari, 2022; Sari et al., 2022). Additionally, the practical implication of studies evaluating smartphone companies' strategies for product quality, electronic word of mouth, brand image, and consumer purchase decisions further emphasizes the significance of smartphone production and its impact on consumer behaviour (Saraswati & Giantari, 2022).

Jardim (2017) aimed to determine the cumulative production of smartphones since the introduction of Apple's first iPhone in 2007, and the result was astonishing—over 7 billion smartphones have been manufactured. In theory, if all these smartphones were still in operation, there would be enough for each person on Earth. However, the reality is different, as the average lifespan of a phone in the United States is slightly over 2 years, even though these devices are capable of functioning for a more extended period (Jardim, 2017). Users often succumb to the temptation of prematurely replacing their phones, driven by factors like the allure of a "free" new phone with a contract renewal or the perceived complexity and costliness of repairing a single failing component, such as the screen or battery. At the current pace, individuals are projected to go through at least 29 phones in their lifetimes. Some key findings include the production of 7.1 billion smartphones since 2007, the use of more than 60 different elements in manufacturing, significant e-waste generation, with only a fraction being recycled, and the substantial energy consumption (968 terawatt hours) in smartphone manufacturing since 2007, equivalent to India's annual power supply in 2014 (Jardim, 2017). Additionally, the challenging disassembly process at the end of a phone's life, often involving shredding and smelting, contributes to inefficiencies in material recovery due to the diverse and minute nature of the components (Jardim, 2017).

Since 2008, the smartphone industry has witnessed consistent growth, expanding in both market size and the variety of models and vendors available. The global shipments of smartphones reached approximately 1.2 billion units in 2022, experiencing a decline from the previous year. As of the end of 2022, 68 percent of the world's population were smartphone users (Statista, 2023b). However, considering that many individuals own multiple smartphones, the actual number of smartphone subscriptions surpasses the count of users. In 2022, smartphone owners utilized an estimated 6.5 billion subscriptions, and this figure is anticipated to rise to nearly eight billion by 2028 (Statista, 2023b).

The increasing adoption of smartphones has implications for various sectors, including the potential for smartphones to be used in motion picture production, as technological improvements in smartphone cameras may lead to a significant increase in movies produced with smartphones (Schulz et al., 2021). However, the widespread use of smartphones also introduces security and privacy challenges, as many smartphone applications produce residual data, highlighting the need for addressing these concerns in the context of smartphone usage (Grispos et al., 2021).

Lithium-ion batteries, commonly used in consumer electronics, possess several lesser-known characteristics that are crucial for users and manufacturers to understand. These facts shed light on various aspects of lithium-ion batteries, showcasing both their advantages and considerations. The finite number of charge cycles before capacity diminishes, sensitivity to temperature extremes, stable voltage throughout most of the discharge cycle, absence of the memory effect, and the presence of built-in Battery Management Systems (BMS) are some of the key features (G. Chen et al., 2023). Additionally, the relatively low self-discharge rate, susceptibility to thermal runaway, cobalt dependency, and trade-offs associated with fast charging are important considerations (Bates, 2000). Ongoing research and development aim to address challenges and improve the overall performance and sustainability of these batteries (Yang et al., 2018).

Lithium-ion batteries have a finite number of charge cycles before their capacity starts to diminish, and extreme temperatures can impact their performance and lifespan. Additionally, they offer a stable voltage throughout most of their discharge cycle and do not suffer from the memory effect. These batteries often include a built-in Battery Management System (BMS) to monitor and manage their performance, preventing issues like overcharging, over-discharging, and overheating (G. Chen et al., 2023). Furthermore, they have a relatively low self-discharge rate, but are susceptible to thermal runaway, which can lead to a fire in extreme cases. Many lithium-ion batteries rely on cobalt for their cathodes, raising ethical and environmental concerns. Fast charging, while convenient, can contribute to increased heat generation, potentially impacting long-term battery health (Bates, 2000).

The ongoing research and development in lithium-ion battery technology aim to enhance energy density, safety, and charging speeds. Emerging technologies, such as solid-state batteries, might eventually replace or complement traditional lithium-ion batteries in consumer electronics (Yang et al., 2018). These advancements are crucial for addressing the challenges associated with lithium-ion batteries and improving their overall performance and sustainability.

Electric Vehicles (EVs)

Lithium-ion batteries have emerged as the dominant energy storage solution for electric vehicles (EVs), owing to their high energy density, lightweight composition, and ability to deliver ample power for extended driving ranges. These batteries serve as the primary energy storage devices in EVs, storing electrical energy in a chemical form that can be converted into electricity to propel the vehicle's electric motor. The high energy density of lithium-ion batteries is critical for EVs, allowing for significant energy storage in a compact and lightweight package, a crucial factor for achieving optimal driving distances.

The stored energy in lithium-ion batteries powers the electric motors that drive EVs, eliminating the reliance on internal combustion engines. These batteries are designed to endure numerous charging and discharging cycles, supporting various charging methods, including standard outlets, dedicated home chargers, and fast-charging stations that have seen improvements in charging speed. Additionally, regenerative braking systems are often integrated into EVs, where the electric motor acts as a generator during braking, converting kinetic energy back into electrical energy and storing it in the lithium-ion battery for increased efficiency.

Sophisticated Battery Management Systems (BMS) are incorporated into EVs equipped with lithium-ion batteries, monitoring and managing aspects such as temperature, state of charge, and voltage to optimize overall battery health and safety. The use of lithium-ion batteries has significantly extended the driving range of modern EVs, addressing a key concern of consumers in the transition to electric mobility.

However, the environmental impact of lithium-ion battery production, involving the mining and processing of materials like lithium, cobalt, and nickel, has raised ethical and sustainability concerns. Efforts are underway to address these challenges, with a focus on reducing the environmental footprint of battery production through sustainable practices.

Ongoing research and development in battery technology aim to enhance lithium-ion batteries used in EVs, with a focus on increasing energy density, reducing charging times, and developing more sustainable and cost-effective battery chemistries. In summary, lithium-ion batteries are pivotal in the electrification of transportation, providing a reliable and efficient energy storage solution for EVs. Continued advancements are anticipated to contribute to the ongoing growth and improvement of electric mobility.

The electric vehicle (EV) battery serves as the primary power source for the electric drive unit and the entire vehicle, functioning as a substantial, high-voltage energy storage block positioned beneath the vehicle, akin to a conventional fuel tank. Traditional EV battery packs consist of multiple smaller module blocks, each containing cells with various shapes such as pouch, prismatic, or cylindrical (Man, 2023).

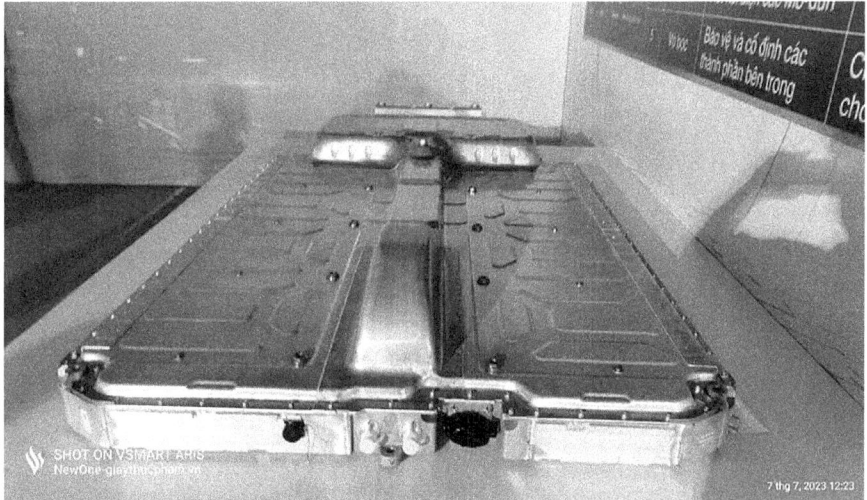

Figure 15: VinFast VF 8 Li-Ion battery. Newone, CC BY-SA 4.0, via Wikimedia Commons.

These cells are composed of a cathode (positive terminal), a separator with liquid electrolyte, and an anode (negative terminal). During charging, charged particles (ions) move from the cathode to the anode through the electrolyte, facilitating the flow of electrons between cathode and anode current collectors, ultimately charging the battery (Zhu et al., 2013). The reverse process occurs during discharging.

A notable innovation in EV battery design is the emergence of 'structural batteries', where cells are directly embedded within the vehicle chassis, eliminating the need for space- and weight-consuming modules in a pack enclosure (Man, 2023). While this design enhances space efficiency, aerodynamic performance, and body stiffness, concerns have been raised regarding repairability in the event of an accident.

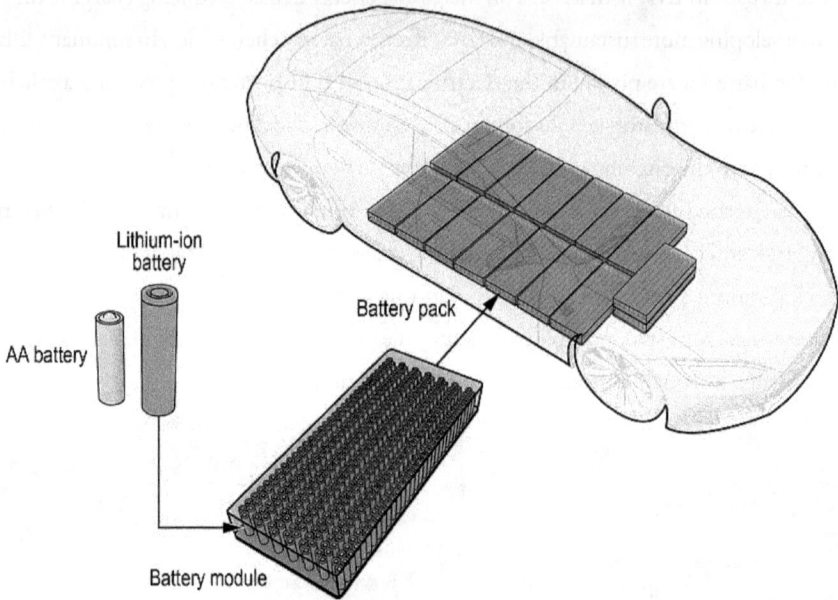

Figure 16: An illustration of a high-voltage, lithium-ion battery in an electric vehicle, showing the location of the vehicle's battery pack, a detail of the battery module, and a size comparison between the lithium-ion batteries in the module and a typical AA battery. National Transportation Safety Board, Public domain, via Wikimedia Commons, NTSB Graphic by Christy Spangler.

In the realm of lithium-ion battery technology, three key types are prevalent, each with distinct characteristics. Lithium-ion (Li-ion) batteries, employing nickel-manganese-cobalt (NMC) cathodes, offer high energy density but are challenged by a shorter life cycle, higher thermal runaway risk, and the use of unsustainable, expensive materials (Man, 2023). Nickel-cobalt-aluminium (NCA) batteries, a variant of NMC, replace manganese with more sustainable aluminium but share similar advantages and disadvantages. Lithium-iron-phosphate (LFP) batteries, featuring a longer life cycle and better safety, are more cost-effective but have lower energy density and are sensitive to temperature variations (Man, 2023).

The cell is a crucial component in the generation of electrical energy within a battery, accounting for approximately 70% of the overall battery pack cost. Typically employing a liquid electrolyte, such as LiPF6, the classification of batteries often centres on the cathode chemistry (Daryanani, 2022). Common cathodes include LCO (lithium cobalt oxide), NCM (lithium nickel cobalt manganese oxide), NCA (lithium nickel cobalt aluminium oxide), and LFP (lithium iron phosphate). NCM, with variations like 8:1:1, is the most prevalent chemistry, but there is a shift toward higher nickel content for increased capacity at the expense of cycle life. NCA, used by manufacturers like Tesla, shares properties with high-nickel NCM cells. LFP, based on iron, is cost-effective and safer but has lower energy densities than NCM and NCA (Daryanani, 2022). The anode, usually graphite, is evolving with silicon nanoparticles for enhanced capacity, while packing methods include modules, packs, and newer approaches like cell-to-pack (C2P) and cell-to-chassis (C2C) for space efficiency. Different cell shapes—cylindrical, pouch, or prismatic—offer varying sizes and packing efficiencies. The Battery Management System (BMS) serves as the battery pack's brain, managing operational and safety functions, including protection, state monitoring, cell balancing, and communication. The cooling system, essential for maintaining optimal temperatures (air-cooled or liquid-cooled), and electrical interconnects, which can vary from ultrasonic bonding to laser welding, contribute to the intricate design and functionality of lithium-ion batteries for electric vehicles (Daryanani, 2022).

Several electric vehicle models utilize nickel-manganese-cobalt (NMC) batteries, including the Tesla Model Y (Long Range and Performance variants) (See Figure 17), Tesla Model 3 (Long Range and Performance variants), Polestar 2, Volvo EX30, and BMW iX3. These models leverage the high energy density of NMC batteries, contributing to longer driving ranges and efficient performance.

Figure 17: Tesla Model Y. Alexander Migl, CC BY-SA 4.0, via Wikimedia Commons.

Various electric vehicle models incorporate nickel-cobalt-aluminium (NCA) batteries, such as the Audi Q8 E-Tron (see Figure 18), older pre-2021 models of the Tesla Model 3, and discontinued models in Australia, namely the Tesla Model S and Tesla Model X. NCA batteries share similarities with NMC batteries, offering advantages like good energy density and charging performance, along with disadvantages such as potential thermal risks. While not as widely adopted by car brands, these models showcase the utilization of NCA technology in the electric vehicle market.

Figure 18: Audi Q8 E-Tron. Alexander-93, CC BY-SA 4.0, via Wikimedia Commons.

A range of electric vehicle models features lithium-ferrous-phosphate (LFP) batteries, including the BYD Dolphin (see Figure 19), MG 4 (specifically the Excite 51 variant), GWM Ora (limited to the Standard Range model), BYD Atto 3, BYD Seal, Tesla Model 3 (available in the Rear-Wheel Drive configuration only), and Tesla Model Y (also exclusively in the Rear-Wheel Drive version). LFP batteries are known for their longer lifespan and improved safety, offering a viable option for those seeking more affordable and enduring electric vehicles. Despite their lower energy density compared to other battery types, these models cater to individuals prioritizing longevity and cost-effectiveness in their electric vehicles.

Figure 19: BYD Dolphin photographed in Zhuhai, Guangdong province, China. User3204, CC BY-SA 4.0, via Wikimedia Commons.

Despite the varied advantages and limitations of these battery chemistries, there is ongoing research into alternative technologies, such as sodium-ion batteries and solid-state batteries, expected to debut in the coming years. Each battery type caters to specific priorities, emphasizing factors like longevity, driving range, charging speed, and environmental impact. The pursuit of advancements in EV battery technology underscores the ongoing evolution of electric mobility (Man, 2023).

Based on the references provided, the total number of electric vehicles sold to date can be estimated. According to Kim et al. (2022), over 1.9 million plug-in electric vehicles

(PEVs), including electric vehicles (EVs) and plug-in hybrid electric vehicles (PHEVs), have been sold in the USA since 2010. Additionally, Tianjin Chen et al. (2020) reported that the total sale number of global EVs reached 3 million by 2017. Furthermore, Ismail and Mulyaman (2021) mentioned that the total electric cars sold worldwide, including all brands, reached 3 million units from 2010 to 2017. Moreover, Xia et al. (2023) stated that by 2021, the holdings of battery electric passenger vehicles in China were about 5.44 million. By synthesizing these references, it can be estimated that the total number of electric vehicles sold globally, including both EVs and PHEVs, is approximately 5.44 million in China, 3 million globally by 2017, and over 1.9 million in the USA since 2010. Therefore, the total number of electric vehicles sold to date is likely to be well over 10 million, considering the sales in other countries and the years following 2017.

Estimating the number of individual lithium-ion battery cells required for electric vehicles involves several variables, including the capacity of each cell and the total capacity needed for each vehicle. The number of cells can vary based on the specific battery chemistry, energy density, and design choices made by manufacturers. However, we can provide a rough estimate using some common parameters.

Let's assume an average electric vehicle battery pack capacity of around 60 kWh, which is a typical capacity for many electric cars. This is a generalization, and actual capacities can vary widely.

Now, to estimate the number of cells, we need to know the capacity of an individual lithium-ion battery cell. Let's assume an average capacity of 3000 mAh (milliampere-hours) per cell, which is a common capacity for cylindrical 18650 cells.

First, convert the pack capacity to ampere-hours (Ah):

60 kWh = 60,000 Wh

60,000 Wh / 3.7 V (average voltage of a lithium-ion cell) = ~16,216 Ah

Now, divide the total ampere-hours needed for one vehicle by the capacity of an individual cell:

16,216 Ah / 3000 mAh = ~5.4 cells

This means that roughly 5.4 individual lithium-ion battery cells would be needed to achieve a 60 kWh battery pack for one electric vehicle.

Now, for 10 million electric vehicles:

5.4 cells/vehicle * 10,000,000 vehicles = 54,000,000 cells

So, a rough estimate suggests that around 54 million individual lithium-ion battery cells would be required for 10 million electric vehicles, given the assumptions and parameters mentioned. Keep in mind that this is a simplified calculation, and actual numbers can vary based on different factors in real-world applications.

The amount of raw materials required to produce a lithium-ion battery cell depends on various factors, including the specific chemistry of the battery, the size and capacity of the cell, and the materials used in its construction. However, I can provide a general overview based on common materials used in lithium-ion battery cells.

Typical components of a lithium-ion battery cell include:

1. Cathode Materials: Common cathode materials include lithium cobalt oxide ($LiCoO_2$), lithium manganese oxide ($LiMn_2O_4$), lithium nickel cobalt manganese oxide (NCM), and lithium iron phosphate ($LiFePO_4$). The quantity of these materials depends on the specific chemistry.

2. Anode Materials: Graphite is commonly used as the anode material, and the quantity depends on the size and capacity of the cell.

3. Electrolyte: The electrolyte is usually a lithium salt dissolved in a solvent. The quantity depends on the cell size.

4. Separator: A separator material is used to keep the cathode and anode apart. Common materials include polyethylene or polypropylene.

5. Binders and Conductive Additives: These materials help hold the electrode materials together and improve conductivity.

6. Metal Foils: Thin foils of metals like aluminium and copper are used as current collectors.

Given that we're estimating for 54 million cells, it's important to note that the size and capacity of these cells can vary. Also, specific formulations and manufacturing processes can influence the quantities of materials used.

For a rough estimate, we can consider a moderate-sized lithium-ion cell with an average cathode material like NCM. In this case, the raw materials needed for cathode, anode, electrolyte, separator, and other components would likely amount to several kilograms per cell.

Let's make a very rough estimation and assume an average of 5 kilograms of raw materials per cell. This is a highly generalized figure, and actual numbers can vary.

54,000,000 cells * 5 kg/cell = 270,000,000 kg or 270,000 metric tons

So, a very approximate estimate suggests that around 270,000 metric tons of raw materials might be required to produce 54 million lithium-ion battery cells, based on the assumptions and generalizations made. Keep in mind that this is a simplified calculation, and actual numbers can vary based on different factors in real-world applications.

To offer a sense of scale for 270,000 metric tons of raw materials, consider these comparisons: the iconic Eiffel Tower in Paris, weighing about 10,100 metric tons, would be roughly equivalent to 26.7 instances of its mass. Alternatively, the Boeing 747-8, a sizable commercial airplane with a maximum take-off weight of approximately 447 metric tons, would find a parallel in 604 such aircraft. The blue whale, Earth's largest animal averaging around 200 metric tons, would be represented by 1,350 of these majestic creatures. To envision the volume, filling an Olympic-sized swimming pool (holding about 660,000 gallons or 2,500 cubic meters of water) with these raw materials would be akin to having approximately 108 such pools. These comparative examples aim to visually convey the magnitude of 270,000 metric tons, emphasizing the substantial quantity of raw materials required for the production of 54 million lithium-ion battery cells. It is important to note that these analogies serve as conceptual aids and may not precisely capture the density or physical attributes of the actual raw materials in battery production.

To compare the raw material requirements of manufacturing a petrol engine car to an electric vehicle, it is essential to consider the differences in the materials and processes involved in their production. The manufacturing of a petrol engine car typically involves the use of traditional materials such as steel, aluminium, and various composite materials (Busarac et al., 2022). On the other hand, the production of an electric vehicle (EV) requires a different set of raw materials, including lithium-ion batteries, electric motors, and lightweight materials such as advanced high-strength steel (AHSS) and aluminium alloys (Busarac et al., 2022).

The raw material requirements for manufacturing a petrol engine car are primarily focused on traditional automotive materials such as steel and aluminium, which are used in the construction of the car body and engine components (Herrmann et al., 2018). Additionally, the production of petrol engine cars involves the use of fuel-related materials such as gasoline and liquefied petroleum gas (LPG) (Jaworski et al., 2019). In contrast, the manufacturing of an electric vehicle involves a significant amount of raw materials for the production of lithium-ion batteries, electric motors, and lightweight materials for the car body and components (Busarac et al., 2022).

The production of lithium-ion batteries for electric vehicles requires a wide range of raw materials and industrial processes, which can result in supply risks and high economic importance in the production chain (Filomeno & Feraco, 2020). Furthermore, the manufacturing of electric vehicle components, such as driving motors, involves specific material trends and manufacturing characteristics that differ from those of traditional petrol engine car components (Kim et al., 2023).

In terms of energy consumption during the raw material preparation stage, the production of traditional automotive materials for petrol engine cars is electricity-intensive, requiring a significant amount of kilowatt hours per ton of raw material (Nalobile et al., 2020). Conversely, the production of lightweight materials for electric vehicles, such as AHSS and aluminium alloys, also requires electricity-intensive processes but may offer potential energy savings in the overall vehicle weight and efficiency (Busarac et al., 2022).

Moreover, the environmental impact of raw material extraction and manufacturing processes is a critical consideration. A life cycle assessment approach has been used to evaluate the environmental performance of battery electric vehicles, which includes the extraction of raw materials, manufacturing of components, assembly, use phase, and end-of-life treatment (Messagie et al., 2010). This approach provides insights into the environmental implications of the raw material requirements for electric vehicles compared to petrol engine cars.

The raw material requirements for manufacturing a petrol engine car differ significantly from those of an electric vehicle. The production of traditional automotive materials and fuel-related components characterizes petrol engine car manufacturing, while electric vehicle production involves a complex supply chain of raw materials for lithium-ion batteries, electric motors, and lightweight materials. Understanding these differences is crucial for assessing the environmental, economic, and technical aspects of transitioning from petrol engine cars to electric vehicles.

Energy Storage Systems

Electrical energy plays a crucial role in various aspects of our lives, contributing to industrial development, urbanization, and economic progress. However, the fluctuating demand for electricity creates imbalances in power generation and utilization. To address issues related to carbon emissions, climate change, and energy supply shortages, the global landscape of electrical energy generation is undergoing significant changes, accompanied by the development of renewable energy sources (Tianmei Chen et al., 2020).

The challenges posed by the instantaneous demand for electrical energy and unpredictable variations in daily and seasonal demand are addressed through grid-level energy storage systems. These systems convert electricity from the generation network into a storable form, facilitating a just-in-time supply system. Grid-level electrical energy storage systems have versatile applications, supporting power generation, transmission, and large-scale electronic devices. For stationary applications, they store excess electrical energy during peak generation periods and provide it during peak load periods, contributing to load leveling and peak shaving (Tianmei Chen et al., 2020). The energy storage system plays a pivotal role in balancing the load and power of the grid network through efficient charging, discharging, and providing regulated power with a fast response time. Moreover, it aids in establishing sustainable and low-carbon electric patterns by efficiently utilizing intermittent renewable energy.

Various energy storage systems, including hydroelectric power, capacitors, compressed air energy storage, flywheels, and electric batteries, have been explored as enablers of the power grid. Among these, electric batteries, particularly lithium-ion batteries (LIBs), exhibit significant potential for grid-level electrical energy storage due to their flexible installation, modularization, rapid response, and short construction cycles. LIBs meet complex requirements for grid energy storage applications, considering factors like capacity, energy efficiency, lifetime, and power and energy densities. Notably, LIBs have become a dominant choice in the USA, constituting 77% of electrical power storage systems used for grid stabilization, emphasizing their high-value market (Tianmei Chen et al., 2020). The global Li-ion Battery for Energy Storage Systems (ESS) market size was valued at USD 20430 million in 2022 and is forecast to a readjusted size of USD 58690 million by 2029 with a CAGR of 16.3% during review period (Market Reports World, 2023).

Energy Storage Systems (ESS) are essential for modern energy infrastructure, particularly in integrating renewable energy sources, enhancing grid stability, and providing backup power. Lithium-ion batteries have emerged as a prevalent technology in ESS due to their high energy density, long cycle life, and reliability (Cheng et al., 2011). The versatile applications of lithium-ion batteries in ESS include grid stabilization, renewable energy integration, peak shaving, frequency regulation, microgrid support, uninterruptible power supply (UPS), load shifting, islanded operation, regenerative energy, and demand response (Hesse, Schimpe, et al., 2017). These applications demonstrate the crucial role of lithium-ion batteries in addressing the challenges of modern energy infrastructure.

Grid stabilization is achieved through the quick and responsive energy injections or absorptions provided by lithium-ion batteries, which help in balancing supply and demand on the grid, especially in addressing fluctuations caused by intermittent renewable sources like solar and wind (Hesse, Schimpe, et al., 2017). Furthermore, lithium-ion batteries store excess energy generated from renewable sources during periods of high availability, ensuring a consistent and reliable power supply during low renewable energy generation or high demand periods (Hesse, Schimpe, et al., 2017). This highlights their significance in renewable energy integration.

Due to its widespread availability, renewable sources have emerged as one of the most economically viable options for power generation across many regions (Tianmei Chen et al., 2020). The significant expansion of variable renewable sources in recent years has spurred the development of electrical energy storage systems, necessitating greater flexibility. Battery energy storage systems play a crucial role in efficiently storing electricity generated from renewable sources, enhancing grid system stability and reliability, and thereby fostering the widespread use of renewable energy (Tianmei Chen et al., 2020).

Wind power, a major contributor to renewable energy, faces challenges due to its seasonal and geographical variability, leading to intermittent power generation. Additionally, mismatches between peak power generation and demand are common. An effective solution involves storing excess energy from wind farms to supply electrical energy during peak demand periods. Tianmei Chen et al. (2020) highlights the future potential of lithium-ion batteries (LIBs) as the primary energy storage system in grid-level power stations integrated with renewable energy sources. Notably, a company installed a LIB energy storage system with a capacity of 32 MW/8 MWh (Laurel Mountain) to support a 98 MW wind generation plant in New York in 2011 (Tianmei Chen et al., 2020). In the UK, a significant LIB energy storage pilot project is underway, deploying a LIB system with a capacity of 6 MW/10 MWh at the primary substation, effectively addressing the intermittency of wind and other renewable energy sources (Tianmei Chen et al., 2020). Toshiba has also launched a project installing a 40 MW/20 MWh LIB system in Tohoku, Japan, in December 2013 (Tianmei Chen et al., 2020). These initiatives collectively contribute to the seamless integration of renewable energy into the grid.

Solar photovoltaic power farms can similarly benefit from integrated LIBs, enabling the storage of electrical energy and smoothing out output power. Intermittency during the night and periods of sunlight blockage pose challenges to solar photovoltaic power generation. The integration with batteries creates an efficient operating system capable of handling high-gradient power spikes and steady-state power requirements. The use of batteries in solar photovoltaic fields has demonstrated stability in output power, especially under conditions of partial shading and varying solar radiation (Tianmei Chen et al., 2020). Zubi, Dufo-López, Carvalho and Pasaoglu (2018) recently emphasized the anticipated continued growth of the LIB market with the integration of power supply systems with solar photovoltaics and wind power, projecting an increase to 2 GWh/year in 2020 and 30 GWh/year in 2030.

The process of grid stabilization with lithium-ion batteries centres on leveraging their rapid response capabilities to rectify imbalances between electricity supply and demand. This becomes particularly critical when addressing the inherent fluctuations associated with intermittent renewable energy sources like solar and wind.

Figure 20: 2.56 mW array in Saratoga Springs, NY. Photo by Greg Johnstone. – U.S. Department of Energy from United States, Public domain, via Wikimedia Commons.

Quick and Responsive Energy Injections or Absorptions: Lithium-ion batteries possess the capability to swiftly charge and discharge energy, enabling them to promptly respond to changes in the electricity grid. Whether there is an excess of electricity due to a sudden surge in renewable energy or a decrease in demand, these batteries can absorb surplus energy rapidly. Conversely, during a spike in demand or a decline in renewable energy generation, stored energy can be injected into the grid promptly. This dynamic responsiveness is integral to maintaining grid stability.

Balancing Supply and Demand: The primary role of lithium-ion batteries in grid stabilization is to harmonize the supply of electricity with its demand. Their instantaneous energy injections or absorptions act as a buffer, smoothing out fluctuations caused by the intermittent nature of renewable energy sources. This contribution enhances the overall stability of the electrical grid, ensuring a balance between the amount of electricity generated and consumed at any given moment.

Addressing Fluctuations from Intermittent Renewables: Renewable energy sources, such as solar and wind, introduce unpredictability due to varying electricity generation based on factors like weather conditions. Lithium-ion batteries play a pivotal role in mitigating these fluctuations. Excess electricity generated during peak times by renewable sources is stored in the batteries. Subsequently, during periods of low renewable energy generation or high electricity demand, the stored energy is released to the grid. This process effectively smoothens the intermittent nature of renewable energy, offering a more consistent and reliable power supply.

Storage of Excess Energy: The statement underscores that lithium-ion batteries store excess energy during periods of high renewable energy availability. This storage function is essential for ensuring a continuous power supply during times when renewable sources are not actively generating electricity. It facilitates the efficient capture and retention of surplus energy, creating a reservoir that can be tapped into when needed.

Consistent and Reliable Power Supply: Due to their energy storage capability, lithium-ion batteries contribute significantly to maintaining a consistent and reliable power supply. When renewable energy generation is low or demand peaks, the stored energy in these batteries can be discharged to meet electricity demand. This proactive measure prevents disruptions in power delivery, ultimately enhancing the overall reliability of the grid.

The use of lithium-ion batteries in grid stabilization involves their rapid response to fluctuations, the balancing of supply and demand, and the storage of excess energy from renewable sources. Collectively, these capabilities contribute to a more stable, resilient, and reliable electrical grid.

Figure 21: Invenergy Beech Ridge Energy Storage System at Beech Ridge Wind Farm in Greenbrier County, West Virginia. Z22, CC BY-SA 4.0, via Wikimedia Commons.

Peak shaving, another important application, involves storing energy during periods of low demand and discharging it during peak hours, thereby reducing strain on the grid during high-demand periods and potentially lowering overall electricity costs (Hesse, Schimpe, et al., 2017). Additionally, lithium-ion batteries respond rapidly to changes in grid frequency, contributing to frequency regulation and ensuring the stability of the electrical grid (Hesse, Schimpe, et al., 2017).

Peak shaving, as a significant application, plays a crucial role in optimizing the utilization of lithium-ion batteries. This process entails the storage of energy during periods characterized by low electricity demand and subsequently releasing it during peak hours. The objective is to alleviate the strain on the grid during high-demand periods, offering a means to potentially lower overall electricity costs. This application aligns with the broader goal of enhancing grid efficiency and managing energy resources more effectively.

Furthermore, the rapid responsiveness of lithium-ion batteries to changes in grid frequency adds another layer of functionality. This characteristic enables these batteries to actively contribute to frequency regulation within the electrical grid. In practical terms, as the demand for electricity fluctuates, the grid frequency may vary. The swift response of lithium-ion batteries allows them to assist in maintaining the stability of the grid by actively adjusting to these frequency changes. This contribution to frequency regulation is essential for sustaining the overall reliability and performance of the electrical grid, ensuring that it operates within the desired frequency range.

In essence, the dual capabilities of peak shaving and rapid response to grid frequency changes highlight the versatility of lithium-ion batteries in addressing dynamic energy demands. These applications showcase their adaptability in not only storing and releasing energy strategically but also in actively contributing to the stability and regulation of the electrical grid.

For frequency regulation services, most projects have been reported with a nominal power exceeding 1 MW and a power/energy ratio of approximately 1:1 (Tianmei Chen et al., 2020). The demands of frequency regulation, necessitating a swift response, high rate performance, and robust power capability for the energy storage system, pose challenges for batteries. Ensuring stable and reliable power in large-scale deployment and islanded applications requires careful consideration of voltage and frequency stability. Energy storage systems play a significant role in maintaining the stability of voltage and frequency when there is a disparity between power generation and utilization, addressing both short-term and long-term applications. Given their notable round-trip efficiency and energy density, lithium-ion batteries (LIBs) exhibit significant potential for such applications (Tianmei Chen et al., 2020). A commercially operated LIB energy storage system achieved a power of 8 MW/2 MWh in 2010, which was later increased to 16 MW in 2011 in New York for frequency regulation services (Tianmei Chen et al., 2020).

In the context of applying LIBs to peak shaving services, power sizes vary widely, ranging from 10 kWh to several MWh, reflecting the diverse requirements of different customers. Mitsubishi Heavy Industries, for instance, installed a 1 MW and 400 kWh battery utilizing a combination of nickel, manganese, and cobalt for peak shaving and load leveling in wind farms and solar-power-connected energy storage systems (Tianmei Chen et al., 2020). Furthermore, LIB energy storage systems have been proposed for use in a newly designed DC line interactive UPS due to the high rate pulse discharge capability of LIBs (up to 10 C for less than 10 s), making them attractive for load leveling in the 150 kVA UPS system of a medical imaging machine. The proposed system consists of two modules, each containing 54 LIBs, providing 170 V and pulsed peak power of 75 kW (less than 5 s) (Tianmei Chen et al., 2020). Boasting high discharge/charge efficiency, high specific energy, and long cycle life, LIBs based on electrochemistry emerge as a highly attractive energy storage technology to meet the needs of grid-level applications.

In microgrid applications, lithium-ion batteries serve as a primary energy storage solution, storing energy generated locally from renewable sources and providing backup power during grid outages, thereby enhancing the resilience and reliability of microgrids (Hesse, Schimpe, et al., 2017). Moreover, they are commonly used in UPS systems for various critical applications, providing seamless transition to backup power in the event of a power outage, thus preventing disruptions (Hesse, Schimpe, et al., 2017).

In the realm of microgrid applications, lithium-ion batteries play a pivotal role as the primary energy storage solution. Their fundamental function revolves around the storage of locally generated energy from renewable sources within the microgrid. This aspect aligns with the broader trend toward decentralized energy systems, enabling microgrids to capture and store power derived from sources such as solar panels or wind turbines. The emphasis on local energy storage significantly bolsters the self-sufficiency of microgrids, diminishing reliance on external power sources and fortifying overall resilience.

A key attribute of lithium-ion batteries in microgrid contexts is their ability to provide backup power during instances of grid outages. This feature becomes particularly crucial in bolstering the resilience and reliability of microgrids, ensuring a consistent power supply even when external grids face disruptions. The capacity to seamlessly transition to stored energy during grid outages serves as a critical safeguard for essential operations within the microgrid, establishing lithium-ion batteries as an indispensable component in scenarios where uninterrupted power is imperative.

Beyond their role in microgrids, lithium-ion batteries find common application in Uninterruptible Power Supply (UPS) systems for various critical functions. These systems are employed in diverse contexts, such as data centres, healthcare facilities, or other mission-critical operations where maintaining continuous power is paramount. Lithium-ion batteries contribute significantly to UPS systems by facilitating a smooth and instantaneous transition to backup power in the event of a power outage. This rapid response mechanism proves instrumental in averting disruptions and ensuring the seamless continuation of critical operations, underscoring the indispensable role of lithium-ion batteries in upholding the reliability and functionality of essential systems.

The utilization of lithium-ion batteries in microgrid applications revolves around their primary function as an energy storage solution, supporting local energy generation and fortifying resilience during grid outages. Additionally, their widespread incorporation into UPS systems highlights their instrumental role in ensuring uninterrupted power for critical applications across various sectors. The multifaceted applications of lithium-ion batteries underscore their versatility and crucial contribution to shaping resilient and reliable energy infrastructures.

Load shifting is facilitated by ESS with lithium-ion batteries, allowing energy consumption to be shifted from high-demand to low-demand periods, thereby optimizing energy usage and cost (Hesse, Schimpe, et al., 2017). Furthermore, lithium-ion batteries enable islanded operation, allowing certain sections of the grid or specific facilities to operate independently from the main grid, which is particularly useful in remote areas or during grid failures (Hesse, Schimpe, et al., 2017).

The concept of load shifting takes centre stage in the realm of Energy Storage Systems (ESS) featuring lithium-ion batteries. This application offers a strategic approach to managing energy consumption by leveraging the storage capabilities of lithium-ion batteries. Load shifting entails the deliberate adjustment of energy usage patterns, allowing the transfer of energy consumption from periods characterized by high demand to those marked by lower demand. This nuanced and dynamic energy management strategy proves instrumental in optimizing both energy usage and associated costs.

At its core, load shifting is made possible by the inherent characteristics of lithium-ion batteries. These batteries excel in storing and releasing energy rapidly, allowing for a responsive and controlled redistribution of power as needed. During periods of low energy demand, excess power can be stored in lithium-ion batteries. Subsequently, during high-demand periods or peak hours, the stored energy is seamlessly released, serving as an additional power source to meet the heightened demand. This sophisticated balancing act not only aids in the efficient utilization of electricity resources but also contributes to cost optimization by reducing reliance on external grids during peak tariff hours.

Moreover, lithium-ion batteries contribute significantly to the concept of islanded operation within the grid infrastructure. This capability allows specific sections of the grid or individual facilities to operate independently of the main grid, even in the absence of external power supply. The practical application of islanded operation becomes particularly evident in remote areas or during instances of grid failures. In scenarios where maintaining a continuous power supply is critical, such as in remote installations or critical facilities, the ability of lithium-ion batteries to enable islanded operation ensures a reliable and autonomous source of energy. This proves especially valuable during grid failures or in regions where connection to the main grid is challenging, underscoring the versatility and resilience of lithium-ion battery-based ESS.

Load shifting with lithium-ion batteries in Energy Storage Systems represents a dynamic approach to energy management, optimizing consumption patterns and reducing costs. Additionally, the capability of lithium-ion batteries to support islanded operation enhances resilience, enabling specific grid sections or facilities to operate independently in remote areas or during grid failures. The combined effect of these functionalities contributes to a more flexible, efficient, and robust energy infrastructure.

Regenerative energy applications involve the use of lithium-ion batteries to capture and store energy during braking in electric vehicles and certain industrial processes, thereby improving overall efficiency (Hesse, Schimpe, et al., 2017). Additionally, lithium-ion batteries contribute to demand response programs by adjusting energy consumption based on signals from the grid, discharging stored energy during periods of high demand or when additional grid support is required (Hesse, Schimpe, et al., 2017).

Regenerative energy applications mark a significant frontier in harnessing the capabilities of lithium-ion batteries to enhance overall efficiency in diverse sectors. One key facet of these applications lies in the realm of electric vehicles (EVs) and specific industrial processes. In this context, lithium-ion batteries play a pivotal role in capturing and storing energy generated during braking. This regenerative process, commonly employed in electric vehicles, involves converting kinetic energy into electrical energy during braking events. Lithium-ion batteries efficiently store this captured energy, providing a valuable reservoir for subsequent use. By incorporating regenerative systems, electric vehicles can optimize their energy utilization, extend driving range, and contribute to a more sustainable transportation ecosystem.

Beyond transportation, lithium-ion batteries find application in industrial processes that leverage regenerative energy capture. Certain industrial setups incorporate regenerative systems to recover and store energy generated during specific operations. Lithium-ion batteries, with their high energy density and rapid charge-discharge capabilities, serve as an effective means to store this harvested energy. The stored energy can then be utilized within the industrial facility, contributing to increased energy efficiency and reduced overall energy costs.

In addition to regenerative applications in electric vehicles and industrial processes, lithium-ion batteries play a crucial role in demand response programs. These programs aim to optimize energy consumption by adjusting it based on signals received from the grid. Lithium-ion batteries excel in this context by offering a responsive mechanism for discharging stored energy during periods of high demand or when additional support is required on the grid. By participating in demand response initiatives, lithium-ion batteries contribute to grid stability and resilience. The ability to adapt energy consumption in real-time aligns with the dynamic nature of energy demand, making lithium-ion batteries an integral component of smart grid solutions.

Regenerative energy applications showcase the versatility of lithium-ion batteries in capturing, storing, and strategically deploying energy. Whether in the context of electric vehicles, industrial processes, or demand response programs, these batteries stand as key enablers of enhanced efficiency and sustainability across diverse sectors. Their role in regenerative systems underscores the ongoing evolution of energy storage solutions toward more dynamic, responsive, and resource-efficient paradigms.

Power Tools

One of the prominent applications of lithium-ion batteries is in power tools, where they are preferred for their ability to discharge at high rates, providing backup power, and offering high energy and power densities (Grabmann et al., 2021; Jiang & Song, 2021). These batteries are crucial for applications that require high power density, such as electric vehicles, hybrid vehicles, and portable power tools (Wang et al., 2010). The high power rate of lithium-ion batteries makes them a significant secondary source for various electronic and high-technology industry applications (Lee & Cheng, 2005).

Figure 22: Bosch 12V Max Lithium Ion Impact Driver.
Charles & Hudson, CC BY-SA 2.0, via Wikimedia
Commons.

The safety and reliability of lithium-ion batteries are also important considerations. While lithium-ion batteries offer advantages such as high energy density, long cycle lifetime, and fast recharging, there are safety concerns related to thermal runaway and the potential for fire and explosion accidents (Kimura et al., 2022; Wang et al., 2005). Research has been conducted to monitor the temperature inside lithium-ion batteries and to develop early monitoring and warning systems for thermal runaway (A. Gao et al., 2022; Lee et al., 2011). Additionally, there is ongoing work on predicting the remaining useful life of lithium-ion batteries and estimating their state of health, which is crucial for ensuring their reliability in power tools and other applications (Gao et al., 2020; Y. Li et al., 2019).

In terms of materials and technology, there is a focus on developing electrode materials that can operate at high currents to supply higher powers, as well as exploring new lithium-intercalation cathodes that can store more energy and deliver it at higher power while being safer and cheaper (Liu et al., 2015; Shaju & Bruce, 2006). Furthermore, research has been conducted on the use of nanomaterials and composites to enhance the performance of lithium-ion battery anodes and cathodes, aiming to improve their energy storage capabilities (Jingjie Liu et al., 2021; Wang et al., 2010).

The cost of cordless tools is significantly influenced by the batteries they use, as these batteries dictate the tool's runtime and, to a large extent, its power output. Power tools typically rely on different types of batteries, each with unique maintenance and usability requirements. It is important to recognize that the effectiveness of a cordless tool is heavily dependent on the quality of its battery (Tool Cobler, 2023).

Voltage serves as a measure of the electrical source's strength at a given current level. Rechargeable batteries for power tools commonly consist of cells with voltages of 1.2V, 1.5V, or 3.6V (Tool Cobler, 2023). Manufacturers leverage these cells to create batteries with voltages ranging from 3.6V to 48V. Generally, higher voltage corresponds to heavier batteries capable of delivering more power.

The voltage of a battery does not necessarily determine the specific power output of the tool. The tool's internal components, especially the transmission quality and design, play a more significant role in this regard. It's not uncommon to find a high-quality 14.4V tool that outperforms a lower-quality 18V tool (Tool Cobler, 2023).

Batteries with voltages between 3.6V and 12V are typically designed for light-duty tools like cordless screwdrivers and small cordless drill drivers. For heavier-duty tools such as cordless hammer drill drivers, cordless angle grinders, and cordless circular saws, batteries with voltages ranging from 14.4V to 36V are manufactured. Among manufacturers, 18V is the most common voltage due to its balance between weight, cost, and potential power output (Tool Cobler, 2023).

The maximum runtime of a cordless tool, once fully charged, is predominantly determined by the battery's amp-hour (Ah) value. Essentially, this value serves as the tool's fuel tank, with a larger fuel tank corresponding to a longer operational period. Amp-hour values typically vary between 1.3Ah and 3.3Ah, depending on the tool's intended applications, whether for light or heavy work. It's important to note that as the capacity increases, batteries become heavier and require longer charging times.

In theory, a 1.5Ah battery in a device drawing a current of 1.5 Amps should last for 1 hour. Therefore, the formula for estimating operation time is:

Operation Time (hours) = Amp-hour Value (Ah) ÷ Current Draw (A).

However, this theoretical approach is not entirely effective for accurately predicting the realistic runtime of a cordless tool, as other crucial factors (detailed below) also play a significant role.

The global market for power tools powered by lithium-ion batteries is anticipated to experience significant growth, increasing from $12.36 billion in 2022 to $21.56 billion by 2029, with a Compound Annual Growth Rate (CAGR) of 8.06% from 2023 to 2029 (Huang, 2023). In 2023, North America is poised to contribute substantially to this market, accounting for over 38% of the global revenue. This region, particularly the United States, boasts a notable presence of service providers. Southeast Asia is projected to exhibit the highest growth rate, with a CAGR of 14.02%, followed by China at 12.45% over the forecast period (Huang, 2023).

Several factors propel the growth of this market, including the rising adoption of cordless power tools, increased demand for fastening tools in industrial settings, and the expansion of the construction industry in emerging economies (Huang, 2023). A noteworthy trend in the power tool market is the gradual transition from corded to cordless tools. While corded tools still maintain significant sales, the shift towards cordless tools is a major industry trend. Rechargeable power tools encompass a diverse range of battery-powered tools such as saws, impact drills, impact wrenches, and impact drivers, proving ideal for applications like welding, grinding, cutting, and metalworking (Huang, 2023).

The longevity of a lithium-ion battery is often measured in charging cycles, but the number of cycles can vary based on charging habits and definitions. Factors like when and how often you charge the battery, as well as the battery's configuration, capacity, and storage conditions, all contribute to its overall lifespan (Boll, 2023).

Manufacturers typically define a charging cycle as each instance the battery is placed on the charger and begins recharging. Although lithium-ion batteries are known for lacking cell memory, excessive charging can still diminish the cells' and the pack's maximum lifespan. Currently, most power tool manufacturers claim that their batteries can endure over 1,000 charge cycles. If you charge your battery once a day, this translates to approximately 2.7 years of use, or 3.8 years for a 5-day workweek (Boll, 2023). Some manufacturers even assert 2,000 charge cycles, potentially doubling these durations.

Medical Devices

Lithium-ion batteries, recognized for their high energy density, lightweight design, and rechargeable characteristics, have become integral components in various medical devices. In the medical field, lithium-ion batteries are prevalent in many portable electronic and implantable medical devices, such as flashlights, watches, digital cameras, cellular phones, hearing aids, and pacemakers (Morse et al., 2018). The use of lithium-ion batteries in medical devices is crucial, and there is ongoing research on the state of health prediction of medical lithium batteries based on multi-scale decomposition and deep learning, which has shown promising results (Liu et al., 2020). These batteries play an essential role in powering a diverse array of medical equipment, offering portable and dependable energy sources that are vital for critical healthcare applications.

The demand for lithium-ion batteries in the healthcare industry is experiencing rapid growth. According to a report by Markets and Markets, the global medical battery market size is projected to reach $8 billion by 2025, with a compound annual growth rate of 7% from 2020 to 2025 (Energy5, 2023). This substantial growth is driven by the increasing adoption of portable and implantable medical devices.

However, it is essential to consider the safety aspects of lithium-ion batteries, especially in medical devices. There have been reports of injuries and burns associated with lithium-ion batteries, emphasizing the need for stringent safety measures in their use (Gibson et al., 2018; Nicoll et al., 2016). Despite the hazards, the potential of lithium-ion batteries in medical devices is significant, and research and development work on lithium-ion batteries for environmental vehicles has highlighted their importance as a fundamental solution to critical issues (Abe et al., 2007).

Portable medical devices, such as infusion pumps (see Figure 23), leverage lithium-ion batteries to deliver medications and fluids to patients. The inherent portability and extended battery life of lithium-ion batteries make them particularly well-suited for these critical applications. Similarly, patients requiring supplemental oxygen benefit from the use of portable oxygen concentrators powered by lithium-ion batteries, providing them with the ability to maintain mobility while undergoing oxygen therapy.

In the realm of diagnostic equipment, handheld or portable ultrasound devices used for diagnostic purposes often rely on lithium-ion batteries to meet their energy needs.

This incorporation enhances the mobility of healthcare professionals during diagnostic procedures. Additionally, many blood glucose monitoring devices, especially those designed for continuous monitoring, utilize lithium-ion batteries to ensure reliable and long-lasting power for continuous monitoring of blood glucose levels.

Monitoring and life support systems, such as patient monitors tracking vital signs and implantable medical devices like neurostimulators and cardiac devices, also rely on lithium-ion batteries. These batteries offer a stable and durable power source for devices crucial in intensive care units and other healthcare settings.

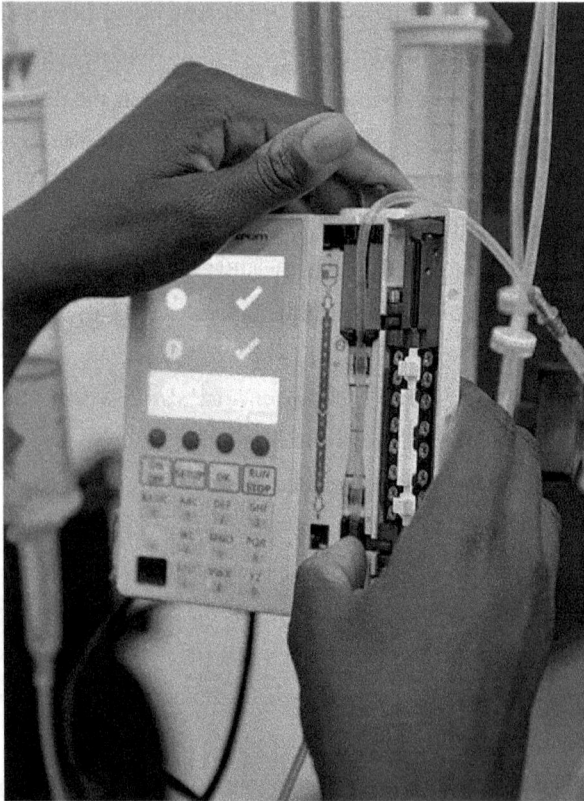

Figure 23: An infusion pump used to infuse fluids, medication or nutrients into a patient's circulatory system. Senior Airman Andrea Posey/U.S. Air Force, Public domain, via Wikimedia Commons.

Lithium-ion batteries play a vital role in powering implantable medical devices such as pacemakers, neurostimulators, and insulin pumps. Their high energy density and extended lifespan make them optimal for these applications. Patients benefit from longer-lasting devices, reducing the need for frequent battery replacement surgeries. Additionally, the compact size of these batteries allows for less invasive implantation procedures.

Rechargeable lithium batteries are a prevalent choice for hearing aids due to their ability to hold a charge for extended periods, lasting multiple years before requiring replacement. This longevity makes lithium batteries a safe option, as users typically do not dispose of them frequently, reducing the risk of being a choking hazard for small children and pets.

In terms of environmental impact, lithium batteries are considered eco-friendly because of their rechargeable nature, leading to less frequent disposal. This not only minimizes hazards associated with improper disposal but also aligns with sustainable practices. Additionally, considering that a significant portion of individuals with hearing loss is over 60, and dexterity issues are common in older adults, the user-friendly aspect of lithium-powered hearing aids is advantageous. Users benefit from not having to regularly handle intricate batteries.

Considering the use of hearing aids, the World Health Organization reported that only one out of five people who could benefit from a hearing aid actually wears one (McCormack & Fortnum, 2013). In the United States, et al. estimated the overall prevalence of hearing aid use among adults 50 years and older with audiometric hearing loss and found that the extent of hearing loss remains untreated, indicating a low prevalence of hearing aid use (Chien, 2012). Similarly, a study in China found that only 0.8% of respondents wore hearing aids despite a high proportion of hearing loss (Heine et al., 2019). These findings suggest a significant gap between the potential benefit of hearing aids and their actual usage.

Furthermore, disparities in hearing aid use were observed among different demographic groups. For instance, a study focusing on Hispanic/Latino adults in the United States revealed that only 13% of older adults from this background reported hearing aid use despite a high prevalence of hearing loss (Arnold et al., 2019). Additionally, the cost of hearing aids was identified as a barrier to access, particularly for low-income older adults, with nearly six million untreated hearing loss cases due to the high cost of hearing aids (Mamo et al., 2016).

Globally, over 1.5 billion individuals currently grapple with hearing loss in at least one ear, necessitating rehabilitation for approximately 430 million people (Wirth, 2023). A noteworthy 13% of adults aged 18 and above encounter hearing difficulties even with the use of hearing aids. Strikingly, those with hearing loss typically wait seven years before seeking assistance (Wirth, 2023). In the United States, approximately 22 million individuals face hazardous noise levels in their workplaces. Alarmingly, almost 60% of hearing loss cases in children result from preventable causes, emphasizing the importance of proper health interventions. Among adults aged 45 to 64, 2.8% of men and 1.9% of women utilize hearing aids, while over 14% of adults aged 65 and above rely on these devices. Moreover, a quarter of individuals worldwide aged 60 and older contend with disabling hearing loss. Projections suggest that an alarming 700 million people globally will grapple with disabling hearing loss by the year 2050.

Lithium-ion batteries have significantly impacted the accessibility of hearing aids due to their high energy density and long lifespan, making them a suitable power source for these devices. The use of lithium-ion batteries in hearing aids has made them more accessible and convenient for users. These batteries offer high energy density and long-lasting power, allowing hearing aids to operate for extended periods without the need for frequent battery replacements (Passerini et al., 2000). Additionally, the lightweight nature of lithium-ion batteries makes them ideal for use in hearing aids, as they do not add significant weight or bulk to the devices, enhancing user comfort and convenience (Ansari et al., 2021).

Furthermore, the rechargeable nature of lithium-ion batteries has contributed to the accessibility of hearing aids by reducing the ongoing cost of disposable batteries. By utilizing rechargeable lithium-ion batteries, users can simply recharge their hearing aids, eliminating the need for frequent battery replacements and reducing long-term costs (Delnavaz & Voix, 2014). This aspect has made hearing aids more economically viable for individuals who require them, thereby increasing accessibility. The high gravimetric capacities and theoretical energy densities of lithium-sulphur batteries, a type of lithium-ion battery, have been under intense scrutiny for over two decades, indicating the potential for even greater advancements in battery technology that could further enhance the accessibility of hearing aids (Ji et al., 2009).

Pacemakers play a critical role in providing essential medical support by regulating the user's heartbeat through electrical pulses. Given the life-saving function of pacemakers, a reliable power source is imperative for their operation.

Lithium batteries are widely employed in pacemakers due to their attributes such as extended lifespan, low drain current, and voltage characteristics. In comparison to previous usage of mercury-zinc batteries for these devices, lithium batteries have surpassed them in numerous benefits. Notably, lithium batteries stand out by not producing any gas, allowing them to be sealed in an airtight manner for enhanced safety and efficiency.

Lithium-ion batteries have significantly impacted the accessibility of pacemakers due to their ability to provide a reliable power source for these life-saving devices. Historically, the dominance of lithium-iodine cells in the 1970s marked a significant advancement in the power source for low voltage, microampere current, single- and dual-chamber pacemakers (Mond & Freitag, 2014). This transition to lithium-iodine cells revolutionized the small pacemaker industry, making pacemakers more reliable and long-lasting. Additionally, primary lithium batteries have been crucial for implantable devices such as cardiac pacemakers, further emphasizing the role of lithium batteries in the field of medicine (Takeuchi & Leising, 2002).

Furthermore, the development of lithium/carbon monofluoride (Li/CFx) batteries has been motivated by the need for pacemakers to become smaller while adding more current-demanding functions, highlighting the continuous evolution of battery technology to meet the specific requirements of pacemakers (Greatbatch et al., 1996). This evolution has been further supported by the practical use of flexible lithium-ion batteries, which has expanded the potential applications of lithium-ion batteries in the medical field (Zeng et al., 2019).

The impact of lithium-ion batteries extends beyond pacemakers, as they have become the mainstream power source for automotive batteries, indicating their widespread adoption and reliability in various critical applications (Haizhou, 2017). While there is a growing interest in sodium-ion batteries for electric vehicles and portable electronic devices, lithium-ion batteries remain the dominant energy storage system, underlining their significance and impact (Jin et al., 2023).

Surgical tools, particularly powered surgical instruments designed for minimally invasive procedures, may incorporate lithium-ion batteries. These tools demand reliable and rechargeable power sources to ensure precision and effectiveness in surgical procedures.

Wearable health technology, exemplified by devices like fitness trackers and health monitors, often integrates lithium-ion batteries due to their compact design and ability to provide sustained power for continuous monitoring of various health parameters.

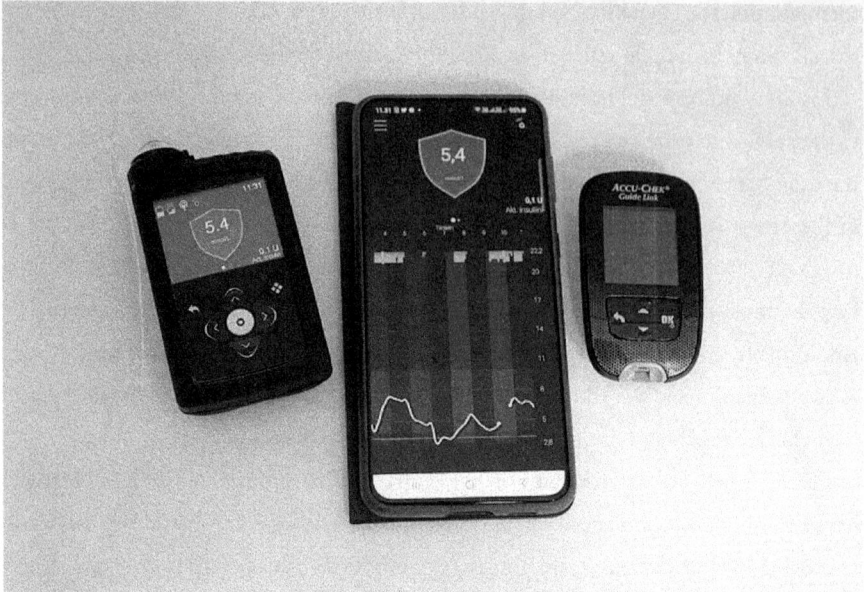

Figure 24: Minimed Mobile APP is in use on a mobile phone (Samsung Galaxy S21). MailariX, CC BY-SA 4.0, via Wikimedia Commons.

Rehabilitation devices, including powered wheelchairs and scooters, utilize lithium-ion batteries to provide users with a reliable and rechargeable power source, contributing to enhanced independence for individuals with mobility challenges.

The integration of lithium-ion batteries into medical devices significantly contributes to improved patient care, heightened mobility for both patients and healthcare professionals, and the advancement of innovative, portable healthcare solutions. Manufacturers place a paramount emphasis on ensuring the safety and reliability of lithium-ion batteries in medical applications to safeguard the well-being of patients and enhance the effectiveness of medical devices.

Aerospace, Marine and Defence

Lithium-ion batteries (LIBs) have emerged as revolutionary power sources, driving innovation across various industries. In the realms of aerospace, marine, and defence, these advanced energy storage systems have catalysed transformative changes, offering lightweight, high-energy-density solutions. Lithium-ion batteries have become increasingly important in various industries such as aerospace, marine, and defence due to their high energy density, long cycle life, and lightweight characteristics (Bugga et al., 2010). The application range of lithium-ion batteries has widened to include electric power, aerospace, military, and other critical fields (L. Zhang et al., 2017). In the aerospace industry, lithium-ion batteries are used in electric and hybrid vehicles, satellite systems, and interplanetary missions (Sone et al., 2010). Similarly, in the marine industry, multifunctional structure-battery composites have been developed using fibre-reinforced marine composites for energy storage and structure function (Thomas et al., 2012). Additionally, in defence applications, lithium-ion batteries are used in various devices and systems due to their high energy density and long cycle life (L. Zhang et al., 2017).

Figure 25: Quad propeller drone with Lidar technology. Jonte, CC BY-SA 4.0, via Wikimedia Commons.

The safety of lithium-ion batteries is a critical concern, especially in aerospace and defence applications. There are regulations related to the safety of lithium-ion batteries during civil aviation transportation, and it is essential to ensure their safe use in these environments (Kimura et al., 2022). Furthermore, to ensure battery safety, evaluation criteria for the safety of lithium-ion batteries have been pioneered by the United States and Japan (Shan, 2017).

The integration of lithium-ion batteries into multifunctional composite structures has been a topic of interest, particularly in the marine industry, where fibre-reinforced composites have been used for both energy storage and structural functions (Thomas et al., 2012). Additionally, the development of energy storage composites containing integrated lithium-ion batteries has been presented for potential applications in aerospace, aircraft, spacecraft, marine, and sports equipment (Galos et al., 2021). This integration of batteries into composite structures allows for efficient use of space and weight, making them suitable for aerospace and marine applications.

The use of lithium-ion batteries in defence applications has also been highlighted, with references to their application in military and critical fields (L. Zhang et al., 2017). The high energy density and long cycle life of lithium-ion batteries make them suitable for various defence applications, including portable electronics and military equipment.

Unmanned Aerial Vehicles (UAVs) and drones have become integral in various applications, such as surveillance, package delivery, agriculture, environmental monitoring, and disaster response. The pivotal role of Lithium-ion Batteries (LIBs) in extending the flight duration of these devices is well recognized due to their lightweight design and high energy density, enabling longer missions and increased payload capacity (Hao et al., 2018). LIBs are widely used in UAVs for military, commercial, and personal applications, and the demand for UAVs with long flight times is increasing exponentially (Jung & Jeong, 2017). Additionally, the use of LIBs for electric vehicles (EVs), drones, and energy storage has been steadily growing, owing to their high specific energy, lightweight, and low self-discharge rate (Zhao et al., 2020).

Efficient thermal management of LIBs is crucial for UAVs, and various techniques have been explored, such as a passive interfacial thermal regulator based on a shape memory alloy and direct evaporative cooling approaches (Jiang & Song, 2021; Türk et al., 2021). These techniques are essential for ensuring the safety and performance of LIBs in UAVs. Moreover, the state estimation of LIBs is critical for the safety of UAVs, and novel equivalent modelling methods have been developed for accurate state estimation (Hao et al., 2018).

In the context of UAVs, it is essential to consider the high discharge rate of batteries for safe landing in emergencies, such as cable faults or high winds (Huang et al., 2023). Furthermore, the development of UAVs powered by electric motors offers advantages such as low noise, high efficiency, and zero emissions, making them suitable for various applications. The integration of electric vehicle-assisted charging mechanisms for UAVs reflects the growing interest in utilizing advanced technologies for extending the operational capabilities of UAVs

The aviation industry is currently undergoing a significant transformation with the emergence of electric aircraft. This paradigm shift is driven by the development of electric propulsion systems, which offer a more energy-efficient alternative to traditional aviation fuels (König et al., 2023). The transition from conventional to electric propulsion systems represents a major shift with substantial potential for the aviation industry, promising reduced emissions and operational costs (S. Li et al., 2021). Furthermore, this shift is expected to lead to weight reductions, lower life cycle costs, reduced environmental impact, and increased reliability of the entire aircraft system (Baldo et al., 2023).

Lithium-ion batteries (LIBs) are at the forefront of this transition, serving as a cornerstone for the development of electric aircraft. The versatility of LIBs in aviation is exemplified by projects such as the eFlyer 2 and Alice aircraft, which are exploring electric aircraft for regional flights (Madonna et al., 2018). These developments align with the industry's pursuit of more electric solutions to optimize performance, decrease operating and maintenance costs, and reduce environmental impact (Schäfer et al., 2018).

The shift towards electric aircraft is also reflected in the technological, economic, and environmental prospects of all-electric aircraft, driven by the need to reduce greenhouse gas emissions and advancements in battery technology (Barzkar & Ghassemi, 2020). As aircraft systems transition from conventional to more electric and all-electric configurations, there is a significant increase in the complexity of electric power systems, presenting new challenges and opportunities for the industry (Nøland et al., 2019).

Safety-critical power conversion systems play a pivotal role in the paradigm shift towards more electric aircraft architectures, emphasizing the criticality of reliable and efficient electrical systems in aviation (Oliveira et al., 2023). Additionally, the development of electric aircraft is not only driven by environmental considerations but also by the potential for performance optimization and cost reduction, aligning with the industry's pursuit of more sustainable and efficient solutions (Chen-Glasser & DeCaluwe, 2022). The transition to electric aircraft is not without its challenges, including technical hurdles related to increasing battery energy density and logistical complexities associated with the lithium supply chain. However, the growing interest in electric propulsion in aviation, along with advancements in battery technology, underscores the industry's commitment to embracing this transformative shift.

LIBs have found their way into space exploration, powering satellite systems, rovers, and probes. The high energy density and long lifespan of lithium-ion batteries make them ideal for sustaining the extended missions required in deep space. These batteries contribute to the reliability of critical space equipment, ensuring consistent power supply for scientific exploration.

Maritime transportation is undergoing a transformation with the adoption of electric propulsion systems. LIBs are at the forefront of this evolution, enhancing the efficiency of marine vessels while minimizing environmental impact. Electric ferries, cargo ships, and even naval vessels are incorporating lithium-ion batteries to reduce reliance on traditional fuels and cut emissions. This shift aligns with the growing emphasis on sustainability in the maritime industry.

Underwater applications present unique challenges for energy storage, and lithium-ion batteries (LIBs) have emerged as a solution due to their compact design and high energy density. LIBs play a critical role in ensuring reliable power sources for underwater exploration and defence operations, enabling extended missions and improved manoeuvrability (Meem et al., 2022). The compact design of LIBs is particularly beneficial for submarines and underwater drones, where space is limited and energy density is crucial for prolonged missions (Trinca, 2022).

In the context of underwater energy storage, all-solid-state lithium metal batteries (ASSLiMB) have been considered as a promising next-generation high-energy storage system, replacing liquid organic electrolytes with solid-state electrolytes (He et al., 2019). Additionally, two-dimensional (2D) materials such as graphene-based silicene, borophene, and phosphorene heterostructures have been identified as new electrode candidates for future energy storage devices, potentially offering advancements in underwater energy storage technologies (Özkan et al., 2020).

Furthermore, the challenges of underwater networking and communication have led to the development of energy-centric green underwater networks, with various techniques and approaches being suggested to enable efficient energy usage in underwater applications (Rathore et al., 2020). Additionally, the potential for underwater compressed air energy storage in offshore wind integration has been highlighted, indicating the relevance of alternative energy storage methods for underwater applications (Pete et al., 2015).

Moreover, the use of energy harvesting technologies, such as autonomous low-power management systems for energy harvesting from piezoelectric transducers, has been explored to address the lifetime and environmental impact of electrical batteries in underwater applications (Diab et al., 2019). This aligns with the need for sustainable and long-lasting energy solutions in underwater environments.

In the context of underwater drones, the employment of 12V and 24V lithium-ion batteries as power sources further emphasizes the significance of LIBs in powering underwater technologies (Meem et al., 2022). Additionally, the development of open-source underwater drone control platforms and the design and analysis of remotely amphibious drones highlight the ongoing efforts to enhance communication systems and structural integrity for underwater applications (Aristizábal et al., 2016; Suresh et al., 2020; Yusof et al., 2021).

In the realm of leisure and recreation, lithium-ion batteries are powering a new generation of electric boats and yachts. The marine industry is witnessing a surge in electric propulsion, driven by the desire for cleaner and quieter boating experiences. LIBs provide the necessary energy density for extended trips, and their rechargeable nature aligns with the convenience sought by recreational boaters.

Lithium-ion batteries have become the preferred choice for enhancing soldier-wearable technology due to their high energy density, lightweight design, and longer lifespan compared to other battery technologies (J. M. Tarascon & M. Armand, 2001). These batteries are widely used in various practical applications, offering mature production technology and relatively high energy density, making them suitable for flexible and wearable devices (Chang et al., 2021; Song et al., 2021). The reliability and energy efficiency of LIBs play a crucial role in improving the mobility and effectiveness of soldiers in the field, as they provide lightweight and portable solutions for communication devices, night vision equipment, and other soldier-wearable technology (J. M. Tarascon & M. Armand, 2001).

Furthermore, the use of LIBs in soldier-wearable technology aligns with the need for high energy density and lightweight energy sources, as highlighted in the literature (Chang et al., 2021; Song et al., 2021). The flexibility and lightweight design of LIBs make them suitable for integration into wearable technology, contributing to the mobility and effectiveness of soldiers in the field. Additionally, the high energy density of LIBs enables the development of compact and portable solutions for military personnel, addressing the requirements for soldier-wearable technology (Chang et al., 2021; Song et al., 2021).

The transition to electric mobility is evident in the defence sector, with military vehicles incorporating lithium-ion batteries. Tanks and armoured vehicles benefit from the high power density of LIBs, providing the necessary energy for propulsion and onboard systems. The adoption of electric military vehicles contributes to reduced fuel dependency and operational flexibility.

Moreover, the widespread commercialization and demand for LIBs since the 1990s have led to an exponential growth in their usage, making them a prominent choice for various applications, including soldier-wearable technology (Gangaja et al., 2021). The continuous advancements in LIB technology, such as the development of flexible and stretchable energy storage solutions, have further contributed to their suitability for soldier-wearable technology (C. Chen et al., 2020; Liu et al., 2016).

In the realm of defence, the reliability and precision of missile systems are paramount. Lithium-ion batteries power these critical systems, ensuring the required energy density and voltage characteristics. The sealed and gas-free nature of LIBs makes them suitable for integration into missile systems, contributing to the effectiveness of defence strategies.

Chapter Five

Challenges and Issues

Lithium-Ion Batteries and Other Energy Sources

Lithium-ion batteries are a widely adopted form of energy storage, amidst a diverse array of energy storage technologies, each characterized by unique attributes, advantages, and drawbacks. A comprehensive examination of these technologies provides insight into their respective features. Lead-Acid Batteries, for instance, employ a chemical reaction between lead dioxide and sponge lead for energy storage. While they offer a lower cost relative to lithium-ion, leverage well-established technology, and are recyclable, they suffer from being heavier, less energy-dense than lithium-ion, and having a shorter cycle life (Nakaya et al., 2014). On the other hand, Nickel-Cadmium (Ni-Cd) Batteries utilize nickel oxide hydroxide and metallic cadmium as electrodes. Appreciated for their reliability, extended cycle life, and performance at high temperatures, they are constrained by the toxicity of cadmium, lower energy density compared to lithium-ion, and a tendency to exhibit memory effect (Nivetha & Saravanathamizhan, 2019). Flow Batteries, which store energy in liquid electrolytes within external tanks, are known for scalability, longer cycle life, and potential for extended-duration storage, but grapple with challenges related to complex design, lower energy density, and often a larger physical footprint (Wagemaker et al., 2021). Solid-State Batteries, which employ solid electrolytes in place of liquid electrolytes, are recognized for higher energy density, potential safety improvements, and extended cycle life, but currently face challenges related to higher manufacturing costs and an ongoing developmental phase (Sakuda et al., 2016).

Hydrogen Fuel Cells, which generate electricity through the chemical reaction between hydrogen and oxygen, offer high energy density, rapid refuelling, and zero greenhouse gas emissions during operation. However, they are hindered by a complex infrastructure for hydrogen production and distribution, coupled with lower efficiency compared to some batteries (Habib & Arefin, 2022). Supercapacitors, which store energy by separating positive and negative charges on their surfaces, facilitate rapid charging and discharging, and exhibit a longer cycle life than many batteries. However, they contend with lower energy density than traditional batteries and may not be suitable for prolonged energy storage (Šimić et al., 2021). Pumped Hydro Storage, which operates by moving water between two reservoirs at different elevations for energy storage, boasts attributes such as large-scale implementation, an established technology, and a long cycle life. However, it is limited by site-specific requirements, environmental impact concerns, and applicability to specific geographical locations (Shudo & Suzuki, 2008). Compressed Air Energy Storage (CAES), which stores energy by compressing air and storing it in underground caverns, offers scalability and potential for large-scale energy storage but encounters challenges associated with site-specific requirements and energy efficiency during compression and expansion processes (Zhang et al., 2010).

Lithium-Ion Battery Limitations

Lithium-ion batteries find widespread use in various electronic devices, electric vehicles, and renewable energy systems, owing to their high energy density and relatively long lifespan. However, these batteries are not without their challenges. Common issues associated with lithium-ion batteries include a limited lifespan. Over time, these batteries degrade, experiencing a reduction in capacity and overall performance. This degradation is influenced by factors such as temperature, charge/discharge cycles, and usage patterns.

Safety concerns also surround lithium-ion batteries. They can be prone to overheating, potentially leading to thermal runaway and, in extreme cases, fires or explosions. This risk is often associated with manufacturing defects, physical damage, or exposure to extreme conditions. Another challenge is the cobalt dependency of many lithium-ion batteries. The use of cobalt in cathodes raises ethical and environmental concerns, prompting efforts to develop alternative chemistries with reduced or no cobalt.

Resource depletion is a significant issue in the production of lithium-ion batteries. Relying on rare metals such as lithium, cobalt, and nickel raises concerns about long-term availability and the environmental impact of mining. The charge/discharge rate poses challenges as well, with rapid cycling causing increased stress on the battery, contributing to degradation and potential overheating issues.

Temperature sensitivity is a factor negatively impacting lithium-ion battery performance and lifespan. Operating these batteries outside their recommended temperature range can lead to accelerated degradation and safety issues. While not as pronounced as in some other types of batteries, lithium-ion batteries can experience a memory effect, resulting in a reduction of effective capacity through repeated partial discharges and recharges.

Capacity fading is another issue, with lithium-ion batteries experiencing a gradual loss of capacity even when not in use. This self-discharge can lead to reduced performance over time. Additionally, the cost of lithium-ion batteries, although decreasing over the years, remains a significant factor, especially in large-scale applications like electric vehicles and grid storage.

Researchers and engineers are actively working to address these challenges through advancements in battery technology, the development of alternative battery chemistries, and improvements in manufacturing processes. As technology evolves, there is a likelihood that some of these issues will be mitigated or overcome.

Limited Lifespan

The degradation of lithium-ion batteries is a well-documented challenge associated with these energy storage devices. This limitation is primarily attributed to several interconnected factors that impact the chemical and physical properties of the battery. Temperature, frequency and depth of charge/discharge cycles, and usage patterns significantly influence the degradation process of lithium-ion batteries (G. Zhang et al., 2022; Zhu et al., 2022). Exposure to high temperatures accelerates the breakdown of the electrode materials and electrolytes within the battery, leading to a decline in performance (H. Zhang et al., 2017). Additionally, the movement of lithium ions between the positive and negative electrodes during charge/discharge cycles can lead to "lithium plating," compromising the integrity of the electrodes and causing capacity fade (A. Wang et al., 2019; Xu et al.,

2014). Furthermore, usage patterns such as deep discharges or consistently keeping the battery at a high state of charge can expedite the wear and tear of the battery's components, contributing to a more rapid decline in its overall health (Chen & Ding, 2015; Liew et al., 2022).

The cumulative impact of these factors results in a reduction in the battery's capacity, leading to decreased energy storage and delivery over time (Ashwin et al., 2016; A. Wang et al., 2019). This diminished capacity not only affects the runtime of devices powered by these batteries but also necessitates more frequent recharging, leading to increased inconvenience for users (Ashwin et al., 2016; A. Wang et al., 2019).

Efforts to address this limitation involve ongoing research and development aimed at enhancing the resilience of lithium-ion batteries to these degradation mechanisms (Z. Chen et al., 2023; Guo & Hu, 2008). Scientists and engineers are exploring new materials, improved electrode designs, and advanced manufacturing processes to mitigate the impact of temperature, cycling, and usage patterns, with the goal of extending the overall lifespan of lithium-ion batteries (Z. Chen et al., 2023; Guo & Hu, 2008).

The degradation of lithium-ion batteries is a complex process influenced by temperature, charge/discharge cycles, and usage patterns. Addressing this challenge requires a multidisciplinary approach involving materials science, electrochemical energy storage, and battery management. Ongoing research and development efforts are focused on improving the resilience of lithium-ion batteries to these degradation mechanisms, with the ultimate goal of extending their lifespan and enhancing their performance.

The lifespan of lithium-ion batteries is influenced by various factors, encompassing battery type, usage patterns, operating conditions, and the specific application in which they are employed. Generally, lithium-ion batteries exhibit a lifespan range of 2 to 20 years (J. M. Tarascon & M. Armand, 2001).

One crucial determinant is the number of charge and discharge cycles a lithium-ion battery can endure, defined as a complete 100% discharge followed by a 100% recharge. For instance, if a battery undergoes one cycle daily and is rated for 500 cycles, its lifespan would be approximately 1.5 years. Advances in battery technology have, however, led to improvements, with some lithium-ion batteries now capable of enduring thousands of cycles.

The depth to which the battery is discharged during each cycle is another significant factor impacting its lifespan. Shallow discharges, utilizing only a small percentage of the battery's capacity, contribute to a longer lifespan compared to deep discharges.

Operating temperatures play a crucial role in lithium-ion battery life. High temperatures accelerate chemical reactions, leading to faster degradation, while low temperatures reduce battery efficiency.

The specific battery chemistry employed also plays a role in determining lifespan. Different lithium-ion battery chemistries, such as lithium cobalt oxide ($LiCoO_2$), lithium manganese oxide ($LiMn2O4$), and lithium iron phosphate ($LiFePO_4$), exhibit varying lifespans, with some known for longer cycles and better performance under specific conditions.

The intended application of the battery further influences its lifespan. Batteries in electric vehicles (EVs), for example, may have a different lifespan compared to those in consumer electronics or grid storage systems.

While lithium-ion batteries offer high energy density, flexible and lightweight design, and longer lifespan compared to other battery technologies (J. M. Tarascon & M. Armand, 2001), they are also subject to reduction in lifespan due to transient stress during acceleration (Liew et al., 2022). The continuous generation of heat in electric vehicle lithium-ion battery packs can lead to significant temperature differences between the battery cells, ultimately deteriorating the performance and lifespan of the batteries (S. Wang et al., 2020).

Research has been conducted to predict the residual life of lithium-ion batteries for electric vehicles based on particle filter methods (An, 2021). Additionally, the life cycle of a lithium-ion battery has been divided into three parts, including the degradation process, capacity regeneration process during rest time, and the degradation process of the regenerated capacity, which impacts the overall lifespan of the battery (Xing et al., 2019). Furthermore, the capacity degradation model of lithium-ion batteries has been studied, indicating that the batteries degrade with a decrease in capacity and an increase in resistance over time (Ouyang et al., 2016).

As a general guideline, consumer electronics like smartphones and laptops typically have a lifespan of 2 to 5 years, electric vehicles range from 8 to 15 years, and renewable energy storage may last 5 to 20 years.

It is important to note that the end of a battery's lifespan does not imply sudden failure; rather, its capacity has significantly diminished. Manufacturers often specify the number of cycles a battery can undergo before reaching around 80% of its original capacity. Ongoing advances in battery technology and research may lead to further improvements in lifespan in the future. Further, the materials used in lithium-ion batteries play a crucial role in determining their lifespan. For example, the use of iron nitride@C nanocubes inside core–shell fibres has been shown to realize high air-stability, ultralong life, and superior lithium/sodium storages, thereby contributing to the longevity of lithium-ion batteries (X. Li et al., 2021). Additionally, the toxicity of the gas released during the combustion of lithium-ion batteries has been studied, highlighting the importance of understanding potential factors that may impact the lifespan of these batteries (Chen et al., 2017).

Safety Concerns

The safety concerns regarding lithium-ion batteries primarily revolve around the potential for overheating, which can escalate to a more severe condition known as thermal runaway, and in extreme cases, lead to fires or explosions. This inherent risk is associated with various factors, including manufacturing defects, physical damage, and exposure to extreme operating conditions. The issue of thermal runaway in lithium-ion batteries has been extensively studied by researchers, with a focus on understanding the causes and developing strategies to mitigate the associated hazards.

Figure 26: Philips BT6000A/12 - Lithium Ion Battery
ICR18650, made by Great Power Battery (ZhuHai) Co, Ltd.
© Raimond Spekking, CC BY-SA 4.0, via Wikimedia Com-
mons.

Research has shown that fast and reliable detection of faulty cells undergoing thermal runaway within lithium-ion batteries is crucial for ensuring passenger safety (Koch et al., 2018). When exposed to abusive electrical, thermal, or mechanical conditions, the active materials within lithium-ion batteries break down exothermically, generating large amounts of heat which can lead to thermal runaway, accompanied by fire and/or explosion (Finegan, Darcy, Keyser, Tjaden, Heenan, Jervis, Bailey, Vo, et al., 2017). Basic research on the battery thermal runaway, mainly the internal exothermic reaction, and the safety performance of lithium-ion batteries, along with overcharge and high-rate charging, are identified as important causes of combustion and explosion (Liu et al., 2017).

Furthermore, studies have investigated the influence of factors such as aerogel felt thickness on the propagation of thermal runaway in lithium-ion batteries, providing valuable suggestions for the aviation safety transportation of these batteries (Kimura et al., 2022). Additionally, the development of microcapsule extinguishing agents with a core–shell structure has been explored to enhance lithium-ion battery fire safety, indicating the ongoing efforts to address safety issues associated with these batteries (W. Zhang et al., 2021).

The safety of lithium-ion batteries has become a major constraint on their wide application, particularly in electric vehicles (EVs) (Jia et al., 2022). Comparative analysis of thermal runaway characteristics of lithium-ion batteries under different conditions has highlighted thermal runaway as one of the greatest potential safety hazards affecting the application of these batteries (Zhu et al., 2021). Moreover, numerical studies have been conducted to explore the inhibition control of lithium-ion battery thermal runaway, emphasizing the wide range of applications of these batteries and the need to reduce environmental pollution (Hu et al., 2020).

The safety issue remains a significant obstacle for the commercialization of high energy density lithium-ion batteries, prompting advances in simulation research to understand and address the thermal runaway of these batteries (Ma et al., 2021).

Manufacturing defects are a significant contributor to safety issues in lithium-ion batteries, as occasional defects during the production process can lead to internal flaws within the battery cells, compromising their integrity and making them more susceptible to thermal instability and the risk of overheating (Ahmed et al., 2016; Ali et al., 2019; Feng et al., 2020; Shan, 2017; H. Zhang et al., 2017; C. Zhao et al., 2021). Despite stringent quality control measures, these defects can still occur, posing a challenge to the safety and reliability of lithium-ion batteries (Ahmed et al., 2016; Ali et al., 2019; Feng et al., 2020; Shan, 2017; H. Zhang et al., 2017; C. Zhao et al., 2021).

Feng et al. (2020) highlighted that the abuse conditions triggering thermal runaway are critical for protecting lithium-ion batteries, emphasizing the importance of understanding and mitigating thermal runaway for battery safety. C. Zhao et al. (2021) also emphasized the frequent occurrence of thermal runaway fire and explosion accidents in lithium-ion batteries due to manufacturing defects, improper use, and unexpected accidents, underscoring the pressing need to address safety concerns. Additionally, (Ahmed et al., 2016)stressed the importance of the maturity of the manufacturing process to reduce the cost of lithium-ion battery packs, indicating the significance of addressing manufacturing defects to enhance safety and reduce costs.

Furthermore, Shan (2017) and H. Zhang et al. (2017) discussed the impact of over-charge, over-discharge, and high-temperature environments on the thermal instability of lithium-ion batteries, leading to fire or explosion, thereby hindering their application and necessitating the resolution of safety issues. Additionally, Khan (2020) highlighted the diverse safety issues associated with lithium-ion batteries, emphasizing the need for a charging interface compatible with their specific requirements to ensure safety.

To understand the impact of physical damage on battery safety, it is crucial to consider the various stages of a battery's life cycle, including manufacturing, transportation, and usage. Physical damage, such as impact or compression, can compromise the internal structure of the battery, leading to potential safety hazards like short circuits and over-heating (Fowler & Fowler, 2020). Mishandling or dropping devices containing lithium-ion batteries can also result in physical damage, posing safety concerns in consumer electronics (Fowler & Fowler, 2020).

During the manufacturing process, cell manufacturing has been identified as a significant contributor to greenhouse gas emissions, accounting for 45% of the emissions (H. C. Kim et al., 2016) Additionally, the eco-efficiency of battery manufacturing can be improved by increasing production capacity and utilizing an electricity mix with low carbon intensity (Philippot et al., 2019). Furthermore, the prediction of contaminations or non-obvious battery damage after manufacturing and during use has been facilitated by the usability of classification algorithms (Höschele et al., 2022).

In the usage phase, various studies have focused on fault diagnosis and monitoring of lithium-ion batteries to ensure safety and performance. For instance, online detection of soft internal short circuits in lithium-ion batteries has been explored, highlighting the importance of monitoring for manufacturing defects and battery misuse during usage (Seo et al., 2020). Additionally, the reliability of battery fault diagnosis has been linked to an accurate estimation of the state of charge and battery characterizing parameters, emphasizing the need for precise monitoring during usage (Zheng et al., 2017).

Overall, the impact of physical damage on battery safety is a multifaceted issue that spans the entire life cycle of the battery. It requires considerations at each stage, from manufacturing to transportation and usage, to ensure the safety and efficiency of battery systems.

Exposure to extreme conditions, both environmental and operational, poses a significant risk. High temperatures, for instance, accelerate the chemical reactions occurring within the battery and can trigger thermal runaway. Overcharging or rapid charging at rates beyond the battery's designed specifications can generate excess heat, leading to thermal instability.

Moreover, the safety of lithium-ion batteries is a critical consideration in electric vehicles (EVs) and large-scale energy storage systems. In these applications, the sheer size and energy capacity of the battery packs amplify the potential consequences of a safety incident. Rigorous safety features and management systems are implemented to minimize these risks, including thermal management systems, overcharge protection, and temperature sensors.

To address safety concerns, ongoing research and development efforts focus on improving the design and manufacturing processes of lithium-ion batteries. This includes the development of advanced materials and technologies that enhance thermal stability, as well as the implementation of safety features and protocols to detect and mitigate potential issues before they escalate. Despite the inherent risks, when handled properly and with appropriate safety measures in place, lithium-ion batteries remain a widely used and effective energy storage solution across various applications.

Cobalt Dependency

The cobalt dependency in many lithium-ion batteries arises from its common use in the cathode, a critical component that plays a key role in the electrochemical processes within the battery. While cobalt enhances the performance of lithium-ion batteries, its extraction and usage have raised significant ethical and environmental concerns, prompting efforts to develop alternative battery chemistries that either reduce the reliance on cobalt or eliminate it entirely.

Cobalt mining is associated with environmental degradation and human rights issues. The majority of the world's cobalt production occurs in the Democratic Republic of Congo (DRC), where concerns have been raised about mining practices, including child labour, unsafe working conditions, and environmental damage (Parker et al., 2017). The extraction process often involves large-scale open-pit mining, leading to deforestation, soil erosion, and pollution of water sources, further exacerbating the environmental impact.

Ethical concerns also arise from the social implications of cobalt mining, particularly the exploitation of vulnerable populations in the mining regions. Child labour and unsafe working conditions in cobalt mines have drawn international attention and criticism. As a result, there is a growing awareness of the need to address these issues and find alternative solutions to reduce the demand for cobalt.

Efforts to address the cobalt dependency in lithium-ion batteries involve research and development initiatives focused on alternative battery chemistries. One approach is to replace or minimize the use of cobalt in cathodes by exploring different materials and formulations. Nickel-based cathodes, such as nickel-manganese-cobalt (NMC) and nickel-cobalt-aluminium (NCA), are being investigated as alternatives to traditional cobalt-containing cathodes (Yi et al., 2022). These alternatives aim to maintain or improve the energy density and performance of lithium-ion batteries while reducing the environmental and ethical concerns associated with cobalt.

Moreover, researchers are exploring entirely cobalt-free chemistries, such as lithium iron phosphate (LiFePO$_4$) and other manganese-based cathodes. These alternatives, while offering lower energy density compared to cobalt-containing cathodes, can provide a more sustainable and ethical solution for certain applications (Yi et al., 2022).

The push for cobalt-free or low-cobalt battery technologies aligns with the broader goal of creating more sustainable and environmentally friendly energy storage solutions (Parker et al., 2017). As these alternative chemistries progress and become more economically viable, the hope is to reduce the environmental and social impact of lithium-ion batteries, making them a more responsible choice for a wide range of applications (Yi et al., 2022).

Resource Depletion

Resource depletion in the production of lithium-ion batteries is a significant concern due to the heavy reliance on rare metals such as lithium, cobalt, and nickel (Vera et al., 2023). The increasing demand for lithium-ion batteries, driven by the rising adoption of electric vehicles, renewable energy systems, and portable electronics, raises apprehensions about the long-term availability of these crucial resources and the environmental impact associated with their extraction and processing (Graham et al., 2021). Most of the world's lithium production is concentrated in a few countries, such as Australia, Chile, and Argentina, which raises concerns about the sustainability of lithium extraction (Stringfellow

& Dobson, 2021). Extracting lithium from brine deposits, as commonly done, can impact local water supplies and ecosystems, while traditional lithium mining methods pose environmental challenges, including habitat disruption and water pollution (Vera et al., 2023).

The ethical and environmental issues associated with cobalt mining, including human rights abuses and environmental degradation, amplify concerns about the sustainability of cobalt supply chains (Harper et al., 2011). Similarly, nickel, present in various cathode chemistries, is also a finite resource, and its extraction and processing contribute to environmental concerns, including substantial energy consumption and habitat destruction (Li et al., 2014).

Memory Effect

The memory effect in lithium-ion batteries, although not as prominent as in other types of batteries, can still lead to a reduction in effective capacity due to repeated partial discharges and recharges (Wang et al., 2005). This phenomenon is a concern as it can impact the overall performance and longevity of the battery. While some sources suggest that lithium-ion batteries do not exhibit memory effects (Udianto et al., 2022), it is important to note that the memory effect, although less pronounced, can still occur in these batteries (Wang et al., 2005).

The memory effect in lithium-ion batteries is a complex issue influenced by various factors such as temperature, anode material, and charging/discharging conditions (Lv et al., 2021). Research has shown that the relationship between capacity and temperature of lithium-ion batteries with different anodes is of great significance, indicating that temperature plays a crucial role in the performance of these batteries (Lv et al., 2021). Additionally, the increase in temperature of lithium batteries is observed with increasing discharge ratio, leading to a rapid temperature increase at higher discharge ratios (Li & Chen, 2020). These findings highlight the intricate interplay between temperature and battery performance, which can contribute to the memory effect in lithium-ion batteries.

The reduction of ethylene carbonate in the electrolyte salts for lithium-ion batteries has been studied, indicating the complexity of the chemical processes involved in battery operation (Parimalam & Lucht, 2018). This chemical aspect is crucial in understanding the memory effect and its underlying mechanisms in lithium-ion batteries. In addition, the development of predictive models for the remaining useful life (RUL) of lithium-ion batteries has been a focus of research, indicating the significance of understanding the degradation mechanisms, including the memory effect, to accurately predict the lifespan of these batteries (Gao et al., 2020). These models take into account the gradual decline in performance with continuous use, which can be attributed to the memory effect and other degradation factors (Gao et al., 2020).

Capacity Fading

Capacity fading in lithium-ion batteries is a significant concern due to its impact on the long-term performance and longevity of these energy storage devices (Peled & Menkin, 2017). This phenomenon is attributed to various interconnected factors that affect the chemical and physical processes within the battery cells (Ramadesigan et al., 2010). One of the primary contributors to capacity fading is the formation of a solid electrolyte interface (SEI) layer on the electrodes during initial charging cycles (Peled & Menkin, 2017). While essential for normal battery operation, the continuous growth of the SEI layer over repeated charge and discharge cycles leads to a gradual loss of capacity, particularly in the early stages of the battery's life (Peled & Menkin, 2017).

Additionally, self-discharge during periods of inactivity also contributes to capacity fading, as the battery undergoes a process of losing stored energy over time, even when not connected to any external device (Ramadesigan et al., 2010). This self-discharge is influenced by factors such as temperature, battery chemistry, and component quality (Ramadesigan et al., 2010). Furthermore, the choice of electrode materials, particularly cathode materials containing cobalt, can significantly impact the rate of capacity fading (Joshi et al., 2014). Research is ongoing to explore alternative electrode materials with improved stability and reduced capacity fading to enhance the long-term performance of lithium-ion batteries (Patil et al., 2021).

Cost

Cost is a pivotal consideration in the widespread adoption of lithium-ion batteries, despite notable reductions in their price over the years. While advancements in technology and economies of scale have contributed to lower costs, the economic factor continues to play a crucial role, especially in large-scale applications such as electric vehicles (EVs) and grid storage.

The cost of lithium-ion batteries is influenced by several factors throughout their lifecycle. The production costs, including raw materials, manufacturing processes, and quality control measures, are significant contributors. Raw materials like lithium, cobalt, and nickel, which are essential components of the battery, can account for a substantial portion of the overall cost. Fluctuations in the prices of these materials and their availability can impact the cost dynamics of lithium-ion batteries.

Manufacturing processes, including cell assembly and packaging, also contribute to the final cost. Improvements in manufacturing efficiency and economies of scale, driven by increased production volumes, have contributed to the gradual reduction in the cost of lithium-ion batteries. Research and development efforts focused on optimizing production techniques and exploring alternative materials aim to further drive down costs.

In large-scale applications like EVs and grid storage, where the demand for high-capacity battery packs is substantial, the cost per kilowatt-hour (kWh) becomes a critical metric. Battery packs in electric vehicles, for example, represent a significant portion of the total vehicle cost, influencing the affordability and market adoption of electric cars.

Grid storage projects, designed to store and manage renewable energy, require extensive battery capacity. The cost of lithium-ion batteries directly impacts the feasibility and economic viability of these projects, especially as grid-scale energy storage becomes increasingly essential for balancing intermittent renewable energy sources.

Despite the ongoing reduction in costs, challenges persist in achieving cost parity with conventional energy storage technologies. Incentives, subsidies, and government policies often play a role in offsetting these costs and incentivizing the adoption of lithium-ion batteries in various applications. As technology continues to evolve, and research and development efforts lead to innovations in battery chemistry and manufacturing processes, it is expected that the cost of lithium-ion batteries will continue to decrease, making them even more competitive and accessible for a broader range of applications.

To address the challenges in achieving cost parity with conventional energy storage technologies, it is crucial to consider the role of incentives, subsidies, and government policies in offsetting these costs and incentivizing the adoption of lithium-ion batteries in various applications. Additionally, as technology continues to evolve, research and development efforts lead to innovations in battery chemistry and manufacturing processes, which are expected to further decrease the cost of lithium-ion batteries, making them more competitive and accessible for a broader range of applications.

The challenges in achieving cost parity with conventional energy storage technologies are multifaceted. Conventional battery technology faces substantial challenges in providing the durability, high power, energy efficiency, and low cost required for grid-scale storage (Wessells et al., 2011). Moreover, the transition from conventional to renewable solar energy also presents colossal cost-parity challenges (Kumar et al., 2023). These challenges are twofold, encompassing the need for advanced, cost-effective electrical energy storage technologies and the critical reliability challenges related to fundamental device characteristics (Lee & Wong, 2013; Palomares et al., 2012).

In the context of lithium-ion batteries, the reduction of costs and achieving cost parity with conventional energy storage technologies is a complex endeavour. Despite substantial cost reductions in recent years, lithium-ion batteries are still significantly more expensive than the Department of Energy target of $125/kWh by 2022 (Ciez & Whitacre, 2016). However, there is a growing interest in hybrid electric vehicles (HEV) and plug-in hybrid electric vehicles (PHEVs) designed to reduce dependence on fossil fuel-derived energy and lower the carbon footprint, compounding the need for advanced, cost-effective electrical energy storage technologies (Jayaprakash et al., 2011).

In addressing these challenges, it is essential to consider the potential for future cost reductions in renewable generation and storage technologies, which hold promise for dramatic changes in the design, cost, and carbon emissions from microgrids (Sandler et al., 2020). Additionally, the "PV-battery" configuration has been identified as a means to achieve parity with the reference configuration sooner, at a 21% cost reduction, highlighting the potential for innovative configurations to drive cost reductions (Grosspietsch et al., 2018). Furthermore, the development of sodium-ion rechargeable batteries and potassium-ion batteries as promising candidates for energy storage due to their low cost and sustainable processes presents an avenue for addressing the cost challenges associated with lithium-ion batteries (Wijesinghe et al., 2019; Xiong et al., 2023). Additionally, the use of abundant, low-cost materials in sustainable processes for high-power and high-energy secondary batteries is crucial for efficiently distributed electrical energy storage from clean and renewable sources (Xu et al., 2015).

Balancing cost considerations with performance, safety, and environmental sustainability is a complex challenge that researchers, manufacturers, and policymakers are actively addressing to accelerate the transition to a more sustainable and electrified future.

After experiencing unprecedented price increases in 2022, the cost of lithium-ion batteries is once again on a downward trajectory over 2023. According to an analysis by research provider BloombergNEF (BNEF), the price of lithium-ion battery packs has decreased by 14%, reaching a historic low of $139 per kilowatt-hour (kWh). This decline is attributed to a combination of falling raw material and component prices, as well as an increase in production capacity across the entire battery value chain. Notably, demand growth did not meet certain industry expectations, contributing to the price drop (Catsaros, 2023; Ritchie, 2021; Statista, 2023a).

Despite the reduction in battery prices, the analysis suggests that demand for batteries in electric vehicles and stationary energy storage is poised for remarkable growth, projected to surge by 53% year-on-year, reaching 950 gigawatt-hours in 2023. However, major battery manufacturers reported lower utilization rates for their facilities, with demand and revenue falling short of expectations. This prompted many electric vehicle and battery manufacturers to reassess their production targets, influencing battery prices.

The growth in battery prices closely correlates with fluctuations in raw material prices (Catsaros, 2023; Ritchie, 2021; Statista, 2023a). Traditionally, falling battery prices were driven by scale learnings and technological innovation. However, the current drop in prices is attributed to a substantial increase in production capacity across the value chain, coupled with weaker-than-expected demand.

The average battery pack prices, encompassing various end-uses such as electric vehicles, buses, and stationary storage projects, were lowest in China at $126/kWh, while packs in the US and Europe were 11% and 20% higher, respectively (Catsaros, 2023; Ritchie, 2021; Statista, 2023a). This disparity is linked to the relative immaturity of these markets, higher production costs, lower volumes, and diverse applications. Intense price competition in China further impacted global battery prices as manufacturers sought to capitalize on the growing demand.

An interesting shift in the industry is the adoption of low-cost cathode chemistry, specifically lithium iron phosphate (LFP). LFP packs and cells recorded the lowest global weighted-average prices at $130/kWh and $95/kWh, respectively. This marks the first year that the analysis found LFP average cell prices falling below $100/kWh. On average, LFP cells were 32% cheaper than lithium nickel manganese cobalt oxide (NMC) cells in 2023 (Catsaros, 2023; Ritchie, 2021; Statista, 2023a).

Looking ahead, miners and metals traders anticipate further easing of prices for key battery metals such as lithium, nickel, and cobalt in 2024. BNEF expects average battery pack prices to decrease again next year, reaching \$133/kWh (in real 2023 dollars). Anticipated technological innovation and manufacturing improvements are projected to drive continued declines in battery pack prices, with estimates reaching \$113/kWh in 2025 and \$80/kWh in 2030.

Localization efforts in battery manufacturing, particularly in regions like the US and Europe, could influence battery pack prices as these industries scale up. Higher energy, equipment, land, and labour costs in these regions compared to Asia, the current hub of battery production, might exert upward pressure on prices. However, local policies such as production tax credits could offset some of these costs. The localization trend adds complexity to regional battery pricing dynamics in the coming years.

The cost of batteries is a significant consideration in the transition to low-carbon electricity, with projections indicating a reduction in battery costs to \$135/kWh by 2050, aligning with the trajectory for battery cost reduction (Trieu et al., 2018). Additionally, the role of batteries in energy storage and their impact on the environment underscores the importance of lightweight and environmentally friendly battery packs for electric vehicles (X. Li et al., 2022). The transition from fossil fuel-powered vehicles to battery electric vehicles is a critical aspect of achieving a low-carbon society, with economic evaluations supporting this transition (Xu et al., 2019).

Challenges and Opportunities for Vehicle-Based Batteries

Various types of electrified vehicles have distinct battery requirements, influencing the challenges and opportunities for Lithium-Ion Batteries (LIBs). Common electrified vehicles include Plug-In Hybrid Electric Vehicles (PHEV) and Electric Vehicles (EV), relying on off-board electrical power for part or all of their traction and on-board power. Hybrid Electric Vehicles (HEV) and Stop/Start Hybrids (S/S) are powered by traditional liquid fuels, with HEVs having the ability to electrically power traction drives (Masias et al., 2021).

The United States Advanced Battery Consortium (USABC), representing Stellantis, Ford, and General Motors, has established battery performance targets for each electrified vehicle type (EV, PHEV, HEV, and S/S). Additional variations of EV goals, such as Low-Cost Fast-Charge and Lithium Metal Based EV goals, have also been introduced. Notably, USABC goals are specified at the end-of-life (EOL) battery pack level, while researchers often focus on beginning-of-life (BOL) cell-level performance (Masias et al., 2021).

The European Council for Automotive Research and Development (EUCAR) and the New Energy and Industrial Technology Development Organization (NEDO) in Japan have set their own battery system goals. A comparison of the goals set by the United States Advanced Battery Consortium (USABC), EUCAR, and NEDO reveals global alignment in energy targets and battery life expectations. However, caution is advised when interpreting cost targets due to variations in publication years and rapidly evolving cost performance.

EUCAR's goals are based on previous automotive industry materials requirements studies and are presented in the context of vehicle-level targets (Masias et al., 2021). The goals set by EUCAR align with the global need for new energy-storage systems to meet the demands of the world's population, given the depletion of fossil fuels and the rapid evolution of the global economy (Gómez-Urbano et al., 2020). Additionally, the US Department of Energy (DOE) and the US Advanced Battery Consortium (USABC) have estimated a specific energy target at the pack level that would enable long-range, high-performance, and affordable electric vehicles (EVs) for widespread market adoption (Ladpli et al., 2019).

NEDO, on the other hand, has initiated projects related to the durability and reliability of solid oxide fuel cell (SOFC) stacks and systems (Hosoi & Nakabaru, 2009; Yokokawa, 2011). These projects reflect NEDO's commitment to advancing energy technologies, which is in line with the global trend of increasing sales of electric vehicles and the need for sustainable energy solutions (Iturrondobeitia et al., 2022). Furthermore, extending the life of end-of-life batteries beyond their intended in-vehicle use could help utilize the remaining capacity while increasing the environmental benefits of EVs (Kamath et al., 2020).

The alignment in energy targets and battery life expectations among EUCAR, NEDO, and USABC reflects the global effort to develop advanced energy-storage systems for automotive applications. The need for new energy-storage systems is driven by the increasing sales of electric vehicles and the global shift towards sustainable energy solutions. However, caution is warranted when interpreting cost targets due to variations in publication years and rapidly evolving cost performance

In the pursuit of advancing electric vehicle (EV) technologies, various organizations worldwide, as outlined above, have set specific performance goals for EV battery packs. A comparison of these goals, outlined below, sheds light on the international alignment in expectations and requirements (Masias et al., 2021).

- USABC

  - Target Year: 2020

  - Specific Energy (Wh/kg): 235

  - Specific Power (W/kg): 470

  - Cost ($/kWh): 125

  - Calendar Life (Years): 10

  - Cycles: 1000

- EUCAR

  - Target Year: 2030

  - Specific Energy (Wh/kg): 288

  - Specific Power (W/kg): 1152

  - Cost ($/kWh): 84

  - Calendar Life (Years): Vehicle Life

  - Cycles: 24 MWh

- NEDO

  - Target Year: 2020s

  - Specific Energy (Wh/kg): 250

  - Specific Power (W/kg): 1500

  - Cost ($/kWh): 190

  - Calendar Life (Years): 10–15

○ Cycles: 1000–1500

Note: The currency exchange rates provided are $1.00 = 1.17€ = 105¥.

These goals, set by the United States Advanced Battery Consortium (USABC), the European Council for Automotive Research and Development (EUCAR), and the New Energy and Industrial Technology Development Organization (NEDO), showcase a shared vision for the end-of-life (EOL) performance of EV battery packs (Masias et al., 2021). While there is notable alignment in specific energy, specific power, and calendar life expectations, caution is warranted in interpreting cost targets due to variations in publication years and the rapidly evolving cost landscape of lithium-ion batteries. This global consensus reflects the collective commitment to advancing energy storage technologies for the sustainable future of electric mobility.

The international consensus on energy goals reflects a universal need to reduce vehicle weight. Both fossil fuel and electric vehicles are subject to the same physics of kinematic weight, aerodynamic drag, tire rolling resistance, and elevation, affecting energy consumption rates. The correlation between weight and energy consumption is evident in a comparison of all electric vehicles and Ford gasoline-powered vehicles (Masias et al., 2021). Long-range EVs, requiring heavy battery packs, can benefit from battery weight reduction, leading to improved energy consumption, extended battery life, and potential cost savings.

The challenges and opportunities in battery life and cost are explored further in subsequent sections. Advances in lithium-ion cell densities, ranging from 2 to 3 g/cm3, contribute to improvements in battery specific energy (Wh/kg) and energy density (Wh/L), translating to significant reductions in battery volume when considering the absolute weight of vehicle battery packs (Masias et al., 2021).

The superior energy density and specific energy of Lithium-Ion Batteries (LIBs) have positioned them as the dominant technology across various markets and applications, surpassing competing battery chemistries like nickel metal hydride (NiMH) and nickel cadmium (NiCd), which rely on aqueous electrolyte systems (Masias et al., 2021). LIBs, being non-aqueous, maintain cell voltage levels approximately three times higher than their counterparts. Despite this advantage, the core electrochemical couple in high-energy LIB designs has remained unchanged since their introduction, resulting in only marginal improvements in nominal voltages. Recent advancements in LIB specific energy primarily stem from engineering enhancements in active material capacity and cell/electrode optimization. However, the rate of performance improvement through these methods has slowed, posing a challenge for future electrification implementations in various vehicle lines (Masias et al., 2021).

To propel future improvements in specific energy, the development of different materials becomes imperative. The automotive industry has established cathode and anode specific energy goals, outlined previously, to guide research with specific targets. USABC goals translate vehicle-level targets to material-level targets, while EUCAR goals reference previous materials requirements studies in the context of their vehicle-level targets. Although the anode goals from USABC and EUCAR exhibit similarity, a closer examination of cathode material performance reveals a lack of practical, mature candidates capable of meeting or surpassing future automotive materials targets (Masias et al., 2021). The dominant cathode materials, such as lithium cobalt oxide and lithium-rich layered oxide (LLO), face challenges related to structural stability and voltage fade, hindering their utility and posing a significant hurdle for further development (Masias et al., 2021).

In the LIB anode market, carbon anodes, historically dominant, fall short of meeting future energy targets. Lithium and silicon, as alternative anode materials, can potentially meet these targets depending on their utilization. Lithium poses technical challenges for automotive applications, necessitating further engineering efforts (Masias et al., 2021). Silicon, however, emerges as a more viable advanced anode material in the near term, with carbon/silicon blends already in commercial use. While pure silicon faces challenges due to volume changes during cycling, silicon as a system, even at reduced capacity, holds promise to meet industrial targets. Strategies accommodating silicon's lithium storage capacity while maintaining electrode structure and stable interfacial chemistry can contribute to overcoming existing challenges and realizing the potential of silicon as an advanced anode material in LIBs (Masias et al., 2021).

The sustained growth of Electric Vehicles (EVs) hinges on the critical requirement of battery durability. Original Equipment Manufacturers (OEMs) often provide warranties on EV battery packs, guaranteeing replacement if capacity degrades beyond specified thresholds. However, this creates engineering challenges as these warranties extend over many years and hundreds of thousands of miles. Battery validation, essential for timely implementation of cutting-edge power and energy density, becomes a complex process, demanding precision in test equipment to prevent uncertainty in derived quantities like battery power and energy.

Accelerated testing is a widely used method to assess long-term battery degradation rates under various operating conditions within a short timeframe (Masias et al., 2021).

Once deviations from End-of-Life (EOL) discharge throughput, power, and energy re-tention targets are identified, two main challenges emerge: selecting a comprehensive set of tests characterizing degradation rates across expected operating conditions, and confirming that degradation rates do not significantly accelerate later in the battery's life, a phenomenon known as "rollover" degradation (Ecker et al., 2012). Intelligent selection of state of charge and temperature windows, along with advanced diagnostics, aids in equat-ing early-life results to long-term degradation rates (Lin & Chung, 2019). The rollover challenge requires either continued testing until EOL conditions or a deep understanding of degradation processes, employing mathematical models to simulate contributions to degradation as the battery progresses (Ecker et al., 2012).

Accelerated aging tests are commonly employed in both academic labs and industry, involving testing at higher temperatures and/or rates to expedite battery degradation (Schauser et al., 2022). To shorten testing time, accelerated testing is performed by in-creasing the loading levels on the products beyond normal conditions (Diao et al., 2018). Furthermore, approaches to overcome rollover failure mechanisms have been presented, such as the use of single-crystal NCM523 materials, showing no "rollover" failure even after 200 cycles (Klein et al., 2020). This highlights the importance of understanding degradation mechanisms and employing suitable materials to mitigate rollover degrada-tion.

The review by provides insights into lithium-ion battery aging mechanisms and esti-mations for automotive applications, emphasizing the importance of techniques, models, and algorithms for battery aging estimation (Barré et al., 2013). It is crucial to consider the impact of external factors such as temperature on battery degradation, as thermal runaway events have been observed in various Li-ion battery powered systems, leading to serious safety concerns (Sharifi-Asl et al., 2019). Additionally, the sensitivity of impedance parameters of Li-ion batteries under different states of health has been studied, revealing the relationship between capacity degradation and the fading law of internal impedance parameters identified from Equivalent Circuit Model (ECM) (Yuan et al., 2019).

Analysis of Coulombic efficiency trends suggests conditions yielding similar long-term degradation performance (Masias et al., 2021). Design of experiments can be streamlined using Coulombic efficiency data, reducing the number of cells tested. However, chal-lenges persist in predicting degradation rates, particularly when cycling cells aggressively and continuously, as in heavy-load towing or autonomous driving scenarios. Three-di-mensional electrochemistry models are likely needed to accurately represent cell and pack designs, considering local degradation effects like lithium plating. Machine learning complements physics-based models when handling large datasets or rapid model execution is necessary.

Looking ahead to next-generation anode and cathode chemistries, challenges in battery life prediction persist (Masias et al., 2021). Increasing silicon concentration in graphite anodes for enhanced energy density may pose mechanical stability issues. Battery failure mechanisms, such as excessive cell swelling and electrode resistance rise, are not adequately predicted by current high-precision coulometry studies. Silicon anodes, metallic lithium anodes, and dendrite formation pose additional challenges requiring novel diagnostic methods to ensure accurate battery life predictions and early failure detection.

The acceptance of Battery Electric Vehicles (BEVs) in the market faces a significant hurdle—the comparatively high cost when juxtaposed with Internal Combustion Engine Vehicles (ICEVs). The battery pack, being the most expensive component in an EV due to material costs and engineering complexity, contributes substantially to this cost disparity. Nevertheless, a substantial 80% cost reduction over the past decade, fuelled by investments in battery manufacturing and reduced cobalt in cathodes, brings total cost of ownership parity between BEVs and ICEVs closer. However, achieving purchase price parity remains a challenge, necessitating further cost reduction for broader marketplace acceptance (Masias et al., 2021).

Opportunities for cost reduction span the entire Lithium-Ion Battery (LIB) value chain and encompass manufacturing process improvements, material innovations, performance enhancements to simplify integration, and closing the value-chain loop through recycling. The effectiveness of cost reduction initiatives should be assessed based on per-kWh cost reduction, considering both upstream requirements and downstream effects, with simulation tools and modelling playing a critical role in validating these effects (Masias et al., 2021).

Significant cost reductions thus far have originated from manufacturing process improvements, targeting energy, time, volume, and capital (Masias et al., 2021). Material production processes and cell assembly procedures still rely on energy-intensive methods, presenting avenues for further cost reduction through innovation. Material selection and innovation, particularly in reducing or eliminating expensive metals like cobalt and nickel, offer additional opportunities. Advanced electrodes that increase energy density can offset higher per-kg costs, and researchers should consider material-level and system-level trade-offs.

Improving performance attributes, such as cycle and calendar lifetime, thermal operation window, and integrated safety features, contributes to pack and system integration, reducing cost per usable kWh. Closing the value chain loop through recycling is another avenue for cost reduction, with the potential to shift raw material economics by reintegrating End-of-Life (EOL) battery materials (Masias et al., 2021).

Beyond manufacturing cost reduction, indirect economic benefits to customers, such as lower fuelling and maintenance costs, may accelerate EV adoption. The advantages of lower scheduled maintenance costs and potentially 60% lower fuelling costs with electricity underscore the potential economic benefits over the vehicle's lifetime (Masias et al., 2021).

While BEVs demonstrate cost advantages over ICEVs, externalized costs related to CO_2 emissions and pollutants persist. Assessing the relative environmental benefits of BEVs is crucial, considering manufacturing emissions and the dynamic coupling of operational emissions with the electricity supply mix. Recognizing the evolving landscape of renewable energy sources further underscores the environmental advantages of BEVs over the fixed operational emissions of ICEVs, reinforcing the primary value proposition for BEVs.

Irrespective of the propulsion energy source, be it gasoline/petrol or electric, the automotive sector demands an elevated standard of safety (Masias et al., 2021). Consequently, Lithium-Ion Battery (LIB)-electrified vehicles are held to the expectation of meeting or surpassing the safety levels offered by fossil-fuel-powered vehicles. To achieve this, vehicle manufacturers implement multiple layers of safety measures, extending from the cell and module to the pack and the entire vehicle. Additionally, a suite of electrical, thermal, and mechanical control and monitoring systems is typically deployed around the electrochemical cell.

The widespread integration of LIB technology in vehicles has led to significant safety research, encompassing thermal, mechanical, electrical, and systems aspects. Various governments, including the UN, ECE, China, and South Korea, have issued regulations addressing battery safety, fostering a comprehensive understanding of safety considerations in LIB applications.

During abuse testing, the reaction of a battery is assessed on a scale developed by EUCAR, ranging from 0 (no response) to 7 (explosion). This rating system serves as an effective tool to summarize complex battery abuse responses, despite not being flawless (Masias et al., 2021).

The current system of layered safety protections at the vehicle, pack, and module levels, all reinforcing the individual cell, has proven highly effective, with minimal incidents to date. Key sensor systems, measuring voltage, current, and temperature at the cell, module, or pack level, play a pivotal role in determining the battery's condition. Exploring alternative sensor techniques, such as pressure, fibre-optics, and acoustics, could present a future research opportunity. These alternative sensors may offer novel insights into LIB safety conditions, potentially reducing cost, weight, and volume compared to existing solutions. As LIB researchers strive for continued energy density improvements, maintaining the current high level of safety remains a paramount challenge.

The duration required for charging is a critical consideration for customers undertaking frequent long-distance trips surpassing their battery's total range. Direct Current Fast Charging (DCFC) enables customers to charge up to 80% State of Charge (SOC) at rates reaching 350 kW, providing several hundred miles of range in approximately 40 minutes. Nevertheless, the desire for even faster charging, aiming to approach parity with Internal Combustion Engine Vehicles (ICEVs), persists, especially for drivers without access to overnight or workplace charging. The U.S. Department of Energy's (DOE) long-term target for extreme fast charging is to deliver 200 miles in just 7.5 minutes, while the U.S. Advanced Battery Consortium (USABC) sets a more near-term goal of achieving $\Delta 80\%$ SOC in 15 minutes by 2023 (Masias et al., 2021).

Lithium-Ion Batteries (LIBs), renowned for their high energy density, face challenges related to faster charging rates, primarily due to the risk of unintended lithium plating resulting in irreversible capacity loss, diminished performance, and heightened risks of short-circuiting and thermal runaway. The proximity of the negative electrode potential to that of lithium metal exacerbates this challenge.

In addressing the challenges related to faster charging rates in Lithium-Ion Batteries (LIBs), it is crucial to understand the risk of unintended lithium plating, which can lead to irreversible capacity loss, diminished performance, and heightened risks of short-circuiting and thermal runaway. The proximity of the negative electrode potential to that of lithium metal exacerbates this challenge. Several studies have focused on the detection, mechanisms, and mitigation of lithium plating in LIBs, as well as the development of optimal charging strategies and thermal management systems to enhance battery performance and safety.

Zhao et al. (2010) discussed the extraction and insertion of lithium during charging or discharging of a lithium-ion battery. This process is crucial in understanding the mechanisms leading to lithium plating and irreversible capacity loss. Additionally, Burns et al. (2015) highlighted the in-situ detection of lithium plating using high precision coulometry, providing insights into detecting and monitoring lithium plating in LIBs. Furthermore, Hein et al. (2020) emphasized the deposition of metallic lithium on the surface of graphite electrodes as a major degradation mechanism in LIBs, shedding light on the mechanisms of lithium plating and its impact on battery performance.

Gao et al. (2019) provided a comprehensive review of charging strategies for commercial lithium-ion batteries, including fast charging and optimal charging strategies. This review is valuable in understanding the challenges and future directions in developing effective charging strategies to mitigate the risk of lithium plating and irreversible capacity loss. Moreover, Nizam et al. (2022) proposed a constant current-fuzzy logic algorithm for lithium-ion battery charging, highlighting the importance of advanced control algorithms in optimizing charging processes and mitigating degradation mechanisms.

Thermal management is crucial in addressing the challenges of lithium plating and fast charging. Madani et al. (2021b) demonstrated the thermal characterizations of lithium-ion batteries and the impact of charge-discharge pulses on heat generation, emphasizing the importance of thermal management for enhancing battery performance and safety. Additionally, (Shao et al., 2022)focused on the design of thermal management systems for lithium batteries at low temperatures, addressing the need for effective thermal control to mitigate the risks associated with fast charging and low-temperature operation (Shao et al., 2022).

Multiple opportunities exist at the cell and battery levels for improving fast charging, including electrode engineering, new electrode chemistry, and thermal management. Optimizing electrode thicknesses and porosities through electrode engineering can reduce voltage polarization, stabilizing the voltage for lithium intercalation. This modification of the power-to-energy ratio (P/E) across a substantial range allows for high power achieved through engineering modifications to the electrode structure while utilizing the same active material. Researchers should prioritize finding high-energy materials, leaving power capability as a secondary goal achievable through cell and electrode engineering (Masias et al., 2021).

At the system level, effective thermal management remains crucial. Continuous improvement in cooling strategies is necessary to accommodate the demand for higher charging rates. Present designs often encounter limitations in available surface area for cooling, as Original Equipment Manufacturers (OEMs) are hesitant to expand surface areas defining heat removal pathways to maximize pack energy density. Repurposing volume within the pack for enhanced heat removal capabilities becomes essential as material energy density increases. Advanced strategies, including brief excursions to high temperatures, could enable extreme fast charging. However, simultaneous improvement in high-temperature durability, encompassing capacity loss, resistance rise, and gas generation, is imperative for regular high-temperature charging throughout the battery's life (Masias et al., 2021).

Chapter Six

Environmental Impact

L ithium-ion batteries are widely embraced for their application in clean technologies such as electric vehicles, owing to their high energy density, compact design, rechargeable nature, and longevity across numerous charge cycles (Crawford, 2022). These batteries play a vital role in current initiatives to replace traditional gas-powered vehicles that contribute to CO_2 and greenhouse gas emissions. Their characteristics also position them as suitable candidates for energy storage within the electric grid. However, the manufacturing process of lithium-ion batteries and their components carries environmental and social considerations, including CO_2 emissions.

The production of lithium-ion batteries for electric vehicles involves a higher material intensity compared to traditional combustion engines, and the demand for battery materials is increasing. According to Yang Shao-Horn, JR East Professor of Engineering at MIT, lithium extraction, particularly from hard rock mines, emits approximately 15 tonnes of CO_2 for every tonne of lithium mined (Crawford, 2022). Furthermore, the extraction and processing of battery materials, including lithium, cobalt, and nickel, are labour-intensive, chemical-dependent, water-intensive processes that may lead to environmental contamination and leave behind toxic waste.

The manufacturing process exacerbates the eco-footprint of these batteries. The synthesis of necessary materials requires temperatures between 800 to 1,000 degrees Celsius, typically achieved cost-effectively through burning fossil fuels, contributing to CO_2 emissions (Crawford, 2022). The amount of CO_2 emitted during battery production varies depending on material choices, sourcing methods, and energy sources in manufacturing. Approximately 77% of the world's lithium-ion batteries are manufactured in

China, where coal is a primary energy source, emitting a significant amount of greenhouse gases (Crawford, 2022).

For instance, the manufacturing of an 80 kWh lithium-ion battery, as found in a Tesla Model 3, can result in CO2 emissions ranging from 2,400 kg to 16,000 kg (Crawford, 2022). Efforts worldwide are underway to develop new manufacturing processes and battery chemistries that use more environmentally friendly materials, but these technologies are not yet widely adopted.

Despite the environmental concerns related to manufacturing, lithium-ion batteries contribute to a more climate-friendly approach compared to alternative technologies. In the United States, the electric grid, consisting of fossil fuels and low-carbon energy sources, is cleaner than gasoline combustion, making electric vehicles more environmentally friendly. Even when considering initial manufacturing emissions, electric cars emit less CO_2 than their gas-powered counterparts over extended distances (Crawford, 2022). This is particularly relevant as the transportation sector is a major contributor to greenhouse gas emissions in the United States.

Moreover, lithium-ion batteries offer significant environmental benefits by aiding energy grid stabilization. As the world shifts toward renewable energy sources like solar and wind power, there is a growing need for efficient energy storage solutions (Crawford, 2022). Batteries can play a crucial role in storing excess renewable energy during optimal conditions, addressing challenges associated with oversupply and shortages during unfavourable weather conditions, and facilitating a transition away from CO_2-emitting fossil fuels.

The environmental impacts of lithium-ion batteries (LIBs) are a complex issue that spans various stages of their lifecycle, from raw material extraction to manufacturing, usage, and end-of-life disposal. Several factors need to be considered when assessing the environmental impact of LIBs, including energy, power, charge-discharge rate, cost, cycle life, safety, and environmental impact (Manthiram, 2017). Life cycle assessment studies of lithium-ion traction batteries have reported widely different results, indicating the need for a comprehensive understanding of the environmental impacts of LIBs (Ellingsen et al., 2017). Temperature has been identified as a significant environmental factor that influences the charge and discharge performance of lithium-ion batteries (Lv et al., 2021). Additionally, the disposal of lithium-ion batteries as waste can pose environmental and health risks due to the presence of heavy metals such as lithium, cobalt, nickel, and copper (Nurqomariah & Fajaryanto, 2018a).

Furthermore, the mechanical integrity of lithium-ion battery modules and their packing density have been highlighted as critical factors, given the widespread use of LIBs in electric vehicles (EVs) due to their high energy and power density, capacity, and long lifecycle (Liu et al., 2018). Life cycle assessments of lithium-ion batteries for plug-in hybrid electric vehicles have emphasized the importance of internal battery efficiency and battery weight during the use phase, indicating the need for a holistic approach to assessing environmental impacts (Zackrisson et al., 2010). The revolutionizing impact of lithium-ion batteries on the portable electronics market and their increasing pursuit for transportation and stationary storage of renewable energies have also been noted (Manthiram, 2011).

Moreover, the thermal management of lithium-ion batteries is crucial for achieving maximum utilization in electric vehicles, highlighting the need for optimum operating temperatures across a range of environmental conditions (Madani et al., 2021a). Research in lithium-ion batteries continues to focus on obtaining higher energy density and developing more environmentally friendly materials (Hidayat et al., 2018). Safety concerns related to lithium-ion batteries, including thermal runaway and the toxicity of gases released during combustion, have been identified as critical environmental and health issues (Chen et al., 2017; Kimura et al., 2022). Additionally, the cost of lithium-ion cells and the market fluctuations in lithium carbonate prices have implications for the overall cost of lithium-ion storage systems (Ciez & Whitacre, 2016).

In addition, the synthesis and characterization of lithium-ion battery materials, as well as the study of their combustion behaviour and thermal runaway phenomena, provide valuable insights into addressing environmental challenges associated with LIBs (Ghorpade, 2019; Xu et al., 2021). The cost implications of lithium-ion battery production and market fluctuations in lithium carbonate prices also play a role in understanding their environmental impact (Ciez & Whitacre, 2016). Furthermore, the potential use of geothermal energy sources for the production of lithium-ion batteries highlights the broader implications of energy consumption and production in assessing their environmental footprint (Tao et al., 2011).

Raw Material Extraction

Lithium, a crucial element in lithium-ion battery (LIB) production, is predominantly sourced through mining operations, utilizing methods such as open-pit and underground mining. Despite its significance in advancing renewable energy and electric transportation, the extraction of lithium raises environmental concerns due to its associated impacts.

Cobalt and nickel are essential components for cathodes in many lithium-ion batteries (LIBs) (Slack et al., 2017). However, their production is associated with ethical concerns, particularly in the Democratic Republic of Congo (DRC) (Z. Lei et al., 2022). The DRC is a significant source of global cobalt production, where issues such as child labour and hazardous working conditions have been raised (Z. Lei et al., 2022). Additionally, nickel mining can lead to environmental issues such as deforestation, habitat disruption, and water pollution (Mudd & Jowitt, 2022).

The concerns about the supply chain for cobalt raw materials needed for LIB production, especially for high-energy LIB types, have been highlighted (Noerochim et al., 2021). Furthermore, the environmental impacts of mining and mineral processing operations, including those related to cobalt and nickel, have been studied extensively (Norgate & Haque, 2010). It has been emphasized that cleaner production is crucial for reducing the environmental impacts of mining operations (Norgate & Haque, 2010).

Cobalt and nickel are widely used in various applications, including super alloys, specialty steel, magnets, and rechargeable batteries (Z. Lei et al., 2022). The importance of these metals to the modern world has been underscored, especially in the context of their use in batteries and other technological applications (Mudd & Jowitt, 2022). However, the environmental and social implications of their extraction and processing cannot be overlooked.

In the context of nickel mining, the environmental pertinence of nickel exploitation has been assessed, highlighting the need to account for biomass carbon emissions when sourcing nickel in the future (Merlo et al., 2021). Additionally, the public perception of the socio-economic and environmental impact of gold mining has been studied, indicating the significance of understanding community perspectives on mining activities (Sehol et al., 2022).

Furthermore, the beneficiation of copper-cobalt-bearing minerals in the DRC has been reviewed, shedding light on the mineral processing aspects related to cobalt extraction (Lutandula et al., 2019). This is crucial in understanding the environmental and social implications of the entire mining process, from extraction to processing.

Graphite, used in the anodes of lithium-ion batteries (LIBs), is primarily sourced through mining, which can have environmental impacts similar to those of other mining activities (Kulkarni et al., 2022). The extraction processes for other materials like copper and aluminium, also found in LIBs, contribute to environmental concerns (Notter et al., 2010). Notter et al. (2010) highlighted that the major contributor to the environmental burden caused by batteries is the supply of copper and aluminium for the production of the anode and the cathode, plus the required cables or the battery management system. Kulkarni et al. (2022) emphasized that graphite is a popular anode material for LIBs due to its electrical properties such as good conductance, chemical inertness, and corrosion resistance. Additionally, low-level exposure to manganese has been implicated in neurologic changes, decreased learning ability in school-aged children, and increased propensity for violence in adults, indicating the environmental and health concerns associated with its extraction and use (Finley, 2004).

In its natural state, open land provides a habitat for various forms of vegetation, including plants, trees, and forests. However, when we alter a portion of this land, resulting in land-use change, it has profound implications for natural forestry (Kilgore, 2023). The removal of trees, a consequence of such land-use changes, disrupts a crucial process – carbon sequestration. As trees are eliminated, the Earth loses a significant carbon sink, leading to an increased release of CO_2 into the Earth's atmosphere (Kilgore, 2023).

Chile, renowned for possessing the world's largest lithium reserves, experienced a substantial loss of approximately 56.8 thousand hectares of its natural forest in 2021 alone (Kilgore, 2023). This equates to emissions of 28.5 metric tons of CO_2 equivalents (CO_2-eq). Australia, ranking second among lithium mining countries, encountered the loss of 231 thousand hectares of its natural forest by 2021. This extensive deforestation contributed to emissions exceeding 90.5 metric tons of CO_2-eq (Kilgore, 2023).

While the adverse environmental impact of lithium mining is evident, it is essential to note that not all forest loss is exclusively attributable to lithium mining. Nevertheless, as lithium mining persists and continues to contribute to land-use changes, the detrimental consequences of deforestation, including increased CO_2 emissions, are likely to persist in these countries and reverberate globally (Kilgore, 2023).

Habitat destruction due to open-pit mining for lithium extraction has significant environmental implications (Notter et al., 2010). Wanger (2011) highlighted that the extraction of lithium for Li-ion batteries has a minimal environmental impact, contributing less than 2.3% to the overall impact. However, Vera et al. (2023) emphasized that lithium extraction is likely to cause substantial water pollution, impacting native diversity and human health. Furthermore, Theobald et al. (2020) stressed the need for continuous monitoring of lithium extraction due to potential long-term environmental impacts. Hartanto et al. (2023) supported these concerns by discussing the extensive landscape changes and high erosion rates associated with open-pit mining, leading to habitat fragmentation and loss of land cover vegetation. Figure 27 shows an open pit mine.

Figure 27: The Palabora open pit mine is almost 2,000 meters across and almost 900 meters deep. This makes it Africa's largest man-made hole. On the northwestern side of the hole, on the left side of this photo, the traces of a major landslide remains. Bjørn Christian Tørrissen, CC BY-SA 4.0, via Wikimedia Commons.

Additionally, Crespo-Cebada et al. (2020) mentioned the improved prospects for lithium open-pit mines, indicating a growing interest in exploiting this resource. However, the environmental consequences, as highlighted by Huang et al. (2021), must be considered alongside economic evaluations. The study by Huang et al. (2021) emphasized the need to assess both the environmental impacts and economic performance of lithium extraction. This aligns with the findings of Notter et al. (2010), which indicated the importance of considering the overall environmental impact of lithium extraction for battery production.

Furthermore, the potential impacts of open-pit mining on groundwater levels and the moisture movement in vadose zones, as discussed by Du et al. (2022) and Lian et al. (2021), respectively, underscore the broader ecological implications of such mining activities. These studies highlight the need to consider the long-term effects on groundwater and moisture movement in open-pit mining areas.

Soil erosion is a significant environmental concern associated with mining activities, particularly in open-pit operations, where the removal of topsoil and vegetation exposes the underlying soil to erosion by forces such as wind and water. This process can lead to the degradation of fertile land, reducing its ability to support vegetation, and may result in sediment runoff that contaminates nearby water bodies, adversely affecting aquatic ecosystems and exacerbating water quality issues (Hartanto et al., 2023). The impact of mining activities on soil erosion has been studied extensively, with a focus on various aspects such as the assessment of erosion rates, the influence of mining on groundwater levels, and the development of erosion zones in open-pit mine dumps.

Several studies have highlighted the increased risk of soil erosion due to mining activities. For instance, research has shown that in tropical countries, soil erosion is often intensified due to factors such as high erodibility of geologically old and weathered soils, intensive rainfall, inappropriate soil management, removal of forest vegetation cover, and mining activities (Wantzen & Mol, 2013). Additionally, open-pit mining has been found to have a high erosion rate due to the breakdown of soil aggregation and the loss of land cover vegetation, leading to increased interaction with rainwater (Wantzen & Mol, 2013).

Furthermore, the impact of mining on groundwater levels has been investigated, with findings indicating that drainage operations during open-pit mining can cause a drop in groundwater levels and the formation of a cone of depression, leading to disturbances in the mining area (L. Chen et al., 2022). This disturbance can extend over a significant radius, affecting a large proportion of the mining area and contributing to changes in the hydrological dynamics of the region.

In addition to soil erosion, mining activities have been associated with water pollution characterized by acid mine drainage, extreme pH conditions, and heavy metal contaminations, which further exacerbate environmental degradation (Sahoo et al., 2021). The contamination of environmental media, including water, soil, and vegetation, has been attributed to various mining operations such as ore extraction, processing, and disposal of mining tailings (Neamtiu et al., 2017).

Moreover, the ecological impact of mining activities has been studied, with research indicating that intensive mining activity has led to considerable environmental deterioration, posing challenges for soil recovery and revegetation in mining areas characterized by the excavation and removal of large volumes of mining waste and ore (Guedes et al., 2021). The difficulty in revegetating slopes or benches with varying inclinations in open-pit mines has been highlighted as a significant challenge in soil recovery efforts.

The extraction of lithium, cobalt, nickel, and graphite can pose a significant threat to local water resources due to various mechanisms. The generation of toxic waste in the manufacturing of lithium-ion batteries is primarily associated with metals like cobalt and manganese. Disposing of these metals into water leads to the contamination of drinking water and fertile land, transforming them into toxic environments (Kilgore, 2023). This contamination results in the depletion of Earth's vital resources, including water reservoirs and natural forests. Consequently, communities residing in close proximity to areas where lithium-ion batteries are discarded bear a significant impact (Kilgore, 2023). Moreover, this process contributes to elevated CO_2 levels in the environment. Recognizing this environmental concern, it becomes imperative to implement effective measures for the recycling of lithium-ion batteries (Kilgore, 2023).

Mining activities can lead to the generation of acidic or alkaline drainage, which alters the pH levels of nearby water bodies, posing harm to aquatic life (Singh et al., 2010). Additionally, the use of chemicals in the extraction process, such as acids and solvents, can result in the leaching of these substances into groundwater or surface water, leading to contamination (Singh et al., 2010). Furthermore, the release of heavy metals and other pollutants during mining operations further compounds the environmental impact on water quality, potentially rendering it unsafe for both ecological and human consumption (Bird et al., 2009).

Studies have shown that mining activities, particularly in the context of shale gas development, can have a substantial impact on water resources, emphasizing the importance of protecting vulnerable ground and surface water resources and promoting more water-efficient technologies (Vandecasteele et al., 2015). Furthermore, assessments of trace element levels in sediment and water in artisanal and small-scale mining localities have revealed significant contamination with elements such as cadmium, lead, iron, zinc, manganese, copper, mercury, and arsenic, highlighting the environmental risks associated with mining activities (Agyarko et al., 2014).

Moreover, the environmental impacts of acid mine drainage-bearing river water in copper mining areas have been estimated and compared, demonstrating the significant effect of mining activities on water quality and environmental sustainability (Adamovic et al., 2021). These findings underscore the critical need for sustainable mining practices and effective environmental management to mitigate the adverse effects of mineral extraction on local water resources.

To understand the environmental impact of mining operations for lithium, cobalt, nickel, and graphite extraction, it is crucial to consider the usage of chemicals in these processes. Chemicals such as acids, leaching agents, and solvents are commonly employed to aid in the separation of these valuable materials from surrounding rock formations. Improper handling, storage, or disposal of these chemicals can lead to environmental pollution, with chemical runoff potentially infiltrating nearby ecosystems and adversely affecting vegetation, aquatic life, and soil health (Hernández et al., 2008).

In the extraction processes, various chemicals are utilized. For instance, in the production of graphene from graphite, liquid-phase exfoliation is employed, which involves the use of pyrrolidinones and spectrum analysis (Hernández et al., 2008). Additionally, the extraction of lithium and cobalt ions from battery components involves the use of organic solvents (Koo et al., 2020). Furthermore, the recycling of spent lithium-ion batteries involves the extraction of graphite and the selective extraction of divalent nickel and cobalt elements using complexing agents (Rothermel et al., 2016).

Moreover, the extraction of cobalt from aqueous solutions requires a thorough understanding of thermodynamics and kinetics, as demonstrated in a study on trace iron removal from cobalt sulphate solutions (Wang et al., 2018). Similarly, the separation of cobalt and nickel by employing solvent extraction has been extensively studied (Lee & Oh, 2004). The recovery of cobalt from deposits involves leaching, solvent extraction, and electrowinning processes (Dreisinger et al., 2012). Furthermore, the synergistic effect on the extraction of nickel and cobalt from synthetic sulphate solutions has been investigated, demonstrating the complexity of the extraction processes (Santanilla et al., 2014).

It is also important to consider the environmental impact of mining operations on soil and vegetation. Studies have shown that mine soil physical properties can be significantly affected, and the revegetation of mined areas plays a crucial role in ecological restoration (Angel et al., 2018; Stumpf et al., 2023). Additionally, the survival of hardwood trees on reclaimed land in mining areas has been studied, highlighting the importance of soil management and revegetation in mitigating the environmental impact of mining operations (Skousen et al., 2009).

To address the environmental challenges associated with lithium, cobalt, nickel, and graphite mining, there is a growing emphasis on adopting sustainable and responsible mining practices. Mitigation strategies include sustainable mining practices, water management, chemical management, and biodiversity conservation. Sustainable mining techniques involve minimizing the size of open-pit mines, employing reclamation efforts post-mining, and adopting best practices to minimize habitat disruption (Wanger, 2011). Effective water management strategies, including the use of containment ponds, sedimentation basins, and water treatment facilities, are crucial to prevent contamination and mitigate the impact on local water resources (Schomberg et al., 2021). Implementing stringent controls and regulations on chemical use, with a focus on minimizing environmental impact, is essential (Liang et al., 2021). Furthermore, engaging in biodiversity impact assessments and conservation efforts to mitigate the effects of habitat destruction is crucial (Agusdinata et al., 2018).

The development and adherence to responsible mining standards can contribute to balancing the demand for lithium with ecological preservation and long-term environmental health (Wanger, 2011). Additionally, the recycling of nickel and cobalt from batteries can save natural resources (Wanger, 2011). It is important to note that the environmental impact of direct lithium extraction from brines is an area of concern, and the data are still limited (Vera et al., 2023). Furthermore, the production of 1 ton of lithium requires approximately 400,000 litres of water, highlighting the significant water footprint of lithium production (Gutiérrez et al., 2022).

Manufacturing

The manufacturing of lithium-ion batteries (LIBs) involves several energy-intensive processes, such as electrode production and cell assembly, which significantly contribute to the environmental footprint of LIBs (Dai et al., 2019). The production of electrodes, including anodes and cathodes, requires resource-intensive steps and high-temperature treatments, leading to substantial energy requirements (Nishijima et al., 2014). Similarly, the assembly of cells demands precision manufacturing and controlled environments, further amplifying the energy consumption during LIB production (Changming Zhang et al., 2022). The environmental impact of this energy consumption is closely tied to the sourcing of energy, with non-renewable sources contributing to greenhouse gas emissions and exacerbating climate change (Gauthier et al., 2015).

Lithium-ion battery (LIB) manufacturing is a highly energy-intensive process, with significant energy inputs required during the production of electrodes and the assembly of cells. The magnitude of energy consumption in these crucial stages plays a pivotal role in shaping the environmental footprint of LIBs. The sourcing and nature of this energy are crucial factors, with non-renewable sources contributing to greenhouse gas emissions and exacerbating climate change.

The production of electrodes, fundamental components of LIBs, involves intricate manufacturing processes. The fabrication of anodes and cathodes necessitates the processing of raw materials, such as metals, along with the application of various coatings to enhance performance. These processes include high-temperature treatments, precision engineering, and resource-intensive steps, collectively contributing to the substantial energy requirements of electrode production.

The assembly of cells, the core units of LIBs, requires precision manufacturing and controlled environments. Integrating electrodes, separators, and electrolytes into a functional unit involves handling materials, forming cell structures, and implementing safety measures—all demanding energy inputs. This amplifies the overall energy consumption of LIB production.

The environmental impact of LIB manufacturing's energy consumption is significantly influenced by the sources of the energy used. If derived from non-renewable sources like fossil fuels or coal, the process contributes to greenhouse gas emissions. These emissions, primarily carbon dioxide (CO_2) and other pollutants, intensify climate change and degrade air quality.

To mitigate the environmental impact, there is a growing emphasis on transitioning to renewable energy sources in LIB manufacturing (Q. Wang et al., 2023). The integration of solar, wind, or hydroelectric energy can significantly reduce the carbon footprint associated with LIB production, aligning with global efforts to combat climate change (Hagemeister et al., 2022). Furthermore, a comprehensive life cycle assessment (LCA) is essential to understand the environmental implications of LIBs, considering factors such as energy consumption, resource depletion, and emissions throughout the battery's life cycle (Keil et al., 2016).

The industry is actively exploring innovative technologies and processes to minimize energy consumption during LIB manufacturing (Nguyen et al., 2018). Research and development efforts are focused on optimizing production methods, improving energy efficiency, and exploring alternative materials that require less energy-intensive processing (Xiao et al., 2021). Collaborative initiatives between manufacturers, researchers, and policymakers aim to establish best practices and standards for sustainable LIB manufacturing (Krauss et al., 2022).

To determine the amount of hydrocarbon-based fuel used to produce one lithium-ion battery, it is essential to consider the energy consumption and environmental impact of the production process. A life cycle analysis conducted by Peters et al. (2017) found that it took 330 kWh and 110 kg CO_2-e to produce 1 kWh of lithium-ion battery storage (Kader et al., 2021). This confirms that the production of lithium-ion batteries involves a significant amount of energy consumption and carbon emissions, which are associated with the use of hydrocarbon-based fuels in the manufacturing processes.

Furthermore, the environmental impacts of the production, use, and disposal of lithium-ion batteries have been highlighted in the literature. pointed out that little is known about the environmental impacts of the production, use, and disposal of lithium-ion batteries (Notter et al., 2010). This indicates that the production process may involve the consumption of hydrocarbon-based fuels, which can contribute to environmental impacts such as carbon emissions and resource depletion.

Moreover, reducing the cost and increasing energy density are two barriers for the widespread application of lithium-ion batteries in electric vehicles (Li et al., 2017). This suggests that the production process may involve significant energy inputs, including hydrocarbon-based fuels, to meet the demand for high-energy-density lithium-ion batteries.

In addition, the study by addressed the environmental burdens of the material production, assembly, and recycling of automotive lithium-ion batteries, indicating that the production process involves energy consumption and air emissions, including greenhouse gases (Dunn et al., 2012). This further supports the notion that the production of lithium-ion batteries may rely on the use of hydrocarbon-based fuels, contributing to energy consumption and greenhouse gas emissions. Therefore, based on the available literature, it can be inferred that the production of one lithium-ion battery involves a significant amount of energy consumption and environmental impact, which may be associated with the use of hydrocarbon-based fuels in the manufacturing processes.

The manufacturing process of lithium-ion batteries involves the use of various chemicals, including solvents and electrolytes. Improper disposal or mishandling of these chemicals can lead to soil and water pollution, impacting the local ecosystem. The production of lithium-ion batteries is a complex process that involves several critical steps. The manufacturing process is based around the slurry tape casting of electrodes followed by the assembly of the dried electrodes into cells with a separator and electrolyte (Gorman et al., 2019). The use of various chemicals, including solvents and electrolytes, is integral to this process. For instance, drying the coated cathode layer and subsequent recovery of the solvent for recycle is a vital step in the lithium-ion battery manufacturing plant and offers significant potential for cost reduction (Ahmed et al., 2016).

Improper disposal or mishandling of the chemicals used in the manufacturing process can have severe environmental consequences, leading to soil and water pollution. As a result, there is a growing emphasis on sustainable practices in the manufacturing of lithium-ion batteries. Research on the manufacture of lithium-ion batteries based on biomass electrodes has prospects for commercial development, offering a sustainable plan for the production of these batteries (Santoso et al., 2023). Additionally, from the viewpoint of cleaner production and green chemistry, the efficient recovery and reutilization of spent lithium-ion batteries are necessary, highlighting the importance of environmentally friendly practices in the battery industry (Liang et al., 2021).

Furthermore, the demand for lithium-ion batteries as a major power source in portable electronic devices and vehicles is rapidly increasing, leading to a significant increase in battery waste (Cerrillo-González et al., 2020). This underscores the importance of developing effective recycling and disposal methods for lithium-ion batteries to mitigate their environmental impact. The price of cobalt and lithium has increased due to the production of electric vehicles and hybrid electric vehicles, making the recycling and extraction of valuable metals from used lithium batteries an economically and environmentally important endeavour (Rahman et al., 2017).

In addition to the environmental impact of the manufacturing process, there is also a focus on the safety and operational aspects of lithium-ion batteries. Aqueous sodium-ion batteries promise increased operational safety and lower manufacturing cost compared to current state-of-the-art lithium-ion batteries based on organic electrolytes, highlighting the potential for safer and more sustainable battery technologies (Kühnel et al., 2017). Moreover, the thermal and electrochemical behaviours of lithium-ion batteries are affected by temperature, ultimately impacting performance and cycle life cost, emphasizing the need for effective thermal management in battery design (Panchal et al., 2017).

Usage

The use of lithium-ion batteries (LIBs) in electric vehicles (EVs) and renewable energy systems has been recognized for its potential to reduce greenhouse gas emissions and promote a more sustainable energy landscape (Ramkumar et al., 2022). However, it is important to consider the energy consumption associated with the charging process of LIBs, as the source of electricity used for charging significantly impacts their overall environmental benefit (Gailani et al., 2020).

Fossil fuel-based electricity generation is associated with the release of greenhouse gases and other pollutants, leading to environmental challenges such as habitat disruption, water pollution, and adverse health effects on local communities (Gailani et al., 2020). Therefore, the reliance on fossil fuel-derived electricity to charge LIBs introduces an environmental trade-off, limiting the overall effectiveness of these batteries in achieving sustainable and low-emission energy systems (Gailani et al., 2020).

Figure 28: EV Charging. Photo by Possessed Photography
on https://unsplash.com/photos/black-and-white-car-door-znCLdh5-Srk?utm_content=cr
editCopyText&utm_medium=referral&utm_source=unsplash, Unsplash.

To maximize the environmental benefits of LIBs, it is essential to transition to cleaner and renewable energy sources for electricity generation (Ramkumar et al., 2022). The adoption of solar, wind, hydropower, and other sustainable energy forms aligns the charging phase of LIBs with broader efforts to reduce carbon emissions and minimize environmental degradation associated with conventional energy production (Ramkumar et al., 2022).

While LIBs contribute significantly to reducing emissions during the operational phase in EVs and renewable energy systems, the environmental implications of their energy consumption during the charging process emphasize the importance of transitioning to cleaner electricity sources (Gailani et al., 2020; Ramkumar et al., 2022). The holistic impact of LIBs on the environment necessitates a comprehensive approach that addresses not only their use but also the energy systems supporting their charging infrastructure (Ramkumar et al., 2022).

The increasing demand for lithium and other materials in lithium-ion batteries (LIBs) due to the rising popularity of electric vehicles has raised concerns about resource depletion and its environmental impacts (Hosaka et al., 2018). The limited abundance of lithium, with less than 20 ppm in the Earth's crust, and its restricted resources in specific geographic regions, such as South America, Australia, China, and the US, has led to intensified extraction efforts across the globe, resulting in habitat disruption, soil erosion, and water contamination (Hosaka et al., 2018; Riofrancos, 2023). Furthermore, the environmental impact of resource depletion encompasses both the direct consequences of extraction processes and the broader implications for ecosystems, including habitat destruction and disruption of the delicate balance of ecosystems (Riofrancos, 2023). The environmental impact of direct lithium extraction from brines has also been a subject of concern, emphasizing the need for comprehensive and proactive approaches to address the environmental impacts of resource depletion (Vera et al., 2023).

Efforts to explore recycling technologies and develop alternative materials that reduce reliance on scarce resources are vital for creating a more sustainable and circular economy for LIB production (Huang et al., 2021; Kang et al., 2013). Additionally, sustainable mining practices, such as minimizing the size of open-pit mines, implementing reclamation efforts, and adopting habitat preservation strategies, can help mitigate the immediate consequences of extraction (Riofrancos, 2023). The socio-environmental impacts of lithium mineral extraction have prompted the need for a research agenda to address the multifaceted challenges associated with resource depletion and environmental sustainability (Agusdinata et al., 2018).

The growing demand for lithium and other materials in LIBs, driven by the surge in electric vehicle popularity, necessitates a balanced approach that considers sustainable extraction practices, explores alternative sources, and prioritizes recycling initiatives to ensure the long-term environmental sustainability of the lithium-ion battery industry (Graham et al., 2021).

In considering the carbon footprint of a lithium-ion battery compared to a car battery, lead-acid battery, and gasoline, lead-acid batteries are known for their affordability in production, making them a widely economical energy source globally. However, when compared to a lithium-ion battery, which boasts a longer life cycle and produces no tailpipe emissions, the use of a lead-acid battery in a gasoline-powered vehicle can result in a carbon footprint that is 13.5 times higher (Kilgore, 2023). This stark difference positions the carbon footprint of a lead-acid battery as more detrimental to the environment than that of a lithium-ion battery.

In contrast, electric vehicles (EVs) with lithium-ion batteries generate no tailpipe emissions, whereas gasoline-powered vehicles, irrespective of the battery type, emit CO_2 during operation. The environmental impact of an EV is contingent upon the energy source used for charging after manufacturing. Conversely, a gasoline vehicle consistently emits CO_2 for every minute it operates on the road (Kilgore, 2023).

Delving deeper into the environmental considerations of emerging electric vehicle technology, an EV's carbon emissions are influenced by the region it operates in and the type of energy prevalent in that area. In contrast, a gasoline vehicle emits tailpipe and CO2 emissions regardless of its location worldwide.

For instance, a 2022 Tesla Model X in New York City exhibits a carbon emission rate of 110g/mi. However, if the same car is evaluated across the U.S., where mixed electricity serves as the primary power source, its carbon emission increases to 140g/mi. In comparison, a new gasoline vehicle, on average, generates total emissions of 410g/mi (Kilgore, 2023). These calculations can be performed for various EV models using a carbon footprint calculator for electric cars versus gasoline. Renting cars with carbon offset options during travel provides an additional avenue for environmental consideration.

Despite causing a notable environmental impact during the manufacturing process, Kilgore (2023) posits that the overall emissions of a lithium-ion battery and the vehicle it powers remain more environmentally friendly. The focus should thus shift towards reducing the environmental impact associated with lithium-ion battery production, rather than reverting to gasoline vehicles powered by lead-acid batteries. This approach envisions a future for lithium mining that is significantly more considerate of the planet and its inhabitants.

End-of-Life

With the increasing utilization of LIBs, there is a growing demand for critical raw materials like lithium, nickel, and cobalt, raising concerns about their uneven distribution and potential geopolitical implications. This has prompted the need for establishing a secondary supply chain through the recycling of spent LIBs to mitigate potential shortages. Additionally, considering that LIBs have an average lifespan of around 10 years, the expected accumulation of over 5 million tons of spent LIBs by 2030 underscores the urgency to address their end-of-life disposal (Yu et al., 2022).

The energy storage and lithium battery market are undergoing rapid expansion. However, the escalating demand in this industry is paralleled by a corresponding surge in waste, prompting critical inquiries into how to effectively manage the emerging waste stream from lithium batteries and establish robust end-of-life (EoL) strategies. As an example of scale, in Australia, the yearly generation of approximately 3,300 tonnes of lithium-ion battery waste emphasizes the enormity of the challenge. A myriad of statistics further illuminates the multifaceted nature of this issue (Boxall, 2023): The global lithium battery market, having reached nearly 250 GWh in 2020, is poised for a substantial tenfold increase by 2030, signifying the magnitude of growth. Electric vehicles and large stationary electrical energy storage systems, notably prevalent in Australia, significantly contribute to the expanding lithium battery market. In 2021, a mere 10% of Australia's lithium-ion battery waste underwent recycling, presenting a stark contrast to the impressive 99% recycling rate for lead-acid battery waste. The volume of lithium-ion battery waste is on an upward trajectory, growing at a rate of 20% per year and potentially surging beyond 136,000 tonnes by 2036. Despite presenting a waste management challenge, lithium-ion batteries contain valuable materials, and recycling efforts have the potential to recover up to 95% of battery components for alternative uses or even the production of new batteries (Boxall, 2023).

Preceding the 2019 national ban on landfill disposal and the subsequent implementation of a battery collection system in 2022, only a minimal percentage of end-of-life lithium batteries were domestically collected. Instead, a substantial portion was shipped overseas for processing (Boxall, 2023). However, the current inadequacy in processing capacity within Australia has resulted in the accumulation of significant volumes in warehouses and scrap yards. This not only poses a substantial fire risk but also heightens the potential for environmental contamination. The limited processing capacity for lithium batteries in Australia exacerbates existing challenges, as substantial volumes remain stored in warehouses and scrap yards, thereby intensifying the risk of fires and environmental hazards (Boxall, 2023).

The recycling of lithium-ion batteries (LIBs) presents several challenges that hinder its widespread implementation. Efficient collection systems for used batteries are crucial, but the lack of standardized collection mechanisms impedes seamless retrieval, limiting overall recycling potential (M. Chen et al., 2019). Additionally, LIBs comprise a complex mix of materials, and the effective separation of these materials is a key challenge in the recycling process, requiring innovative technologies to enhance precision and efficiency (M. Chen et al., 2019).

Cost-effective and environmentally friendly recycling technologies are critical, as many existing methods involve high costs or environmentally impactful processes (Zheng et al., 2023). Inadequate infrastructure and the absence of comprehensive regulatory frameworks and standardized recycling practices further hinder effective LIB recycling (Shahjalal et al., 2022). Addressing these challenges requires collaborative efforts from governments, industries, researchers, and environmental organizations, emphasizing the need for investment in research and development, innovation, and supportive policies (Zheng et al., 2023).

The end-of-life (EoL) stage for an EV battery is typically reached when its capacity drops to around 80% of its original nominal capacity or power. At this point, spent EV batteries have two main disposal options: repurposing and recycling (Yu et al., 2022). Repurposing involves analysing individual modules or cells of a battery pack to reconfigure new packs with specific health for use in applications with similar or lower power requirements. However, repurposing faces technical challenges such as evaluating the state of health (SoH) of spent batteries and market difficulties like consumer acceptance and safety concerns (Yu et al., 2022).

If batteries are no longer suitable for repurposing, recycling becomes the primary method for safe EoL disposal. The recycling process includes industrial steps like pre-discharging, disassembling, sorting, shredding, purifying, and re-manufacturing (Yu et al., 2022). Cathode materials, containing precious metals like lithium, cobalt, and nickel, make up a significant portion of the battery's total value. Therefore, recycling efforts often focus on cathode recycling for profit. However, challenges persist in recycling anodes, electrolytes, and current collectors, necessitating further technological advancements for increased profitability and efficiency. While repurposing and recycling present viable disposal pathways, breakthroughs in various technologies are required to overcome existing obstacles and establish a sustainable and effective system for handling spent LIBs (Yu et al., 2022).

In considering recycling methods for Lithium-Ion Batteries (LIBs), Yu et al. (2022) outlines three key approaches: pyrometallurgical, hydrometallurgical, and direct recycling processes. While the first two have attained commercialization, direct recycling remains in the laboratory developmental stage.

Pyrometallurgy, extensively used in Europe, the US, and Japan, efficiently produces mixed metal slags. Nevertheless, it poses challenges such as harmful emissions and difficulties in recovering valuable elements like Li and Al. Hydrometallurgy, prevalent in China, exhibits lower energy consumption than pyrometallurgy but requires higher material input. Challenges in this method include chemical pollution and increased costs (Yu et al., 2022).

Pyrometallurgy, a widely adopted method in Europe, the US, and Japan, has played a pivotal role in efficiently producing mixed metal slags from spent Lithium-Ion Batteries (LIBs). This method, involving high-temperature processes, aims to extract valuable metals from the complex composition of battery materials. While pyrometallurgy offers notable advantages in terms of mixed metal slag production, it is imperative to acknowledge the associated challenges and consider advancements to enhance its sustainability and address environmental concerns.

Pyrometallurgy stands out for its ability to efficiently produce mixed metal slags, which are crucial raw materials for subsequent industrial processes (Lu et al., 2021). The high temperatures involved in pyrometallurgical operations facilitate the breakdown of battery components, leading to the extraction of diverse metals, including cobalt, nickel, and manganese. The resulting mixed metal slags serve as valuable inputs for downstream applications, contributing to the circular economy by reintegrating extracted metals into new manufacturing processes.

Despite its efficiency, pyrometallurgy presents environmental challenges that necessitate ongoing advancements. The high-temperature processes generate emissions that can contribute to environmental pollution (Khaliq et al., 2014). Harmful substances, such as fluorine compounds and toxic organic compounds, may be released into the atmosphere during the smelting of LIBs. These emissions underscore the importance of developing cleaner and more environmentally friendly pyrometallurgical methods to minimize the ecological footprint associated with metal recovery from spent LIBs.

While pyrometallurgy efficiently extracts a range of metals, challenges persist in the recovery of specific valuable elements, notably lithium (Li) and aluminium (Al) (Mutafela et al., 2020). Li, a critical component of LIBs, is essential for the battery industry and renewable energy applications. However, the recovery of Li through pyrometallurgical processes may not be as efficient as desired. Similarly, the extraction of aluminium, which is a key component in battery casings, faces hurdles in achieving optimal recovery rates.

Advancements in pyrometallurgical processes are indispensable to overcome these challenges and enhance the overall sustainability of metal recovery from spent LIBs (Roy et al., 2021). Researchers and industry experts are actively exploring innovative techniques to improve the efficiency of Li and Al recovery during pyrometallurgy. This may involve the development of novel smelting methodologies, optimized temperature control, and the integration of advanced separation technologies.

The quest for sustainability in metal recovery from spent LIBs involves striking a balance between efficiency, environmental impact, and the comprehensive retrieval of valuable elements (Qiu et al., 2023). Ongoing research initiatives focus on creating pyrometallurgical processes that not only maintain or improve efficiency but also minimize harmful emissions and maximize the recovery rates of critical elements.

Hydrometallurgy has gained attention for its lower energy consumption compared to pyrometallurgy, aligning with the global pursuit of energy-efficient processes (Larouche et al., 2020). However, challenges such as higher material input and chemical pollution necessitate focused research and development efforts to refine and optimize hydrometallurgical processes (Larouche et al., 2020). Chemical processes in hydrometallurgy selectively dissolve and recover metals from LIB components, but they bring about challenges related to environmental pollution if not managed effectively. The release of toxic substances during metal extraction poses environmental risks, necessitating stringent waste treatment and disposal measures. To enhance sustainability, researchers are exploring ways to optimize chemical processes, reduce environmental footprint, and enhance metal recovery efficiency. Strategic steps involve fostering interdisciplinary collaborations to optimize chemical processes, mitigate environmental impacts, and reduce material input, aiming to establish hydrometallurgy as a technologically advanced and environmentally responsible approach to metal recovery from spent LIBs.

Electrochemical approaches have been discussed for selective recovery of critical elements in hydrometallurgical processes, focusing on rare earth elements and other key materials for the circular economy. This highlights the potential for innovative solutions to address the challenges of material input and environmental impact in hydrometallurgy. Additionally, the use of deep eutectic solvents for recycling valuable metal elements from spent LIBs has attracted extensive research, indicating a shift towards more sustainable and efficient recovery methods.

The direct recycling method, emerging as a potential solution, seeks to restore spent cathode materials electrochemically or physicochemically for direct reuse or as precursors.

Challenges associated with this method encompass reliance on manual labour, safety concerns, and the imperative need for automation (Yu et al., 2022). Moreover, direct recycling demonstrates promise in upcycling, allowing the modification of recycled cathode materials based on market demands, leading to improved energy density and performance. Overcoming challenges, such as repetitive manual procedures, standardization issues, safety in disassembly, and adherence to industrial material requirements, is crucial. Encouraging collaboration between industry, universities, and laboratories is essential to facilitate research and large-scale experimentation, addressing technical barriers and working towards achieving practical direct recycling in the industry (Yu et al., 2022).

The advancements in sustainable metal recovery methods for lithium-ion battery recycling are crucial for addressing the challenges associated with pyrometallurgy, hydrometallurgy, and direct recycling. Research and development efforts focused on improving the efficiency, sustainability, and environmental impact of these methods are essential for the successful implementation of LIB recycling on a larger scale. Collaboration between industry, academia, and research institutions is vital for driving innovation and overcoming the challenges associated with each method. By addressing these challenges and leveraging the potential of sustainable metal recovery methods, the recycling of lithium-ion batteries can contribute to a more sustainable and circular approach to battery production and end-of-life management.

Improper disposal of Lithium-Ion Batteries (LIBs) has also emerged as a pressing environmental concern, introducing significant risks and consequences that extend beyond the end of a battery's functional life. Whether these batteries find their way into landfills or are subjected to incineration, the implications for the environment are profound and multifaceted.

When lithium-ion batteries (LIBs) are disposed of in landfills, they pose a significant threat to soil and water ecosystems due to the intricate composition involving various materials, including metals and chemicals. If left unmanaged, these batteries can undergo processes leading to environmental contamination, releasing toxic chemicals and heavy metals into the soil, thereby directly risking the surrounding ecosystem and potentially harming plants, animals, and microorganisms that are integral to the natural balance (Kader et al., 2021). The challenges of recycling spent LIBs include contamination, complicated processes, massive chemical/energy consumption, and secondary pollution (B. Zhang et al., 2023). Furthermore, the improper disposal and recycling of LIBs have attracted increasing interest due to the purposes of resource conservation and pollution reduction (Liang et al., 2021). The risks associated with lithium-contaminated waste disposal raise concerns about potential harm to food and fodder crops, emphasizing the need for proper management and recycling of LIBs (Hayyat et al., 2021).

Current recycling technologies for spent LIBs include pyrometallurgy, direct recycling, and wet chemical extraction, highlighting the need for improved recycling systems and resource reuse (Wang, 2022; Zhu et al., 2023). Additionally, it is crucial to minimize heavy metals' composition before disposal by recovering valuable metals from waste LIBs (Nurqomariah & Fajaryanto, 2018a). Educational and community-based programs about appropriate disposal and recycling of lithium-ion batteries are also needed to reduce lithium-contained trash in the environment (Chow, 2022). Moreover, environmentally safe, "green" disposal processes are required, including disassembly of batteries into component materials for recycling.

The rapid growth of LIB manufacturing in certain regions is expected to increase pollution, which is inconsistent with long-term ecological sustainability (Filomeno & Feraco, 2020). Life cycle assessments of LIBs have revealed significant greenhouse gas emissions during battery production and disposal, emphasizing the environmental impact of LIBs throughout their life cycle (Mahmud et al., 2019). Additionally, the environmental impacts of the production, use, and disposal of LIBs are areas that require further understanding and research (Notter et al., 2010; Yu et al., 2014).

The leaching of hazardous substances from LIBs can percolate through the soil layers, ultimately reaching groundwater reservoirs. Once these contaminants infiltrate the water supply, they jeopardize the quality of drinking water and contribute to the deterioration of aquatic habitats. Heavy metals such as cobalt, manganese, and lithium are particularly notorious for their persistence in the environment, and their release into water sources can lead to long-term ecological consequences.

Incineration, often considered as an alternative to landfill disposal, presents a different set of challenges. While it might seem like a method to reduce physical waste volume, the incineration of lithium-ion batteries (LIBs) comes with its own environmental risks. The high temperatures involved in the incineration process can cause the release of toxic fumes and gases into the atmosphere. This poses a significant challenge as hazardous heavy metals may release from incineration residues, causing harm to the environment (Li et al., 2023). Additionally, incineration of LIBs is associated with the release of toxic fumes and gases into the atmosphere due to the high temperatures involved in the process (Li et al., 2023). Furthermore, incineration poses challenges related to waste minimization and recycling, as well as alternate incinerator treatment technologies (Ghazali et al., 2022).

The challenges associated with incineration are further highlighted in the context of waste-to-energy (WTE) incineration technologies. While incineration is considered an effective solution for sustainable and efficient municipal solid waste (MSW) disposal, there are potential health risks associated with incinerators, leading to considerable interest in alternative WTE options (Almanaseer et al., 2020). Moreover, the risk factors associated with public-private partnership waste-to-energy incineration projects vary from one country to another, indicating the complex nature of incineration as a waste management method (Cui et al., 2019).

In the specific context of LIBs, the challenges of incineration are compounded by the need for efficient and sustainable recycling of metal resources from spent LIBs. The overwhelming dependency on chemical and energy in the current prevailing recycling methods presents a significant challenge (Jiang et al., 2022). Additionally, the release of toxic gases during the combustion of lithium-ion batteries poses a threat to both human health and the environment (Chen et al., 2017).

The environmental risks associated with incineration are further underscored by the potential impact on human health and the environment. Studies have shown that incineration of hazardous waste can lead to the increase in heavy metals content in ash, posing risks to the surrounding population and ecosystems (Ferré-Huguet et al., 2005; Latosińska & Czapik, 2020; Mari et al., 2007). Furthermore, the characterization of particulate air pollution exposure from municipal solid waste incinerator emissions highlights the potential health risks associated with incineration, emphasizing the need for comprehensive assessment and management of incineration-related environmental impacts (Ashworth et al., 2013).

The combustion of LIBs releases not only heavy metals but also other potentially harmful substances, including fluorine compounds and volatile organic compounds. These airborne pollutants can contribute to air pollution and, in turn, have detrimental effects on respiratory health and overall air quality. Additionally, the dispersion of these contaminants into the atmosphere can lead to their deposition over larger geographical areas, affecting ecosystems far from the original point of disposal.

The consequences of improper LIB disposal extend beyond environmental ecosystems, posing a dual threat to both the natural world and human health. Contaminated soil and water directly impact the health of plants and animals, disrupting ecological balance and biodiversity. Furthermore, as these contaminants move up the food chain, there is a risk of bioaccumulation, where the concentration of hazardous substances increases with each trophic level.

Human exposure to toxic elements from improperly disposed LIBs can occur through contaminated drinking water, crops grown in polluted soil, or inhalation of airborne pollutants from incineration. Chronic exposure to heavy metals, such as cobalt and manganese, has been associated with various health issues, including neurological disorders, respiratory problems, and developmental issues, making the improper disposal of LIBs a matter of grave concern for public health.

Addressing the disposal challenge of LIBs is important for mitigating environmental risks and safeguarding ecosystems and human health. Implementing proper recycling and disposal mechanisms is imperative to ensure that the revolutionary benefits of lithium-ion technology do not come at the cost of irreparable environmental damage.

A Comparison of the Use of Fossil Fuels Against the Production, Maintenance and of Renewable Energy Through Solar, Wind with Lithium-Ion Battery Storage

The environmental impact of energy sources is a complex issue that necessitates a comprehensive assessment of the entire life cycle of each energy production method. Fossil fuels, including coal, oil, and natural gas, have well-documented environmental costs associated with their extraction, processing, combustion, and waste production (Nayak et al., 2019). These processes result in habitat disruption, water and air pollution, greenhouse gas emissions, and the generation of waste materials that can contaminate the environment (Nayak et al., 2019). On the other hand, renewable energy sources such as solar and wind power, as well as lithium-ion batteries, also have environmental impacts, particularly during their manufacturing, land use, and end-of-life stages.

The production of solar panels and wind turbines involves resource depletion, energy consumption, and waste generation, while the extraction of raw materials for lithium-ion batteries can lead to habitat disruption, soil degradation, and water pollution.

The production of solar panels and wind turbines, integral components of renewable energy systems, entails several environmental considerations. The environmental impact associated with their manufacturing includes resource depletion, energy consumption, and waste generation (Alblawi et al., 2019; Alibaba et al., 2020; Dupraz et al., 2011; Khan et al., 2021; Pavlopoulos et al., 2023; Rahman et al., 2023; Tan et al., 2022).

The manufacturing of solar panels relies on materials such as silicon, silver, aluminium, and other metals, contributing to resource depletion if not sourced sustainably (Dupraz et al., 2011). Additionally, the production of solar panels involves energy-intensive processes, including the extraction, refining, and manufacturing stages, which can offset the environmental benefits of the solar panels over their lifetime if non-renewable energy sources are used (Barron-Gafford et al., 2016). Furthermore, the manufacturing process generates waste, including silicon dust, slurry, and other by-products, which can have environmental impacts (Pavlopoulos et al., 2023).

Similarly, wind turbines, predominantly made of steel, copper, and rare earth metals, contribute to resource depletion through mining and processing these materials (Alibaba et al., 2020). The production of wind turbines also requires significant energy inputs, and the manufacturing processes, transportation, and installation contribute to the overall embodied energy of wind turbines (Alblawi et al., 2019).

The production of lithium-ion batteries involves raw material extraction through mining, leading to habitat disruption, soil degradation, and water pollution if not carried out responsibly (Khan et al., 2021). Additionally, energy-intensive processes from mining and processing raw materials to manufacturing the battery cells contribute to the environmental impact (Tan et al., 2022). Improper disposal of lithium-ion batteries can lead to environmental contamination, and recycling efforts need to address the recovery of valuable materials (Puspitarini et al., 2020).

When evaluating the overall environmental impact, it is essential to consider the full life cycle of each energy source. While fossil fuels have significant environmental costs, renewable energy sources and lithium-ion batteries are generally considered more environmentally friendly due to their lower greenhouse gas emissions during operation and the potential to reduce dependence on finite fossil fuel resources. Efforts are ongoing to improve the sustainability of renewable energy technologies and address the environmental challenges associated with their production and disposal.

The environmental impact of energy sources can be comprehensively assessed through a full life cycle assessment, which considers the entire life cycle from raw material extraction and manufacturing to operation and end-of-life disposal (Gauthier et al., 2015). Fossil fuels, such as coal, oil, and natural gas, have extensive extraction processes that lead to habitat disruption, soil degradation, and ecosystem destruction. Additionally, the transportation of fossil fuels involves energy consumption and environmental risks, such as oil spills and pipeline leaks. The combustion of fossil fuels releases greenhouse gases and pollutants, contributing to air and water pollution. On the other hand, renewable energy sources and lithium-ion batteries, while involving resource extraction and energy-intensive manufacturing processes, generally have lower ongoing environmental impacts once operational, contributing to lower greenhouse gas emissions compared to fossil fuels.

Efforts are ongoing to develop efficient recycling technologies for components like solar panels and batteries, aiming to recover valuable materials and reduce waste. Ongoing research aims to enhance the efficiency of renewable energy technologies, reduce their manufacturing footprint, and identify alternative materials with lower environmental impact. Governments and organizations worldwide are implementing policies to encourage the adoption of renewable energy and enforce responsible disposal and recycling practices. The concept of a circular economy, where materials are reused, recycled, and repurposed, is gaining traction in the renewable energy sector to minimize waste and resource consumption. Certification programs and standards are being established to ensure that renewable energy technologies adhere to environmentally sustainable practices throughout their life cycle.

It is important to note that the environmental impact of energy sources is a topic of ongoing research and debate. Some studies have raised concerns about the environmental impact of renewable energy sources, particularly in terms of land use and resource consumption. Additionally, the transition to renewable energy sources may pose challenges for certain industries and economies, as evidenced by the scepticism and resistance from the fossil fuel industry in some regions. Furthermore, the widespread exploitation of bioenergy has raised questions about its sustainability and potential adverse social and environmental impacts.

The assessment of environmental impacts related to energy sources is a dynamic and evolving field marked by continuous research and ongoing debates. Various studies contribute to the understanding of the complexities associated with different energy technologies, offering insights into both positive and negative aspects. Concerns about renewable energy include land use and resource consumption, as well as transition challenges from fossil fuels to renewable energy sources. Several studies have expressed concerns about the environmental impact of renewable energy technologies, particularly in terms of land use and resource consumption. Large-scale solar and wind farms may require substantial areas, leading to habitat disruption and altered ecosystems (Alola &

Alola, 2018). Additionally, the extraction of raw materials for renewable technologies may raise concerns about resource consumption and environmental degradation (Alves et al., 2023).

The transition from fossil fuels to renewable energy sources can pose challenges for specific industries and economies. Resistance and scepticism from the fossil fuel industry, observed in certain regions, highlight the economic and geopolitical complexities associated with this shift (Jebli & Youssef, 2018). Furthermore, the impacts of renewable energy consumption on economic growth have been a subject of active research, with studies employing dynamic heterogeneous panel data modelling approaches to analyse the relationship (Can & Korkmaz, 2019). In addition, the correlation between renewable energy consumption and economic growth has been a topic of investigation, with studies aiming to understand the causality relationship in different regions (Marinaş et al., 2018). Furthermore, the impacts of renewable and non-renewable energy consumption on environmental pollution have been examined, particularly in the context of an environmental Kuznets curve setting (Karasoy & Akçay, 2019).

The Kuznets Curve, a theoretical framework proposed by economist Simon Kuznets, suggests a relationship between economic development and income inequality over time, following an inverted U-shape trajectory. As a country undergoes economic development, income inequality initially increases, peaks, and then decreases (Sharif et al., 2020). The stages of the Kuznets Curve encompass the early stage, transition stage, peak inequality, and late stage of economic development, each characterized by distinct economic and social features (Sharif et al., 2020).

In the context of environmental pollution, the impact of renewable and non-renewable energy consumption on carbon emissions and environmental degradation has been extensively studied. While some research has focused on the bidirectional causal association between energy consumption and carbon emissions, other studies have attempted to disaggregate the impacts of renewable and non-renewable energy consumption on environmental pollution (Iorember et al., 2022; Pata, 2020; Sahoo & Sahoo, 2020; Shafiei & Salim, 2014).

These studies have revealed that non-renewable energy consumption has a positive and statistically significant effect on CO_2 emissions, contributing to environmental degradation, while the impact of renewable energy consumption varies across different contexts (Iorember et al., 2022; Pata, 2020; Sahoo & Sahoo, 2020; Shafiei & Salim, 2014).

Furthermore, the relationship between economic growth, energy consumption, and environmental pollution has been a subject of investigation. Studies have explored the effects of renewable and non-renewable energy consumption on CO_2 emissions, air pollution, and ecological footprint, highlighting the significance of transitioning towards renewable energy sources to mitigate environmental degradation (Rofiuddin et al., 2019; Shen et al., 2020; Turedi & Turedi, 2021; Zaidi et al., 2018). Additionally, the role of renewable energy consumption in promoting economic growth and environmental sustainability has been examined, with findings indicating a positive impact on economic growth and a potential for reducing environmental pollution (Can & Korkmaz, 2019; Inglesi-Lotz, 2016; Joseph & Charles, 2021).

It is important to note that the Kuznets Curve, while providing a theoretical framework for understanding income inequality in relation to economic development, does not directly address the specific impacts of renewable and non-renewable energy consumption on environmental pollution. However, the studies referenced provide valuable insights into the complex dynamics of energy consumption, economic growth, and their implications for environmental sustainability.

The potential impacts of emission concerned policies on power system operation with renewable energy sources have also been studied, highlighting the need for sustainable energy planning with respect to resource use efficiency (Verma & Kumar, 2013). The significance of foreign direct investment and renewable energy consumption in mitigating environmental degradation has also been explored, emphasizing the need for sustainable practices in energy consumption (Warsame, 2023).

The sustainability of bioenergy has become a topic of concern due to its potential environmental and social impacts. Large-scale biomass production for energy purposes has raised questions about its sustainability, particularly in relation to deforestation, biodiversity loss, and ecosystem disruption (Schleussner et al., 2016). The extensive cultivation of bioenergy crops may lead to adverse social and environmental effects, including land-use conflicts, changes in agricultural practices, and potential disruptions to local communities (Calvin et al., 2021). These concerns have prompted adjustments in policies and regulations to ensure the sustainable deployment of renewable energy technologies (W. Wu et al., 2019). Policymakers face the challenge of balancing environmental sustainability, economic considerations, and social impacts when formulating energy policies, which is crucial for achieving a transition to cleaner energy sources (Raj et al., 2021).

Research findings have highlighted the need to address the genuine concerns related to the large-scale deployment of bioenergy, irrespective of the presence of mitigation policies (Searle & Malins, 2014). Studies have also emphasized the importance of considering the trade-offs and effectiveness of renewable energy policies in achieving dual decarbonization goals, as well as the need for a reassessment of global bioenergy potential to ensure sustainable deployment (Wei et al., 2022). Furthermore, the role of community acceptance in sustainable bioenergy projects has been recognized, with air pollution from bioenergy plants identified as a major concern for local communities (Eswarlal et al., 2014).

In addition, the potential impacts of emission concerned policies on power system operation with renewable energy sources have been studied, highlighting the need to evaluate policies to increase electricity generation from renewable energy (Verma & Kumar, 2013). Moreover, the role of bioenergy in enhancing energy, food, and ecosystem sustainability based on societal perceptions and preferences has been explored, reflecting the social and economic concerns in various countries (Acosta et al., 2016). Overall, the sustainability concerns associated with bioenergy have prompted a significant body of research aimed at understanding the environmental, social, and economic implications of its widespread exploitation. This research has contributed to the ongoing discourse on the development of policies and regulations to ensure the sustainable deployment of bioenergy as a renewable energy source.

To address the environmental impacts of energy production, a holistic approach is essential. This approach involves considering all stages of energy production, from raw material extraction to end-of-life disposal, to comprehensively evaluate environmental impacts (Verma & Agarwal, 2022). Technological innovation plays a crucial role in addressing environmental concerns associated with renewable energy. Ongoing research focuses on developing more efficient, sustainable, and eco-friendly technologies (Muzhikyan & Farid, 2015). Acknowledging the ongoing research and debates in renewable energy is crucial for informed decision-making, policy adjustments, and technological advancements, contributing to a more sustainable and balanced energy future (Johnstone et al., 2008).

The need for a holistic approach in renewable energy solutions is emphasized in literature. It is suggested that a holistic solution is necessary to address interwoven challenges related to fossil energy and climate change (Finocchi, 2021). Furthermore, the successful implementation of sustainable power supply in emerging countries requires a comprehensive methodology, emphasizing the importance of a holistic approach (Peñalvo-López et al., 2019). Similarly, a holistic retrofit approach is recommended for considering renewable energy as an energy supply (Alkhateeb & Abu-Hijleh, 2019). Evaluating a variety of energy sources and complementing it with economic analysis demonstrates the benefits of a sustainable methodology based only on renewable sources, highlighting the importance of a holistic approach (Ramos et al., 2022).

In the context of sustainable design of energy conversion systems, a multi-criteria method is proposed to guide reflections and imagine sustainable solutions, emphasizing the need for a holistic approach in the design process (Mallard et al., 2020). Additionally, the emergence of renewable energy technologies at the country level is linked to relatedness, international knowledge spillovers, and domestic energy markets, emphasizing the interconnected nature of renewable energy solutions (D. Li et al., 2020). The impact of sustainable construction and knowledge management on sustainability goals is reviewed, highlighting the complexity of construction projects and the need for a holistic approach to achieve sustainability (Pietrosemoli & Rodríguez-Monroy, 2013).

The literature also emphasizes the importance of integrating renewable energy solutions for rural communities using a nexus approach to promote efficient resource management while considering interdependencies between water, waste, and energy (Piippo & Pongrácz, 2020). Furthermore, the impact of technology integration on sustainable development is quantified using a holistic sustainable development approach, emphasizing the importance of passive energy buildings and energy conservation technologies in reducing carbon footprint (Hasan, 2014). An integrated approach to supply all energy requirements in a sustainable manner is highlighted as highly required for better indoor environmental conditions (Hejazi, 2016).

In conclusion, the environmental impact of energy sources is a complex and evolving field of study. While fossil fuels have well-documented environmental costs, renewable energy sources and lithium-ion batteries offer potential benefits in terms of reduced greenhouse gas emissions and decreased reliance on finite resources. However, ongoing research and efforts to improve the sustainability of renewable energy technologies are essential to address the environmental challenges associated with their production and disposal.

Chapter Seven

Economic Impact

Lithium-ion batteries, introduced to the commercial market over 35 years ago and initially used primarily in consumer electronics, have witnessed a significant surge in production scale in the past decade, notably driven by the proliferation of electric vehicles (EVs). This surge has led to the emergence of challenges related to the end-of-life (EoL) management of large quantities of lithium-ion battery (LIB) packs and potential negative environmental impacts. While several LIB mega-factories have been announced globally, viable recycling solutions for the substantial EoL LIBs remain limited, raising concerns about long-term environmental consequences (Gonzales-Calienes et al., 2023).

The escalating demand for LIBs has put considerable strain on raw material supply chains, impacting market prices for crucial materials like lithium, cobalt, and nickel. Despite the presence of numerous technologies and startups aiming to address these challenges, policy development, research and development (R&D), and commercial solutions are progressing at a slower pace (Gonzales-Calienes et al., 2023). The complexity is further compounded by the diverse array of battery technologies and material formulations.

Various studies have explored alternative battery technologies, such as solid-state Li-metal batteries and organic batteries, showcasing potential improvements in performance and sustainability. However, challenges persist, particularly in optimizing these technologies and designing them for effective end-of-life management, necessitating further research and development (Gonzales-Calienes et al., 2023).

To address these issues, collaborative efforts at the international and national levels are crucial, focusing on policies and regulations for EoL LIB management. The recent focus on LIB recycling in Canada, particularly in the upstream supply chain, highlights the need for integrated economic and environmental sustainability assessments tailored to local contexts. This study specifically delves into the economic and environmental impact assessments of recycling spent LIBs, concentrating on the supply chains of battery cathode active materials (Gonzales-Calienes et al., 2023).

Economic Factors

The global economic impact of lithium-ion batteries is multifaceted and encompasses various aspects such as production, recycling, and reuse. The increasing demand for lithium-ion batteries, particularly in electric vehicles (EVs) and energy storage systems, has led to significant economic implications. Firstly, the production of lithium-ion batteries has led to the depletion of natural resources and increased environmental concerns due to the disposal of end-of-life batteries (Kader et al., 2021). Additionally, the growth in demand for lithium-ion batteries has led to the need for sustainable recycling technologies to recover valuable materials, which has economic significance (Y. H. Wang et al., 2023). Furthermore, the economic analysis of centralized battery energy storage systems with reused lithium-ion batteries has been studied to evaluate the economic benefits, indicating the economic potential of reusing batteries (Meng, 2021).

Moreover, the economic impact extends to the recovery of metals from used lithium-ion battery cathode materials, which is of both environmental and economic importance (Y. H. Wang et al., 2023). The economic implications are also evident in the development of high-energy batteries through the introduction of high-capacity lithium-storage materials, fostering advancements in battery technology (Cheng et al., 2011). Additionally, the economic value of carbon-based materials in producing electrodes for lithium-ion batteries has been highlighted, indicating the economic significance of material choices in battery manufacturing (Ingried et al., 2022).

Furthermore, the economic impact is evident in the thermal management of lithium-ion batteries, as the working temperature significantly affects battery performance, thereby influencing the economic viability of battery systems (Madani et al., 2020a). Additionally, the safety considerations and evaluation criteria for lithium-ion batteries, pioneered by countries such as the United States and Japan, have economic implications in ensuring the safe and reliable use of these batteries (Shan, 2017).

The widespread integration of lithium-ion batteries has initiated a profound shift across various industries, facilitating advancements in electric vehicles, smartphones, and renewable energy storage systems. As the world increasingly emphasizes sustainability and cleaner energy alternatives, the demand for lithium-ion batteries has experienced unprecedented growth. This article delves into the economics of lithium-ion battery production, examining the market trends that have defined the industry's trajectory and its future prospects.

The production of lithium-ion batteries is influenced by several key factors:

- Raw Materials: Lithium-ion batteries depend on specific raw materials, such as lithium, cobalt, nickel, and graphite. The cost and availability of these materials significantly impact battery production costs, with fluctuations in cobalt and nickel prices influencing overall expenses.

- Manufacturing Process: The intricacies of the manufacturing process, covering electrode production, cell assembly, and quality control, contribute to production costs. Continuous advancements in battery manufacturing technology and automation have the potential to reduce production costs over time.

- Energy Consumption: The energy-intensive nature of battery production adds to the economic equation, where the source and cost of electricity during manufacturing directly affect the final battery price.

- Scale of Production: Economies of scale play a crucial role, with large-scale production facilities benefiting from reduced costs per unit, enhancing affordability as demand rises.

- Research and Development: Investment in research and development is essential for improving battery performance and reducing production costs. Technological advancements result in more efficient and cost-effective production methods.

Several market trends have shaped the growth and development of the lithium-ion battery industry:

- Electric Vehicles (EVs): The surge in EV demand has been a primary driver for the lithium-ion battery market. EV manufacturers seek batteries with extended ranges, faster charging times, and lower costs to encourage consumer adoption.

- Renewable Energy Storage: Lithium-ion batteries play a vital role in storing energy generated from renewable sources, supporting the global transition to cleaner energy and contributing to grid stability.

- Consumer Electronics: The thriving consumer electronics market, including smartphones, laptops, tablets, and wearables, fuels the demand for high-performing and durable lithium-ion batteries.

- Energy Storage Solutions: Large-scale battery energy storage solutions are gaining traction to support grid stability, peak shaving, and emergency backup power.

- Recycling Initiatives: With a growing emphasis on sustainability, battery recycling has become a significant trend, aiming to recover valuable materials, reduce waste, and decrease the demand for new raw materials.

The lithium-ion battery industry has witnessed significant cost reductions and technological advancements over the years:

- Decline in Battery Prices: Consistent declines in lithium-ion battery prices have made electric vehicles and energy storage solutions more accessible. This trend is a result of improved production processes, economies of scale, and advancements in battery chemistry.

- Energy Density Improvements: Ongoing research explores battery chemistry innovations to enhance energy density, directly influencing EV range and performance.

- Solid-State Batteries: Solid-state batteries, with potential benefits in energy density and safety, represent the next frontier in battery technology. Commercialization of these batteries could lead to further cost reductions and performance improvements.

- Recycling and Circular Economy: Efficient recycling processes aid in recovering valuable metals like lithium, cobalt, and nickel, promoting a circular economy in the battery industry.

While the future outlook for the lithium-ion battery industry is promising, certain challenges need addressing:

- Supply Chain Security: The availability and geopolitics of critical raw materials, such as lithium and cobalt, can impact battery production and prices. Diversification of the supply chain and investments in resource exploration are crucial for stable access.

- Environmental Concerns: Despite environmental benefits, concerns persist regarding the impact of mining and battery disposal. Sustainable practices and recycling initiatives are imperative to address these concerns.

- Advancing Technology: Continuous research and development efforts are vital to stay competitive. Technological advancements will drive cost reductions and further enhance battery performance.

The economics of lithium-ion battery production are undergoing a dynamic transformation, driven by technological advancements, the growing demand for electric vehicles and renewable energy storage, and decreasing production costs. The lithium-ion battery industry plays a pivotal role in facilitating the global transition to cleaner and more sustainable energy solutions. As research and development continue to push the boundaries of battery technology, significant improvements in performance and affordability are expected, making electric vehicles and renewable energy storage solutions more accessible worldwide. Embracing these trends and addressing associated challenges will contribute to a greener, more sustainable future powered by lithium-ion batteries.

The global lithium-ion battery recycling market is anticipated to witness substantial growth, reaching USD 22,805 million by 2030, driven by the increasing demand for electric vehicles, depletion of critical minerals, and stringent government policies (Gonzales-Calienes et al., 2023). Despite the positive outlook, challenges such as safety concerns and high capital and operating costs need to be addressed for sustainable growth.

Sustainability indicators play a vital role in assessing the long-term viability of electric vehicles. While a comprehensive sustainability assessment involves multiple economic, environmental, and social indicators, this study focuses on global warming potential (expressed as greenhouse gas emissions) and financial cost as key indicators (Gonzales-Calienes et al., 2023).

However, there is a notable gap in current sustainability assessments related to EV battery cars, particularly in comparing the sustainability of virgin and recycled battery materials using a bottom-up approach. This study aims to fill this gap by providing a detailed and consistent integrated economic and environmental assessment along the full supply chain of recovered LIB cathode materials (Gonzales-Calienes et al., 2023). The focus is on a practical case study within the province of Quebec, Canada, considering both recycled and virgin materials.

As the global demand for LIBs continues to rise, addressing the challenges of recycling and sustainability becomes imperative. The economic impact of lithium-ion batteries extends across industries, influencing energy storage, electric vehicles, consumer electronics, job creation, and investment opportunities. However, achieving a balance between economic growth and environmental sustainability requires ongoing research, policy development, and collaborative efforts to optimize recycling processes and mitigate the environmental impact of LIB production and disposal (Gonzales-Calienes et al., 2023).

To achieve net-zero emissions and cost-effective power production, recent analyses propose that photovoltaic (PV) capacity in the United States could surpass 1 terawatt (TW) by 2050, accompanied by substantial energy storage capacity, mainly from batteries (Heath et al., 2022). In comparison, the total U.S. utility-scale power capacity from all energy sources in 2020 stood at 1.2 TW, with solar contributing approximately 3%. This monumental scale of deployment raises concerns about material supply challenges, end-of-life management, environmental impacts, and economic costs (Heath et al., 2022). A set of solutions revolves around transitioning to a circular economy, departing from the linear take-make-waste economic model to one that maximizes the value of materials and products, recovering them at the end of life for reintroduction into the economy.

Despite limited global experience, scholars and practitioners are increasingly exploring circular economy pathways, especially in fast-growing sectors like renewable energy. This critical review aims to consolidate the expanding literature, identifying key insights, gaps, and opportunities for research and implementation of a circular economy for two pivotal technologies facilitating the shift to a renewable energy economy: solar PV and lithium-ion batteries (LIBs).

Through a systematic literature review of over 3,000 publications, Heath et al. (2022) assess the state of the circular economy for solar PV and LIBs. While neither industry has fully embraced a circular economy, both are progressing toward increased circularity. The literature highlights the need to shift research emphasis beyond recycling technology development. It advocates for a more comprehensive investigation of various circular economy strategies, considering economic, environmental, and policy aspects.

Responding to common questions, the review concludes that a circular economy for LIB and PV has not been fully realized yet but is in progress. The awareness and research efforts suggest active endeavours to define challenges' scope and develop technical and policy solutions. Industries and government policies are contributing to this shift, though challenges like economic sustainability and consistent regulatory frameworks persist.

The review by Heath et al. (2022) underscores the imperative to advance research beyond recycling, urging a focus on diverse circular economy pathways that offer superior environmental and social benefits. Additionally, it advocates for supporting technology deployment through comprehensive economic, environmental, and policy analyses, recognizing the crucial role of non-technological factors in the success of circular economy initiatives (Heath et al., 2022). The review also highlights the importance of leveraging digital information systems to enhance circularity, emphasizing the exploration of digital pathways alongside material approaches. Furthermore, it stresses the need to enhance recycling technologies by addressing challenges associated with integrated processes, costs, and scalability. Lastly, the review suggests a concentrated effort to study and design circular economy-related aspects of markets for lithium-ion batteries (LIBs) and photovoltaic (PV) systems. This involves the regular collection of data to facilitate efficient capital allocation and informed decision-making in these evolving markets.

While acknowledging the journey toward a circular economy is ongoing, the review suggests that the benefits can be realized incrementally, with continuous updates to mark progress and align with sustainability goals. The recommendations provided can inform and complement governmental action plans, promoting a circular economy for LIB and PV, and can be extended to similar products as the world strives to minimize the material impacts of technological transitions.

To combat climate change and transition to a fossil fuel-free economy, there is a global consensus on the urgent and substantial reduction of greenhouse gas (GHG) emissions. Human activities have been responsible for approximately 50 billion tons of GHGs annually at the global level since the mid-2010s, with the electricity and heat sector being the primary contributor, followed by transport, manufacturing, and agriculture (Hiroshi Kawamura et al., 2021).

Lithium-ion batteries (Li-ion) present a promising clean technology to replace traditional fossil fuel-powered devices, playing a crucial role in the two major GHG-emitting sectors: electricity generation and transport. In the electricity sector, affordable Li-ion batteries facilitate the integration of more renewable energy capacity from solar and wind sources into grids. These batteries address the variability of solar and wind power by storing excess energy during surplus generation and distributing it during deficits, reducing the need for maximum power plant capacity and associated construction costs (Hiroshi Kawamura et al., 2021).

In the transport sector, Li-ion batteries drive the revolution in electric vehicles (EVs), with global manufacturers planning to introduce nearly 100 purely electric vehicle models by the end of 2024. The demand for Li-ion batteries has experienced significant growth, from 19 gigawatt hours (GWh) in 2010 to 285 GWh in 2019, with projections reaching 2,000 GWh in 2030, constituting about 8% of the world's energy supply (Hiroshi Kawamura et al., 2021). The decreasing costs and improved portability of Li-ion batteries have contributed to their widespread adoption.

As the energy density of batteries has increased and prices have dropped, Li-ion batteries have become integral to electrifying various types of vehicles, from bicycles to buses and trucks. The electrification of transportation has seen a surge in sales of new passenger electric vehicles, reaching over 3 million in 2020. The adoption of electric buses (E-buses) and the potential electrification of larger trucks indicate a growing trend in sustainable transport (Hiroshi Kawamura et al., 2021).

While electric vehicles can reduce CO_2 emissions in the transportation sector, the overall impact is influenced by factors such as the carbon intensity of power generation and the manufacturing processes of vehicles and batteries. The electrification of transport is expected to have a substantial positive effect on emissions reduction, with estimates suggesting up to 50 billion tons of CO_2 prevented from being emitted into the atmosphere between now and 2050 in an optimistic scenario (Hiroshi Kawamura et al., 2021).

In the energy sector, Li-ion batteries play a vital role in energy storage, particularly in regions with low access to modern energy, like sub-Saharan Africa. Utility-scale stationary energy storage is increasingly in demand due to the growth of renewable energy sources, and the declining costs of Li-ion batteries make them a preferred choice for such projects. The flexibility and scalability of Li-ion batteries contribute to their dominance in new utility-scale battery storage installations (Hiroshi Kawamura et al., 2021).

Battery storage enables renewable generators to integrate seamlessly with existing grids, storing excess generation and ensuring a stable energy distribution. As the costs of renewable energy generation and battery storage become competitive with conventional sources, the transition to renewable electricity sources can accelerate, reducing carbon emissions by 50–80% (Hiroshi Kawamura et al., 2021). Investment in energy storage and transmission infrastructure is crucial for grid stability and economic feasibility, making renewable energy investments more profitable and displacing traditional coal and gas generators.

Li-ion batteries are pivotal in addressing climate change and facilitating the transition to a sustainable, fossil fuel-free future. Their applications span electricity generation, transportation, and energy storage, contributing significantly to the global efforts aimed at mitigating GHG emissions and achieving a cleaner, more sustainable energy landscape.

The transition to a low-carbon future hinges on a crucial yet challenging technology — lithium-ion rechargeable batteries. These batteries, already ubiquitous in laptops and smartphones, are poised to become central to electric vehicles, renewable energy grids, and more. Despite their pivotal role, the lithium-ion technology comes with downsides for both people and the planet, particularly in the extraction of raw materials like lithium and cobalt, which demand significant energy and water resources and often involve hazardous working conditions, including child labour (Nature, 2021).

While countries recognize the imperative for responsible mining, some policies, especially in battery recycling, risk adverse environmental impacts. The European Union, for instance, mandates the collection and recycling of batteries, with targets set to increase (Nature, 2021). However, these requirements may lead to premature disposal of batteries that still have usable life, and the inclusion of recycled material in batteries could exacerbate material shortages and result in increased imports, potentially from China and South Korea, with significant carbon footprints.

Battery reuse emerges as a viable solution, yet it lacks prominence in current proposals. Many batteries are retired not due to complete depletion but inefficiency for certain uses. Without incentivizing battery reuse and repurposing, economically preferable options such as incineration or overseas recycling persist. A shift in mindset is essential, urging scientists to design materials with considerations for recycling, reuse, and repurposing (Nature, 2021).

On one hand, the growth of the battery industry contributes to economic expansion, job creation, and industrial development, fostering innovation in technology and energy storage solutions (Majeau-Bettez et al., 2011). However, there are also challenges and costs associated with lithium-ion batteries. These include resource dependence, environmental and social costs, supply chain risks, investments in research and development (R&D) and infrastructure, and long-term sustainability considerations (Notter et al., 2010).

The expansion of the battery industry, largely fuelled by the widespread adoption of lithium-ion batteries, has yielded several positive impacts on the economy. However, this growth is accompanied by challenges and costs that warrant careful consideration.

The growth in the battery industry has been shown to have a significant impact on overall economic expansion, creating new markets and opportunities (Hanson et al., 2017). The heightened demand for lithium-ion batteries not only stimulates production but also results in the establishment of manufacturing facilities, contributing substantially to economic growth (Hanson et al., 2017). This growth in the battery industry has also been linked to the associated job creation, spanning from skilled positions in research and development to manufacturing jobs within battery production plants (Hanson et al., 2017). Lessons from the pharmaceutical industry's commercialization successes can be identified and applied to the battery industry to potentially improve its startup success rates, which in turn can contribute to job creation and economic growth (Hanson et al., 2017).

The relationship between firm size and employment growth has been found to be sensitive to various issues, indicating that the size of firms can have an impact on job creation and economic expansion (Haltiwanger et al., 2013). Additionally, the gross creation and destruction of jobs and the rate at which jobs are reallocated across plants have been shown to have implications for job creation and economic growth (Davis & Haltiwanger, 1992). These findings suggest that the dynamics of job creation and destruction within firms can significantly influence overall employment growth and, by extension, economic expansion.

Furthermore, the agglomeration of industries has been found to play a role in promoting economic growth (Li, 2020). Industrial agglomeration and industrial upgrading have been shown to better promote economic growth, indicating that the spatial concentration of industries can contribute to overall economic expansion (Li, 2020). Additionally, the development of the insurance industry has been linked to regional economic growth, suggesting that the growth of specific sectors, such as the insurance industry, can have implications for economic expansion (Xie, 2022).

The pivotal role of the battery industry in industrial development is evident in countries prioritizing technological advancements and sustainable energy solutions. The growth of this industry fosters a more diversified and technologically advanced economy, with sectors linked to battery manufacturing experiencing significant expansion. The continual drive for innovation in battery technology, propelled by the demand for improved energy storage solutions, positively influences various industries beyond the battery sector (I. C. Chen et al., 2019; Kittner et al., 2017; Y. Lei et al., 2022; Liu & Guan-jun, 2018; Mallard et al., 2020; Novas et al., 2021; Piippo & Pongrácz, 2020; Sharp & Miller, 2016).

Addressing the challenges posed by resource dependence, particularly in the context of lithium imports for lithium-ion batteries, is crucial for countries heavily reliant on this finite resource. Fluctuations in lithium prices can have profound effects on economies, especially those with burgeoning battery industries, leading to cascading impacts on production costs, employment, investments, and trade balances (Bode & Wagner, 2015). Disruptions in the global lithium supply chain due to geopolitical tensions or unforeseen events in major lithium-producing regions can create vulnerabilities in resource-dependent countries, affecting various sectors and the interconnected global economy (Bode & Wagner, 2015). This underscores the need for countries to diversify their sources of lithium to reduce the risk of over-reliance on a single supplier or region (Bode & Wagner, 2015).

Efforts to diversify lithium sources and invest in alternative battery technologies represent strategic responses aimed at mitigating these risks, enhancing resource security, and promoting long-term economic sustainability (M. Li et al., 2018). Countries can explore new lithium reserves domestically and internationally, thereby reducing their vulnerability to disruptions in a single region (M. Li et al., 2018). Additionally, investments in research and development (R&D) aimed at exploring alternative battery technologies that rely on more abundant or geographically diverse materials can further contribute to reducing resource dependence (M. Li et al., 2018).

Investing in R&D to develop alternative battery technologies not only offers a potential solution to the challenges associated with lithium dependence but also fosters innovation within the battery industry (M. Li et al., 2018). Technologies that use alternative materials, such as sodium-ion or solid-state batteries, have garnered attention as potential successors to traditional lithium-ion batteries, offering the potential to diversify resource bases and insulate economies from uncertainties linked to lithium availability (M. Li et al., 2018).

Efforts to address resource dependence in the context of lithium-ion batteries involve recognizing the economic risks associated with relying on a finite resource and proactively mitigating these risks through diversification of lithium sources and investment in alternative battery technologies (M. Li et al., 2018). These strategic responses are essential for enhancing resource security and promoting long-term economic sustainability, especially for countries heavily reliant on lithium imports for their battery industries (M. Li et al., 2018).

The environmental and social costs associated with the extraction and processing of raw materials for lithium-ion batteries are multifaceted and extend beyond immediate economic considerations. These costs are intertwined with the entire lifecycle of battery production, impacting ecosystems and communities in various ways (Notter et al., 2010). The extraction of raw materials, including lithium, cobalt, and nickel, involves significant environmental consequences, such as habitat disruption, deforestation, and water usage (Zackrisson et al., 2010). Mining activities can lead to the destruction of natural habitats, disrupting ecosystems and biodiversity, and transforming landscapes with long-term consequences for local flora and fauna (Zackrisson et al., 2010). Water usage in mining operations, particularly in regions with existing water scarcity, exacerbates challenges and can lead to competition for water resources between mining operations and local communities, impacting agriculture, drinking water supplies, and overall ecosystem health (Zackrisson et al., 2010).

Furthermore, alterations to landscapes, such as open-pit mining, can leave scars on the land, affecting the aesthetic value of surrounding areas and potentially impacting tourism and local economies (Zackrisson et al., 2010). The social costs associated with lithium-ion battery production are equally impactful, including community displacement, conflicts over resource access, and ethical concerns tied to labour practices in mining regions (Zackrisson et al., 2010). Mining activities, if not managed responsibly, can lead to community displacement, disrupting established communities and leading to social tensions and challenges in resettlement (Zackrisson et al., 2010). Additionally, conflicts over resource access intensify as demand for raw materials increases, leading to conflicts between different stakeholders, including local communities, mining companies, and governmental bodies (Zackrisson et al., 2010). Ethical concerns tied to labour practices, particularly the use of child labour in mining, raise ethical concerns and contribute to the perpetuation of poverty and human rights abuses in affected communities (Zackrisson et al., 2010).

Addressing these costs requires industry stakeholders, policymakers, and local communities to collaborate on developing responsible and ethical practices that minimize environmental impact, ensure the well-being of affected communities, and promote the long-term sustainability of battery technologies (Notter et al., 2010). It is essential to consider the entire supply chain and lifecycle of battery production to mitigate these environmental and social costs effectively (Notter et al., 2010)

Geopolitical factors play a crucial role in the supply chain risks associated with the lithium-ion battery industry, primarily due to the concentration of lithium production in specific regions such as the Lithium Triangle covering parts of Argentina, Bolivia, and Chile (Calisaya-Azpilcueta et al., 2020). Political instability, regulatory changes, trade disputes, or other geopolitical events in these lithium-producing regions can disrupt the extraction, processing, and transportation of lithium, thereby affecting the entire supply chain (Calisaya-Azpilcueta et al., 2020). Changes in government policies in lithium-producing countries can result in shifts in export regulations, export quotas, or taxes on lithium exports, impacting the cost of lithium and creating uncertainties in the lithium market (Um & Han, 2020). Additionally, trade tensions or conflicts between nations can pose risks to the supply chain, leading to disruptions in the flow of lithium and related materials, ultimately affecting the production and availability of lithium-ion batteries globally (Um & Han, 2020).

To mitigate these supply chain risks, various strategies are adopted by countries and industry stakeholders. These strategies include diversification of supply sources, global collaboration and strategic alliances, investment in domestic lithium production, strategic planning and risk management, exploration of alternative materials, and storage and stockpiling (Calisaya-Azpilcueta et al., 2020). Diversifying sources of lithium, establishing strategic alliances, investing in domestic production, and developing contingency plans are crucial measures to enhance the resilience of the lithium-ion battery supply chains (Calisaya-Azpilcueta et al., 2020).

Efforts to expand lithium mining have been observed to be less successful in certain regions, such as Chile, the United States, and Europe, compared to Australia (Graham et al., 2021). This highlights the importance of diversifying supply sources and investing in domestic production capabilities to ensure a more reliable and resilient supply chain, reducing dependence on external sources (Graham et al., 2021). Furthermore, research and development efforts focus on exploring alternative materials or battery technologies that may be less dependent on lithium, providing flexibility in the face of geopolitical uncertainties (Xu et al., 2015). This diversification of materials can contribute to mitigating the risks associated with geopolitical events in lithium-producing regions.

Investments in Research and Development (R&D) and infrastructure are crucial for the growth, innovation, and sustainability of the lithium-ion battery industry, especially in its intersection with the electric vehicle (EV) and renewable energy sectors. R&D investments are essential for technological advancements, safety improvements, and environmental sustainability of lithium-ion batteries (Dunn et al., 2011). These investments aim to enhance energy density, improve safety, and explore greener materials for battery components. Furthermore, infrastructure development, including charging stations, grid upgrades, and energy storage facilities, is pivotal for supporting the widespread adoption of electric vehicles and renewable energy solutions (Dunn et al., 2011). The synergy between R&D and infrastructure investments not only drives technological advancements but also creates an enabling environment for market competitiveness and regulatory support (Dunn et al., 2011).

R&D investments play a crucial role in advancing battery technology, particularly in enhancing energy density, which is essential for improving the performance and range of electric vehicles and optimizing the efficiency of renewable energy storage systems (Dunn et al., 2011). Safety improvements through R&D efforts focus on minimizing the risk of overheating, fires, and other safety concerns associated with battery use, while also addressing environmental sustainability by exploring greener and more sustainable materials for battery components (Dunn et al., 2011). These efforts are integral to ensuring the continued growth, innovation, and sustainability of the lithium-ion battery industry, especially in its intersection with the electric vehicle and renewable energy sectors.

Infrastructure development, including the creation of a widespread network of charging stations, grid upgrades, and energy storage facilities, is essential for alleviating concerns about range anxiety and promoting the adoption of electric vehicles (Dunn et al., 2011). Moreover, modernizing grids enables the seamless integration of renewable energy sources and facilitates the efficient distribution of power, contributing to a more sustainable and energy-efficient future (Dunn et al., 2011). Large-scale energy storage solutions, such as battery energy storage systems, contribute to grid stability and support the integration of intermittent renewable energy sources like solar and wind (Dunn et al., 2011). These infrastructure investments are fundamental for ensuring a reliable and resilient energy grid capable of balancing supply and demand fluctuations.

Furthermore, R&D investments lead to the development of new and improved battery technologies, expanding the range of applications beyond electric vehicles and renewable energy storage (Dunn et al., 2011). Continuous innovation not only improves the performance of lithium-ion batteries but also fosters economic growth by attracting investments, creating high-tech jobs, and driving industrial development (Dunn et al.,

2011). Additionally, regulatory support and incentives, including financial incentives and subsidies provided by governments, play a crucial role in encouraging R&D activities in the battery sector (Dunn et al., 2011). Establishing a supportive regulatory framework is also essential for fostering R&D initiatives, providing a conducive environment for companies to invest in the development of advanced battery technologies (Dunn et al., 2011).

Critical to minimizing environmental impact and fostering a circular economy, proper disposal and recycling of lithium-ion batteries are essential. Sustainable practices throughout the entire lifecycle of lithium-ion batteries, from raw material extraction to end-of-life management, are vital for long-term environmental and economic sustainability.

Countries heavily reliant on lithium imports may face challenges related to resource dependence, as fluctuations in lithium prices or disruptions in the global lithium supply chain can impact the economic stability of importing nations (Zackrisson et al., 2010). Additionally, the extraction of lithium and other components of lithium-ion batteries can have environmental and social consequences, leading to habitat disruption, water usage concerns, and negative impacts on local communities, potentially resulting in social conflicts (Filomeno & Feraco, 2020).

To understand the economic vulnerability of importing nations due to fluctuations in the global lithium market, it is essential to consider the impact of geopolitical events, demand, and technological advancements on lithium prices. The concentration of lithium suppliers and regions intensifies the risk in the supply chain, leading to cascading effects on the importing nation's economy during disruptions caused by geopolitical tensions, trade policies, or natural disasters in lithium-producing countries. The global lithium market is subject to price fluctuations influenced by demand, geopolitical events, and technological advancements (J. Gao et al., 2022). These fluctuations have repercussions on industries heavily reliant on lithium-ion batteries, leading to economic instability in importing nations. The concentration of lithium suppliers or regions intensifies the risk in the supply chain, exacerbating the impact of disruptions caused by geopolitical tensions, trade policies, or natural disasters in lithium-producing countries (Montané et al., 2017).

Geopolitical risks and events have a significant impact on the global economy and financial markets (Demiralay & Kılınçarslan, 2019). The study by highlights the development of a global geopolitical risk index, emphasizing the influence of geopolitical tensions and events on various sectors, including tourism and politics. This underscores the relevance of geopolitical events in shaping economic vulnerabilities.

Furthermore, the impact of oil price fluctuations on different sectors of the economy is well-documented (Nandha & Brooks, 2009). found a positive association between energy stock returns and oil and gas prices, indicating the far-reaching effects of energy price fluctuations on economic sectors. Additionally, the study by Hsiao et al. (2019) emphasizes the significant impact of international oil prices on stock price fluctuations, further underlining the interconnectedness of energy markets and the broader economy.

The vulnerability of economies to external shocks, including those related to commodities such as lithium, is a subject of extensive research (Kuek et al., 2020). discuss the construction of a financial vulnerability indicator in China, highlighting the adverse impact of financial vulnerability shocks on economic growth (Kuek et al., 2020). This underscores the potential repercussions of economic vulnerabilities on overall economic performance.

Moreover, there are supply chain risks, including geopolitical risks and market price volatility, which can affect the supply and pricing of lithium-ion batteries, impacting industries and consumers (Shi et al., 2021). Investments in R&D and infrastructure are essential for enhancing battery performance, safety, and sustainability, but they can strain national budgets, particularly in countries aiming to transition to a low-carbon economy (X. Li et al., 2020). Furthermore, the proper disposal and recycling of lithium-ion batteries pose challenges, incurring costs but essential for mitigating environmental impact and promoting a circular economy (Meng, 2021). It is crucial for countries to consider a balanced and sustainable approach, taking into account social, economic, and environmental factors to maximize the benefits of lithium-ion battery technology (Kader et al., 2021).

Quantifiable Aspects

To provide economical statistics related to lithium-ion batteries, it is essential to consider their dominance in the market due to their higher energy density compared to other rechargeable battery systems, enabled by the design and development of high-energy density electrode materials (Manthiram, 2017). The recycling of waste lithium-ion batteries has become the best choice, providing significant social and economic benefits and global sustainability (Fei et al., 2021). The demand for lithium-ion batteries is expected to further increase, with batteries projected to account for 66 percent of global lithium production (Tadaros et al., 2020). Additionally, the production of lithium-ion batteries has significantly increased since their introduction to the market in 1991 due to their good performance, associated with high specific energy, energy density, specific power, efficiency, and relatively long life.

The shift toward electric vehicles (EVs) across light-, medium-, and heavy-duty sectors is a pivotal element in the United States' endeavours to diminish greenhouse gas emissions (Freeman et al., 2023). Establishing a resilient battery supply chain stands as a crucial factor in fostering the expansion of the domestic EV market. A report by Freeman et al. (2023) details the economic advantages that will arise from a robust domestic EV battery supply chain, presenting a comprehensive evaluation of the jobs, labour compensation, and gross domestic product (GDP) generated from mining battery materials through battery recycling. The assessment spans a 20-year period, from 2021 to 2040, presuming the continuity of existing federal policies, specifically the Inflation Reduction Act, for the next decade until 2032.

Figure 29: Electric Bus. Photo by Mario Sessions on Unsplash.

The analysis projects the economic contribution of the U.S. EV battery industry under two alternative market growth scenarios. In the Moderate EV adoption scenario, it is assumed that annual new sales of light-duty vehicles (LDVs) and medium- and heavy-duty vehicles (M/HDVs) will transition to 75 and 45 percent electric, respectively, by 2040 (Freeman et al., 2023). The High EV adoption scenario is more ambitious, reaching 100 percent for LDVs in 2038 and 74 percent for M/HDVs in 2040.

Utilizing REMI PI+ economic forecasting software, the analysis concentrates on five components of a cradle-to-grave assessment of the domestic EV battery life cycle: EV battery metal and mineral production, EV battery manufacturing, EV manufacturing, internal combustion engine vehicle manufacturing, and EV battery material recycling. By 2040, contingent on the EV adoption scenario (Freeman et al., 2023):

- Annual U.S. battery capacity produced is projected to surge from approximately 44 gigawatt-hours (GWh) per year in 2021 to 1,500 GWh/year (Moderate scenario) or 2,100 GWh/year (High scenario), signifying a more than a hundred-fold increase on the high end.

- Jobs supported are anticipated to grow by 570,000 to 740,000 jobs, constituting a 0.2 to 0.3 percent increase over the expected employment level in 2040 without the modelled transition to EVs.

- Labor compensation is expected to rise by $40 to $50 billion (2020$), representing a 0.2 to 0.3 percent increase.

- GDP is forecasted to increase by $110 billion to $150 billion (2020$), reflecting a 0.4 to 0.5 percent increase.

The widespread adoption of electric vehicles (EVs) necessitates significant quantities of metals and minerals, some of which are presently facing shortages. The key materials essential for battery manufacturing encompass lithium, cobalt, nickel, copper, graphite, iron, manganese, phosphorus, and aluminium. Although evolving battery technologies, cost considerations, and evolving customer preferences may influence the specific battery chemistry and materials utilized for EVs, several raw materials are deemed critical and anticipated to be indispensable (Freeman et al., 2023).

Currently, the United States relies on external sources for many of these materials, but the surge in demand, coupled with provisions in the Inflation Reduction Act (IRA), is making domestic mining and production more economically viable, leading to an expected growth in domestic supply (Freeman et al., 2023). The IRA also influences battery chemistries, encouraging the sourcing of metals and minerals from domestic or fair-trade agreement sources. This shift in the market dynamics is pushing towards chemistries that are more accessible and less costly to source from the United States or IRA-allowed countries. However, the extent of U.S. production growth is contingent on the available reserves of the various materials.

Despite minimal lithium production in the United States in 2021, numerous mining facilities are anticipated to become operational in the near future. Based on projections from Fitch Solutions Mining Forecast, U.S. lithium production is estimated to reach 150,000 metric tonnes (MT) per year in 2030, and by 2040, it is expected to reach 300,000 MT/year.

The sole U.S. cobalt mine commenced operations in October 2022, projecting an annual cobalt production of nearly 2,000 MT (Freeman et al., 2023). Given the limited cobalt reserves in the United States, the analysis assumes that U.S. cobalt production will grow to 2,000 MT/year in the next few years and remain constant until 2040.

In 2021, U.S. nickel production totalled 18,000 MT, with initiatives like Talon Metals Tamarack aiming to expand the domestic nickel supply. However, U.S. nickel reserves are currently meagre, amounting to less than 1 percent of the world's supply. Anticipating a limited increase in U.S. nickel mining, the analysis assumes production will rise to 25,000 MT/year in the next five years and remain unchanged until 2040 (Freeman et al., 2023).

While the United States presently does not mine natural graphite, planned natural graphite mines in Alabama and Alaska, along with synthetic graphite production, are expected to contribute to a total U.S. production of 600,000 MT/year by 2040.

Considering the assessments of potential U.S. metal and mineral production, the analysis assumes that current extraction technology remains relatively unchanged, and existing reserve levels are accurate. While advancements or discoveries may alter the mining potential of these materials, predicting such events is inherently uncertain (Freeman et al., 2023).

For more abundant materials like copper, iron, aluminium, and phosphorus, which are extensively used across various industries, the demand for EV batteries is not expected to significantly impact U.S. production. However, IRA provisions may redirect domestically produced materials from their current applications to EV battery use, satisfying the IRA's domestic content requirements. Consequently, these materials are not explicitly modelled as a separate industry but are integrated into the overall EV battery supply chain (Freeman et al., 2023).

Utilizing a modelling tool, researchers compared lithium requirements for achieving zero transport emissions in personal vehicles (cars, trucks, and SUVs) across various scenarios (Lakhani, 2023). The study forecasts future lithium demand by considering factors such as car ownership, battery size, city density, public transit, and battery recycling, correlating these with potential harms.

In all scenarios, the United States attains zero-emission transportation by 2050, necessitating some additional lithium mining. The report emphasizes that the current policy decisions will significantly impact economic prosperity, public health, environmental justice, ecosystems, and communities across the entire supply chain for decades to come (Lakhani, 2023).

In the most optimistic scenario, where EV battery size increases and U.S. car dependency remains stable, ambitious public transit, city density, and recycling policies could reduce lithium demand by 92% (Lakhani, 2023). The size of EV batteries, akin to a fuel tank's size, influences range—the distance a vehicle can travel before recharging.

Even if Americans continue their reliance on large lithium batteries, enhancing metropolitan density and investing in mass transit could reduce cumulative lithium demand by 18% to 66% (Lakhani, 2023). Additionally, restricting the size of EV batteries alone has the potential to decrease lithium demand by up to 42% by 2050 (Lakhani, 2023).

The most significant reduction is expected by transforming urban transportation—reducing the number of cars and promoting more walking, cycling, and public transit, facilitated by denser city planning. Subsequently, downsizing vehicles and implementing battery recycling contribute to further reductions.

Global cities have already initiated efforts to curtail car use, aiming to enhance air quality, road safety, and overall quality of life. Paris witnessed a nearly 30% decline in car use from 2001 to 2015, while London experienced a nearly 40% reduction (Lakhani, 2023).

Apart from the financial challenges confronting Original Equipment Manufacturers (OEMs) and their belated realization of raw material supply issues, governments are awakening to the fact that geopolitical factors have deeply permeated the battery industry. This shift is unsurprising, given China's substantial role throughout the supply chain, coupled with the deterioration of U.S.-China relations (Mehdi & Moerenhout, 2023). While China's involvement in upstream mining is relatively small but noteworthy, it dominates the midstream and downstream segments of the battery supply chain. Currently, Chinese companies refine approximately 60 percent of the world's lithium, 69 percent of nickel, 75 percent of cobalt, and the entirety of the world's natural graphite (Mehdi & Moerenhout, 2023).

China's stronghold in battery cell production is bolstered by its control over cathode and anode production, coupled with the economies of scale achieved by major players like CATL and BYD (Mehdi & Moerenhout, 2023). China contributes to over 75 percent of the global supply of cathodes and anodes, along with about 78 percent of the global battery cell supply. Notably, the most significant dependency lies in LFP cells, where China dominates around 99 percent of LFP cathode output (Mehdi & Moerenhout, 2023).

Lithium emerges as arguably the most geopolitically sensitive material in the battery supply chain. The complexities of introducing new lithium supply, with average project delays of 2–3 years, and limited substitution risk from alternative technologies underscore its critical role. In essence, lithium serves as a key element in realizing net-zero targets (Mehdi & Moerenhout, 2023).

While China holds a considerable role in lithium refining capacity, its domestically mined lithium output, mainly comprising low-grade micas and lepidolite grades, is high-cost. China has heavily relied on raw material supply from key countries over the past decade, notably Australia (a major provider of hard-rock spodumene) and Chile (known for using brine to produce lithium carbonate, convertible into hydroxide at a cost) (Mehdi & Moerenhout, 2023). Over the years, China has secured offtakes or equity stakes in high-quality deposits globally.

Despite the presence of some U.S. companies (e.g., Albemarle) with robust refining positions in China and Australia's dominance in hard rock production acting as a geopolitical counterbalance to China's midstream dominance, the reality remains that China has enjoyed a decades-long lead over the West (Mehdi & Moerenhout, 2023). Amid concerns about the Ukraine conflict and the security of supply, scenarios involving potential disruptions in cell supply, imposition of tariffs, or curtailment of chemical supply have gained attention. However, US policies such as the Inflation Reduction Act (IRA) are not only responses to potential scenarios but also signify government support to scale raw material supply. In an energy landscape where supply chains are no longer regime-agnostic, striking a balance between addressing issues such as cost, emission intensity of transport and production, and geopolitics becomes increasingly challenging—a fact already evident in the discussions surrounding the IRA in Washington (Mehdi & Moerenhout, 2023).

China's investment spree in the lithium battery industry is now causing significant challenges for its smaller battery manufacturers, who had poured billions into capitalizing on the increasing demand and government subsidies. By 2025, these manufacturers are anticipated to produce a staggering 4,800 gigawatt-hours (GWh) of batteries, a quadruple of the demand projected by the country's electric vehicle (EV) makers (Chaudhury, 2023). Unfortunately, this oversupply is poised to lead to the closure of smaller-sized battery manufacturers.

China dominates the processing of essential materials for battery production, accounting for 65% of global lithium, 74% of cobalt, 100% of graphite, and 42% of copper processing units (Chaudhury, 2023). Despite its leadership role in the global battery industry, the production capacity at China's battery factories is estimated to surpass demand levels by more than twice, intended for approximately 22 million EVs (Chaudhury, 2023).

The oversupply issue has been exacerbated by a series of expansion projects announced, totalling around USD 20 billion by 2020 and a subsequent tripling of investment amounts in the following two years (Chaudhury, 2023). By 2022, Beijing's businesses captured about 57% of the global demand for lithium-ion batteries (Chaudhury, 2023).

While China is the largest EV market globally, the oversupply of lithium products is anticipated to pose a significant economic burden. Analysts predict that numerous smaller companies will shut down, while larger ones will continue to grow (Chaudhury, 2023). However, Chaudhury (2023) indicates that the Chinese government has yet to take action against the over-expansion, providing massive subsidies to allay concerns of mass business closures. This has led to frequent announcements of battery plant constructions, contributing to overcapacity, expected to be four times what the country needs by 2027.

The excess supply is anticipated to exert downward pressure on battery prices, impacting the profitability of manufacturers, especially smaller firms (Chaudhury, 2023). These smaller companies are likely to accumulate unsold inventory, tying up capital and resources, and leading to significant environmental costs associated with resource-intensive lithium and battery production processes (Chaudhury, 2023).

Large battery manufacturers have already begun offering discounted prices, intensifying pressure on smaller manufacturers lacking the capital to compete. Industry experts predict that, in the next three years, top battery producers' capacity utilization will remain around 50%, while smaller producers will fall below 20% (Chaudhury, 2023).

The increase in demand in recent years led to a fourfold increase in the price of lithium carbonate, a crucial raw material. As a result, major EV battery manufacturers like CATL, BYD, and CALB have started offering discounted products, intensifying competition for smaller, second-tier companies and jeopardizing the future of many in the industry.

The oversupply issue is further exacerbated by the significant imbalance between the 545 gigawatt-hours of batteries produced by China and the 294 GWh installed by EVs for commercial purposes (Chaudhury, 2023). The decrease in EV sales and overinvestment in production facilities contribute to this growing disparity.

Chaudhury (2023) posits that China's flawed economic policies are expected to lead to significant revenue and resource losses, particularly affecting its small and medium-sized industries. The repercussions of these policies will not only impact the domestic industry but are likely to influence global demand and supply dynamics in the EV markets. With an oversupply of lithium batteries, Chinese firms may resort to dumping their products in countries with moderate demand, further complicating the global EV landscape (Chaudhury, 2023).

Economic Impacts of Shifting to Battery Storage Renewable Energy

The shift to battery storage for renewable energy significantly influences various sectors of the economy. One key consideration is the enhancement of grid stability and reliability. Battery storage positively impacts the electrical grid by storing excess energy during high renewable generation periods and releasing it during peak demand or low generation, contributing to a more reliable and resilient grid. This, in turn, can reduce the economic costs associated with power outages and grid instability.

Shifting to battery storage for renewable energy has significant economical impacts. The incorporation of demand-side management (DSM) and battery storage systems can reduce peak power demand, increase renewable self-consumption, enhance grid stability, and decrease distribution system losses (Pholboon et al., 2016). Additionally, reducing embodied energy costs, increasing efficiency, and improving the energetic performance of batteries can further enhance the economical impacts of battery storage for renewable energy (Barnhart et al., 2013). Battery storage systems increase the flexibility of power systems and enable the integration of renewable energy sources, contributing to their economic feasibility (Ramelan et al., 2021).

Figure 30: Renewable Energies: Biogas (fermenter), wind power and photovoltaics on a farm in Horstedt (Schleswig-Holstein/ Germany). Florian Gerlach (Nawaro), CC BY-SA 3.0, via Wikimedia Commons.

Another crucial aspect is the integration of renewable energy sources. Battery storage facilitates the smooth integration of intermittent sources like solar and wind into the grid, enhancing overall energy system efficiency. By smoothing out fluctuations in power generation, it reduces the need for backup power plants, optimizing the use of renewable resources.

The reduced need for peak power plants is an additional positive impact. Battery storage mitigates the requirement for expensive peaker plants designed for peak demand, leading to cost savings in infrastructure investments and operational expenses, ultimately impacting electricity prices positively. Battery storage has become fundamental in renewable energy systems, providing instantaneous response to changes in demand, clean electricity to customers, and integration of intermittent power sources, thus contributing to the economic viability of renewable energy (Quann & Bradley, 2017). Additionally, the economic feasibility of energy storage systems, particularly lithium-ion batteries, has been demonstrated in peak load management and addressing the supply instability of renewable energy (M. Kang et al., 2019). Microgrids equipped with energy storage, such as batteries, enable the utilization of local renewable energy resources, thereby contributing to the economic feasibility of renewable energy (Khemir et al., 2022).

Demand management and energy arbitrage also contribute positively. Battery storage enables storing energy during low-demand periods and discharging it during high-demand periods, resulting in cost savings for consumers and utilities by avoiding the use of more expensive power sources during peak times.

The role of energy storage in conjunction with renewable electricity generation is crucial for grid stability and the integration of intermittent power sources (Denholm et al., 2010). Economic optimization of component sizing for residential battery storage systems, particularly those coupled with rooftop-mounted photovoltaic generation, is gaining attention and market penetration, indicating their economic viability (Hesse, Martins, et al., 2017). Furthermore, a comparative analysis highlights the practical benefits of renewable energy generation and battery energy storage under different generation and storage capacities, emphasizing their economic significance (Karimi & Kwon, 2021).

The decreased reliance on fossil fuels is a significant benefit. Battery storage reduces dependence on fossil fuel-based power plants for meeting peak demand, contributing to a decrease in greenhouse gas emissions and external costs associated with air pollution and climate change. This positively impacts public health and environmental quality.

Moreover, battery storage systems contribute to energy independence and security by providing a decentralized power source. This reduces vulnerability to disruptions in centralized power generation, enhancing energy security for regions and nations.

The use of second-life batteries in combination with renewable energy sources may generate additional economic benefits in certain areas, further emphasizing the economic potential of battery storage in renewable energy systems (J. Li et al., 2022). Moreover, the installation of energy storage technologies, such as solar batteries, facilitates the management of electricity flow from renewable resources, contributing to their economic viability (Azahar et al., 2021). The intermittent nature of renewable energy systems necessitates the integration of energy storage, such as batteries, within microgrids, highlighting their economic significance (Jayawardana et al., 2019).

Job creation and industry growth are additional positive impacts. The growth of the battery storage industry creates job opportunities in manufacturing, installation, maintenance, and research and development, stimulating economic growth in these sectors.

Reduced transmission and distribution costs also play a positive role. Battery storage can reduce the need for extensive investments in transmission and distribution infrastructure, alleviating congestion on the grid and leading to potential cost savings in grid expansion and maintenance.

The increasing demand for battery storage technologies presents investment opportunities, attracting capital and fostering innovation in the energy storage sector. However, challenges related to initial costs, affordability, and supply chain considerations must be addressed to realize the full economic potential of energy storage technologies. The upfront costs of large-scale battery storage systems can be significant, posing affordability challenges for smaller utilities and regions. Government incentives and subsidies may be required to make these technologies more accessible. Additionally, ensuring a secure and sustainable supply chain for critical materials like lithium, cobalt, and nickel is essential to prevent potential bottlenecks and price fluctuations.

While the shift to battery storage for renewable energy offers positive economic impacts, overcoming challenges related to initial costs, affordability, and supply chain considerations is crucial for realizing its full economic potential. The transition to battery storage for renewable energy undoubtedly brings forth a multitude of positive economic impacts, revolutionizing the energy landscape and fostering sustainability. However, to fully unlock its economic potential, there are critical challenges that must be addressed.

The substantial initial costs associated with implementing large-scale battery storage systems pose a significant barrier to entry for smaller utilities and regions with limited financial resources. Overcoming this challenge necessitates strategic financial planning, innovative financing models, and, in many cases, government incentives or subsidies to make these technologies more financially accessible (Gissey et al., 2018). The lack of transparency and adequate data for risk assessment, high upfront costs, heterogeneity, complexity, and the presence of a large number of parties, and the lack of a clear benchmark for measuring investment performance were identified as the severest challenges in exploring innovative energy infrastructure financing (Kukah et al., 2021). Additionally, the most important regulatory barrier is the current classification of storage as a generation asset, despite it being unable to provide a positive net flow of electricity, which justifies double network usage charges (Gissey et al., 2018).

Innovative financing models and government incentives are crucial for addressing the high costs associated with the development, installation, and integration of advanced energy storage technologies (Kukah et al., 2021). Green finance has been shown to have a positive impact on the transformation of energy consumption structure, but its effectiveness may decline beyond a certain threshold (Gu et al., 2023). Furthermore, the increase of investment flows by private investors, institutional investors, or banks is hampered by various green investment barriers, leading to the so-called green finance gap, which describes the current lack of investments required for the realization of a green trajectory (Hafner et al., 2021).

Affordability is a key consideration tied to the initial costs. While the long-term benefits of battery storage, such as reduced transmission costs, grid stability, and minimized reliance on fossil fuels, contribute to economic advantages, the immediate financial burden can deter widespread adoption. Finding ways to make battery storage solutions economically viable and ensuring that the costs align with the long-term benefits is crucial for fostering a sustainable transition.

Another pivotal aspect is the intricate supply chain considerations associated with battery storage technologies. Critical materials like lithium, cobalt, and nickel play a central role in the production of batteries. Ensuring a secure, sustainable, and ethical supply chain for these materials is paramount. Challenges may arise from geopolitical factors, fluctuations in commodity prices, and ethical concerns related to mining practices. Addressing these issues requires a concerted effort from policymakers, industry stakeholders, and environmental advocates to establish transparent and responsible supply chains.

Furthermore, research and development efforts are essential to drive innovation and reduce the overall costs of battery storage technologies. Advancements in battery chemistry, manufacturing processes, and materials can lead to more cost-effective solutions, making them more accessible to a broader range of users. Collaboration between governments, research institutions, and private sector entities is crucial to accelerate technological advancements and streamline the production processes.

Public awareness and education campaigns can also play a significant role in overcoming barriers to adoption. By informing businesses, communities, and policymakers about the long-term economic benefits of battery storage, there is a greater likelihood of garnering support and commitment to overcome the initial challenges.

While the positive economic impacts of transitioning to battery storage for renewable energy are evident, the journey towards realizing its full potential requires a comprehensive approach to tackle challenges related to initial costs, affordability, and supply chain considerations. Through a combination of innovative financing, research and development initiatives, ethical supply chain practices, and increased awareness, societies can pave the way for a sustainable energy future that leverages the economic benefits of battery storage.

The feasibility of all countries shifting to Battery Storage Renewable Energy (BSRE) is a complex and nuanced issue that depends on various factors, including economic capacity, technological readiness, and geopolitical considerations. While the potential benefits of BSRE are substantial, not all countries may find it economically or logistically feasible to make a swift and complete transition.

Economic Capacity is a critical factor in this consideration. Some developed countries with robust economies may find it easier to invest in the infrastructure required for BSRE. On the other hand, developing nations may face challenges in mobilizing the necessary funds for large-scale deployment.

Technological Readiness varies among nations and plays a crucial role in the transition. Countries with advanced research and development capabilities and existing renewable energy infrastructure may find it easier to adopt BSRE. However, others may need to invest in building technological capabilities to support this transition effectively.

Resource Availability, particularly of key materials like lithium, cobalt, and nickel, is essential for the production of batteries for energy storage. Countries with access to these resources may have a strategic advantage, while others may face challenges in securing a stable supply.

Infrastructure Development is a significant consideration for shifting to BSRE. Countries with well-established grids and supportive infrastructure may find the transition smoother, while those lacking such infrastructure may face higher implementation costs.

Government Policies and Incentives play a vital role in supporting the transition to BSRE. The commitment of governments to policies, incentives, and regulatory frameworks that encourage investment in BSRE can significantly influence the pace of adoption.

Geopolitical Considerations, including international relations and dependencies, can impact a country's ability to transition. Global cooperation and geopolitical stability are essential for the smooth functioning of the global supply chain for key materials required for BSRE.

Public Awareness and Acceptance also play a role in the transition. Countries with populations supportive of renewable energy transitions may experience smoother adoption as public acceptance is crucial for the success of such initiatives.

Climate and Environmental Factors, such as the urgency of addressing climate change and environmental concerns, can motivate countries to invest in BSRE. Regions facing severe environmental challenges may prioritize renewable energy solutions for sustainability.

While many countries have the potential to shift to BSRE, the ability to afford and implement such a transition varies widely. Affordability, technological readiness, available resources, and supportive government policies all contribute to the feasibility of adopting BSRE. Global collaboration, financial assistance to developing nations, and advancements in technology can contribute to making BSRE more accessible on a global scale. However, each country's unique circumstances will shape its capacity to embrace and afford such a transition.

The division in renewable energy installation capacity between the Global South and the Global North underscores a significant disparity. In 2019, Africa contributed a mere 1.3% (586,434 MW) to the global installed solar capacity, and in Asia (excluding China, India, Japan, and South Korea), the figure stood at only 5.4% (330,786 MW) of the total installed solar capacity (Babayomi et al., 2022). The wind energy sector exhibits similar trends, with Africa possessing only 0.9% of the global installed wind capacity in 2019, and in Asia (excluding specific nations), the total installed wind capacity constituted 2.0% of the global sum (Babayomi et al., 2022). These figures highlight the sluggish pace of the energy transition in the Global South.

The gradual nature of the clean energy transition is profoundly influenced by the ease or difficulty of accessing finance and the associated costs. Most renewable energy patents are filed in China, the United States, the European Union, Japan, and Korea, and a similar pattern is observed in renewable energy manufacturing. For instance, China supplies 70.0% of the photovoltaic cell production market (Babayomi et al., 2022). Consequently, many developing countries become consumers of clean technology rather than contributors to innovation or manufacturing.

Furthermore, developing economies (excluding China and India) receive considerably fewer capital flows for clean energy infrastructure than their developed counterparts. Attracting investment is contingent on various factors, including local environmental conditions, macroeconomic strength, state capacity, economic and regulatory governance gaps, policy uncertainty, and regulated power tariffs. The increased financial burden associated with building clean infrastructure contributes to the slowing progress in lower-income economies (Babayomi et al., 2022).

Several peculiarities characterize developing countries that need consideration in formulating viable low-carbon energy transition pathways (Babayomi et al., 2022). These include highly inefficient grid infrastructure, inadequate grid capacities to supply total demanded baseload power, high costs of capital and country-risk factors, a high rural population with a low electrification rate, and a significant youth population facing challenges of migration and limited opportunities.

The clean energy transition in developing countries faces challenges due to inefficiencies in grid infrastructure, resulting in high transmission and distribution losses. Additionally, many developing nations lack sufficient generation capacity to meet baseload demand, leading to frequent power outages. The high cost of capital and various risk factors further hinder progress in these regions (Babayomi et al., 2022).

The substantial rural population in developing countries, particularly in Sub-Saharan Africa and South Asia, poses challenges in achieving universal energy access. Limited electrification in rural areas contributes to lower income levels and hampers economic growth (Babayomi et al., 2022). Furthermore, the high youth population in developing countries, coupled with inadequate education and skills, may lead to migration to urban areas, putting pressure on already strained electricity infrastructure.

In considering the clean energy transition, policymakers should navigate the delicate balance between achieving sustainable human development and economic growth. Human development and economic growth should be complementary inputs, fostering a self-sustaining cycle of growth (Babayomi et al., 2022). However, challenges such as the affordability of clean energy technologies and the need for strategic plans to integrate economic growth with rural human development should be addressed for a successful transition.

The head of the International Energy Agency, Fatih Birol, has asserted that Europe's rising energy prices are not linked to the continent's shift toward renewables (Toh, 2021). Despite this claim, recent findings from economists at the University of Chicago suggest that a higher penetration of renewables contributes to increased energy prices (Toh, 2021). The widespread belief that wind and solar energy lead to cheaper and cleaner power contradicts the current global energy crisis, occurring shortly after the addition of record amounts of new renewable capacity.

The reality is that the affordability of wind and solar energy is prominent in the early stages of transition, relying on a substantial base of fossil fuel generation (Toh, 2021). However, as renewable penetration grows, depending on fossil fuels to support intermittent renewables becomes more costly and risky. Europe's predicament, exacerbated by soaring natural gas prices, results from the unpredictable and variable nature of wind and solar energy, necessitating increased reliance on gas (Toh, 2021).

Transitioning away from fossil fuels requires effective storage solutions for excess electricity generated during peak renewable output. While pumped-storage hydro is a viable option, it is limited by specific geographical and water resource requirements. Battery storage, often promoted as a low-carbon solution, must witness a significant cost reduction, estimated at 90% by MIT researchers, to replace fossil fuels, a development unlikely within the next decade (Toh, 2021). Green hydrogen, another touted solution, lags even further behind in development.

Moreover, the best locations for wind and solar installations are quickly depleting, necessitating the expansion of transmission grids to reach remote areas with favourable conditions. Estimates indicate that the United States' transmission grid needs to expand by at least 50%, considering the increased demand for electricity by 2050 (Toh, 2021).

While upfront investment from large companies and financiers can cover a portion of the costs, public spending and higher energy bills are inevitable (Toh, 2021). It is crucial to prepare the public for rising prices and growing public debt to avoid unrest, as witnessed in events like the Yellow Vest protests in France. Transparency about the substantial investments required, estimated at $30.3 trillion by 2030, is essential to address the challenges of climate change effectively (Toh, 2021). Misleading narratives about an easy transition to a green future may offer short-term momentum but risk backlash as the world enters more complex phases of the transition. Leaders must communicate honestly about the challenges ahead to garner support and prevent resistance when time is even more critical.

Figure 31: Yellow vests. Taken at Place de la République, Paris, France. On the 3rd of August, 2019. Late afternoon. The 'Yellow Vests' protesters are hitting the streets of Paris for the 38th week in a row, to demonstrate against the French government's policies. Elekes Andor, CC BY-SA 4.0, via Wikimedia Commons.

Battery Energy Storage Systems

Battery energy storage systems operate by storing and releasing energy through electro-chemical processes. Key requirements for effective battery storage include high energy density, power capacity, long life in terms of charge-discharge cycles, efficient round-trip efficiency, safety, and cost competitiveness (World Nuclear Association, 2022). Factors such as discharge duration and charge rate also play a role, necessitating trade-offs among these criteria and highlighting the limitations of battery energy storage systems (BESS) compared to dispatchable generation sources. The issue of energy return on energy in-vested (EROI) becomes crucial, particularly concerning the duration a battery remains in service and how well its round-trip efficiency holds up over time.

Figure 32: The 230-megawatt Desert Sunlight Battery Energy Storage System is now fully operational. The project is on 94 acres of BLM-managed public lands near Desert Center in Riverside County. Bureau of Land Management California, Public domain, via Wikimedia Commons.

Figure 32: The 230-megawatt Desert Sunlight Battery Energy Storage System is now fully operational. The project is on 94 acres of BLM-managed public lands near Desert Center in Riverside County. Bureau of Land Management California, Public domain, via Wikimedia Commons.

To integrate batteries into an ordinary AC system, a power conversion system (PCS) with an inverter is essential, adding approximately 15% to the base battery cost (World Nuclear Association, 2022). Several large-scale projects have demonstrated that batteries are well-suited for mitigating the variability of power from wind and solar systems over short durations, such as minutes and hours. They exhibit faster and more precise responses compared to conventional resources like spinning reserves and peaking plants. As a result, large battery arrays are increasingly favoured as the preferred technology for stabilizing grids during the integration of short-duration renewables, primarily due to their power-related functions. The demand for this capability is considerably lower than that for energy storage, with the California ISO estimating its peak frequency regulation demand for 2022 at just over 2000 MW from all sources (World Nuclear Association, 2022).

Some battery installations act as virtual synchronous machines, replacing spinning reserve for short-duration backup and utilizing grid-forming inverters. Smart grids are often associated with discussions about battery storage, as they optimize power supply by leveraging information on both supply and demand through networked control functions (World Nuclear Association, 2022).

In 2015, lithium-ion batteries accounted for a significant portion of newly-announced energy storage system (ESS) capacity and deployed ESS power capacity (World Nuclear Association, 2022). With a 95% round trip direct current efficiency, lithium-ion batteries have become popular for distributed energy storage systems. According to the International Energy Agency (IEA), lithium-ion batteries are the fastest-growing storage technology globally, with projections indicating a substantial increase in battery storage capacity by 2030 and 2050.

At the household level, battery storage, especially lithium-ion batteries, is being promoted, particularly in conjunction with solar photovoltaic (PV) systems. In Germany, a notable percentage of new solar PV installations in recent years have been equipped with backup battery storage. Government support, such as low-interest loans and payback assistance, has contributed to the growth of residential storage in Germany.

Different types of batteries, including sodium-sulphur batteries and redox flow cell batteries, offer varying advantages and challenges. Sodium-sulphur batteries, while established and having a long service life, operate at high temperatures and are relatively expensive. Redox flow cell batteries, such as vanadium redox flow batteries (VRFB), are known for their long cycle life and low cost per kWh, making them suitable for large stationary applications.

Cost reductions have been observed in lithium-ion battery prices over the years, driven by advancements in technology and increased adoption in the automotive sector. However, power conversion system costs have not followed the same downward trend (World Nuclear Association, 2022). The ongoing evolution of battery technologies and their integration into energy systems will likely play a crucial role in the transition to more sustainable and resilient power grids.

Lithium-ion batteries can be categorized based on the cathode chemistry, leading to distinct characteristics for each battery type. Lithium nickel cobalt aluminium oxide (NCA) batteries showcase a specific energy range of 200-250 Wh/kg, high specific power, and a lifespan of 1000 to 1500 full cycles. This battery type is often favoured in high-end electric vehicles, such as those produced by Tesla, despite being relatively more expensive compared to other chemistries.

The Lithium nickel manganese cobalt oxide (NMC) battery offers a specific energy range of 140-200 Wh/kg and a lifespan of 1000-2000 full cycles. Widely employed in electric and plug-in hybrid electric vehicles, NMC batteries, although having lower energy density than NCA, make up for it with longer lifetimes.

The Lithium iron phosphate (LFP) battery features a specific energy range of 90-140 Wh/kg and a lifespan of 2000 full cycles. While its lower specific energy may limit its suitability for long-range electric vehicles, LFP batteries could find preference in stationary energy storage applications or vehicles where battery size and weight are less critical. Notably, they are recognized for being less prone to thermal runaway and fires (World Nuclear Association, 2022).

The Lithium manganese oxide (LMO) battery displays a specific energy range of 100-140 Wh/kg and a lifespan of 1000-1500 cycles. Its cobalt-free chemistry is considered advantageous, making it suitable for use in electric bikes and certain commercial vehicles (World Nuclear Association, 2022).

Battery systems worldwide are experiencing significant growth and diversification, with developments observed across various regions, with summaries below based on World Nuclear Association (2022) on data to 2022.

Europe: The total installed non-hydro storage capacity in Europe reached 2.7 GWh by the end of 2018 and is expected to be between 2.8 and 3.3 GWh in 2022. This encompasses household systems, constituting over one-third of the additions in 2019-20. EDF has ambitious plans to achieve 10 GW of battery storage across Europe by 2035. Notably, Total launched a substantial 25 MW/25 MWh lithium-ion battery project in Mardyck near Dunkirk, aiming to be the largest in France.

STEAG implemented the first of its six planned 15 MW lithium-ion units in a €100 million, 90 MW program in June 2016 at the Lünen coal-fired site in Germany. Rigorous criteria, including a rapid response to automated calls within 30 seconds and a minimum 30-minute feed-in capability, must be met for commercial operation.

In Germany, RWE invested €6 million in a 7.8 MW/7 MWh lithium-ion battery system at its Herdecke power station site near Dortmund, operational since 2018. Additionally, a 10 MW/10.8 MWh lithium-ion battery storage system was commissioned in 2015 at Feldheim, Brandenburg, supporting a local 72 MW wind farm and contributing to grid stabilization. Future plans in Germany include RWE's 45 MW lithium-ion battery at Lingen and a 72 MW unit at Werne Gerstein power plants by the end of 2022, primarily for Frequency Control Ancillary Services (FCAS). Siemens is also planning a substantial 200 MW/200 MWh battery at Wunsiedel in Bavaria for energy storage and peak management.

Eneco and Mitsubishi, operating as EnspireME, collaborated to install a 48 MW/50 MWh lithium-ion battery in Jardelund, northern Germany, supplying primary reserve to the grid and enhancing stability in a region with abundant wind turbines.

German operators participating in the primary control reserve market reported an average price of €17.8/MWh over an 18-month period to November 2016.

In Spain, Acciona commissioned a wind plant in May 2017, incorporating two Samsung lithium-ion battery systems, contributing to grid stability and frequency response.

Fortum in Finland contracted Saft to supply a €2 million megawatt-scale lithium-ion battery energy storage system for its Suomenoja power plant in May 2016. This project aims to offer frequency regulation and output smoothing to the Transmission System Operator (TSO).

The UK witnessed significant growth, with battery storage reaching 2.4 GW/2.6 GWh by the end of 2022, and an additional 20.2 GW approved for development. RES, a renewable energy company, provides 55 MW of dynamic frequency response from lithium-ion battery storage to the National Grid.

In March 2020, Finland's Wartsila secured a contract to supply two 50 MW lithium-ion batteries to EDF's Pivot Power for a 2 GW storage program supporting grid-scale batteries for ancillary grid services and electric vehicle charging.

AES completed a 10 MW/5 MWh energy storage array at its Kilroot power station in Carrickfergus, Northern Ireland. It is the largest advanced energy storage system in the UK and Ireland and operates at transmission scale.

The Orkney Islands house a 2 MW/500 kWh lithium-ion battery storage system using Mitsubishi batteries, storing power from wind turbines.

Cranborne Energy Storage in Somerset operates a 250 kW/500 kWh Tesla Powerpack lithium-ion storage system associated with a 500 kW solar PV setup.

Statoil commissioned the design of a 1 MWh lithium-ion battery system, Batwind, for onshore storage at the 30 MW offshore Hywind project in Peterhead, Scotland.

North America: In November 2016, Pacific Gas & Electricity Co (PG&E) reported on an 18-month technology demonstration project involving battery storage systems in California's electricity markets. Operational control proved complex, emphasizing the challenges faced even with a 20-year battery life assumption. The Yerba Buena battery was used for another technology demonstration in 2017, exploring coordination with third-party distributed energy resources using smart inverters and battery storage.

In August 2015, GE was contracted to build a 30 MW/20 MWh lithium-ion battery storage system for Coachella Energy Storage Partners in California, aiding grid flexibility and increasing reliability.

San Diego Gas & Electric operates a 30 MW/120 MWh lithium-ion BESS in Escondido, supplying evening peak demand and partly replacing the Aliso Canyon gas storage.

Southern California Edison is undertaking large battery installations, including a 100 MW/400 MWh battery system scheduled for commissioning in 2021.

The Tehachapi 8 MW/32 MWh lithium-ion battery storage project, coupled with a 4500 MWe wind farm, is a significant project by Southern California Edison.

A large-scale battery system with 256 Tesla 3 MWh Megapack units is approved for Vistra's gas-fired Moss Landing power plant in California.

In West Virginia, the 98 MW Laurel Mountain wind farm uses a 32 MW/8 MWh grid-connected BESS for frequency regulation, grid stability, and arbitrage.

In December 2015, EDF Renewable Energy commissioned a 40 MW flexible (20 MW nameplate) capacity BESS project on the PJM grid network in Illinois, contributing to regulation and capacity markets.

E.ON North America is installing two 9.9 MW short-duration lithium-ion battery systems for its Pyron and Inadale wind farms in West Texas.

SolarCity employs 272 Tesla Powerpacks for its 13 MW/52 MWh Kaua'i Island solar PV project in Hawaii.

In Ontario, Powin Energy and Hecate Energy are building two projects totaling 12.8 MW/52.8 MWh for the Independent Electricity System Operator.

A 4 MW sodium-sulfur (NaS) battery system enhances reliability and power quality for the city of Presidio in Texas.

East Asia: China's National Development and Reform Commission (NDRC) calls for multiple 100 MW vanadium redox flow battery (VRFB) installations by the end of 2020. China connected 100 MW of VRFB capacity in September 2022, with plans for further expansion. Rongke Power installed a 200 MW/800 MWh VRFB in Dalian, China, claiming it to be the world's largest.

In Japan, Hokkaido Electric Power contracted Sumitomo Electric Industries to supply a 17 MW/51 MWh vanadium redox flow battery (VRFB) for a wind farm.

In Australia, the Hornsdale Power Reserve, a Tesla 150 MW/194 MWh lithium-ion system, contributes significantly to grid stability and system security in South Australia.

Neoen is building the 300 MW/450 MWh Victorian Big Battery near Geelong, Victoria, with Tesla as the system supplier.

AGL has contracted Wärtsilä to supply a 250 MW/250 MWh lithium iron phosphate (LFP) battery at Torrens Island near Adelaide for use from 2023.

Australia's first utility-scale flow battery, supplied by Invinity, is being built at Neuroodla, with 2 MW/8 MWh capacity to provide evening peak supplement and ancillary services.

Other Countries: In Rwanda, 2.68 MWh of battery storage from Germany's Tesvolt is contracted to provide backup power for agricultural irrigation, off-grid, using Samsung lithium-ion cells. Tesvolt claims 6000 full charge cycles with 100% depth of discharge over 30 years of service life.

A Comparison of the Use of Nuclear Energy Against the Production, Maintenance and of Renewable Energy Through Solar, Wind with Battery Energy Storage Systems

To compare battery storage renewable energy and nuclear energy, it is essential to understand the distinct approaches and challenges associated with each. Battery storage renewable energy relies on renewable sources such as solar, wind, or hydropower, storing energy in batteries to address intermittency. On the other hand, nuclear energy relies on nuclear fission, providing a continuous and stable power source unaffected by weather conditions.

In the context of battery storage renewable energy, the use of energy storage technologies is crucial for addressing intermittency and ensuring a reliable power supply from renewable sources. These technologies play a significant role in the configuration and operation of energy storage systems, especially in multi-scenario settings. Furthermore, the influence of different energy storage modes on the confidence capacity of renewable energy is a critical aspect that needs to be studied to ensure the reliability of renewable energy power systems. Additionally, the feasibility analysis of innovative energy storage systems, such as vacuum pipeline magnetic levitation energy storage, contributes to the advancement of energy storage technologies.

Moreover, the challenges associated with renewable energy sources, including their intermittent nature and unpredictability due to natural conditions, highlight the need for effective grid-connected power flows and comprehensive energy-optimized operation strategies based on energy storage models. The capacity value of energy storage, considering control strategies, is also a crucial factor in ensuring the effectiveness of energy storage systems. Furthermore, energy cooperation in cellular networks with renewable-powered base stations demonstrates the potential benefits of integrating renewable energy sources with energy storage technologies.

Figure 33: Olkiluoto 1 nuclear power plant in Eurajoki, Finland. Marabu, CC0, via Wikimedia Commons.

In contrast, nuclear energy presents its own set of challenges and opportunities. Financial challenges and public opposition have been identified as significant obstacles to the development of nuclear energy projects. Additionally, the dilemma between perception and reality regarding nuclear power emphasizes the need for a comprehensive understanding of the potential risks and benefits associated with nuclear energy. Furthermore, the analysis of the high-quality development of nuclear energy under the goal of peaking carbon emissions and achieving carbon neutrality highlights the importance of addressing environmental concerns associated with nuclear power.

According to Matthews (2022), nuclear energy exhibits several advantages over renewables, particularly concerning reliability, greenhouse gas (GHG) emissions, land utilization, and waste management. The reliability, or dispatchability, of nuclear power surpasses that of renewable sources like wind and solar. Nuclear plants consistently produce energy, unaffected by fluctuations in wind or sunlight (Matthews, 2022).

Furthermore, nuclear power stands out as one of the cleanest energy sources. Recent research in the Journal of Cleaner Production indicates that GHG emissions and natural resource usage associated with nuclear power are comparable to those of renewable energy (Matthews, 2022). The European Commission's analysis suggests that, considering full-cycle production, nuclear emissions are approximately equivalent to those of wind, and some studies propose that nuclear may even be cleaner than solar. According to Orano, nuclear power generates four times fewer GHGs than solar (Matthews, 2022).

In terms of land use, nuclear energy requires substantially less space compared to wind and solar. Assessments indicate that nuclear demands 1/2,000th as much land as wind and 1/400th as much land as solar. U.S. government data reveals that a 1,000-megawatt wind farm occupies 360 times more land than an equivalently powered nuclear facility, while a solar plant requires 75 times more area (Matthews, 2022).

Figure 34: Solar power plant (Serpa, Portugal). Ceinturion, CC BY-SA 3.0, via Wikimedia Commons.

While concerns exist about nuclear waste, similar concerns surround renewable waste (Matthews, 2022). Wind and solar energy production result in various chemical wastes, including toxic heavy metals like cadmium, arsenic, chromium, and lead. Although nuclear waste can remain radioactive for thousands of years, the hazardous metals associated with renewables remain perilous indefinitely. Crucially, the volume of nuclear waste is a fraction of renewable waste, constituting 1/10,000th of solar waste and 1/500th of wind waste (Matthews, 2022).

The comparison between battery storage renewable energy and nuclear energy underscores the diverse technological, operational, and environmental aspects associated with each approach. While battery storage renewable energy focuses on addressing intermittency and unpredictability through energy storage technologies, nuclear energy provides a continuous and stable power source. Understanding the distinct advantages and challenges of these energy sources is essential for developing sustainable and reliable energy systems.

In the context of environmental impact, both Battery Storage Renewable Energy and Nuclear Energy have their respective advantages and challenges. Battery Storage Renewable Energy is generally considered environmentally friendly due to its reliance on clean, renewable sources (Stoppato et al., 2021). However, concerns arise from the production and disposal of batteries (Faisal et al., 2019). On the other hand, nuclear energy boasts low greenhouse gas emissions during electricity generation (Kadiyala et al., 2016), but faces challenges related to nuclear accidents, radioactive waste disposal, and uranium mining (Ramana & Rao, 2010).

The environmental impact of Battery Storage Renewable Energy is a topic of interest, particularly in assessing the environmental impact of storage systems integrated with energy plants powered by renewable sources (Stoppato et al., 2021). Additionally, the development of efficient fuzzy logic control systems for charging and discharging battery energy storage systems in microgrid applications is a focus area (Qays et al., 2020). Furthermore, the simulation and techno-economic analysis of on-grid battery energy storage systems in various regions, such as Indonesia, provides insights into the potential environmental impacts of such systems (Ramelan et al., 2021).

In contrast, the environmental impact of nuclear energy has been extensively studied, with a focus on the quantification of the lifecycle greenhouse gas emissions from nuclear power generation systems (Kadiyala et al., 2016). Additionally, the examination of the Indian experience in the environmental impact assessment process for nuclear facilities sheds light on the complexities and challenges associated with nuclear energy (Ramana & Rao, 2010). Moreover, the efficient recovery of radioactive materials and their transmutation into stable nuclides using fast reactors is proposed as a method to reduce the environmental impact of nuclear energy (Albrecht-Schönzart et al., 2022).

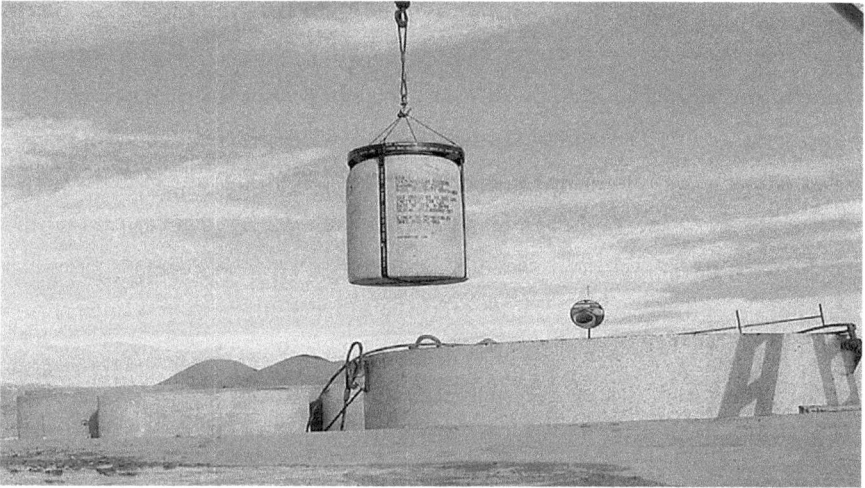

Figure 35: Class A radioactive waste disposal at the Clive, Utah disposal facility in Tooele County, Utah. U.S. National Nuclear Security Administration, Public domain, via Wikimedia Commons.

It is evident that both Battery Storage Renewable Energy and Nuclear Energy have implications for environmental policy and stringency. The impact of environmental policy on important economic outcomes, such as innovation, productivity, competitiveness, and energy efficiency, is a subject of solid tests (Galeotti et al., 2020). Furthermore, the architecture design of battery energy storage coordinated control systems based on multi-agent mechanisms highlights the role of battery energy storage in promoting the efficient consumption of new energy and the safe and stable operation of the power system (Qiu et al., 2022). Similarly, the application of hybrid renewable energy systems is seen as a solution to environmental challenges, integrating renewable energy along with traditional power plants (Al-Ammar et al., 2011).

The environmental impact of both Battery Storage Renewable Energy and Nuclear Energy is a complex and multifaceted issue. While Battery Storage Renewable Energy is generally considered environmentally friendly, challenges related to battery production and disposal persist. On the other hand, nuclear energy offers low greenhouse gas emissions but faces significant challenges related to nuclear accidents, radioactive waste disposal, and uranium mining. Both forms of energy have implications for environmental policy and stringency, and ongoing research and technological advancements aim to mitigate their environmental impact.

In terms of reliability and stability, the intermittent energy production of battery storage renewable energy presents a challenge, requiring effective storage systems to address fluctuations (Suchet et al., 2020). Renewable energy sources such as solar and wind power are inherently intermittent, relying on suitable wind speeds and sufficient sunlight for consistent energy production (Deutzmann et al., 2022). This intermittency poses a challenge to the balance of the electricity grid (Drobinski & Tantet, 2021). On the other hand, nuclear energy provides a stable and continuous power supply, demonstrating high reliability (Y. Wu et al., 2019). However, challenges may arise during maintenance or unforeseen incidents, which could impact its stability (Y. Wu et al., 2019).

The variability of climate induces fluctuating or intermittent production of renewable energy, further challenging the stability of the electricity grid (Drobinski & Tantet, 2021). This intermittency in renewable energy production requires means to balance demand and supply, which strongly depends on the energy mix as a whole and cannot characterize a specific source (Suchet et al., 2020). Additionally, the sequestration of nuclear waste is highly desirable for energy sustainability and environmental safety (Di et al., 2023).

Nuclear energy has been one of the most contentious energy sources, provoking highly active anti-nuclear groups mobilizing at different scales around the world (Şahin & Ün, 2021). Safety challenges in the nuclear industry have been a topic of concern, and experts have highlighted the importance of addressing these challenges (Germán et al., 2016). Furthermore, the potential interruption of energy supply from nuclear facilities poses a threat to national security and global stability (Burger et al., 2013).

Land use considerations play a significant role in determining the feasibility and impact of different energy generation technologies. Battery storage for renewable energy, such as solar or wind farms, requires significant land, depending on the scale of energy production. Similarly, nuclear power plants, while requiring less land, must account for safety buffer zones (Hernandez et al., 2014). The environmental impacts of utility-scale solar energy and wind farms underscore the potential landcover and land-use change impacts relative to carbon-intensive energy and other renewable energy sources (Hernandez et al., 2014). Additionally, the impacts of wind farms on land surface temperature and the spatial and temporal variation of offshore wind power production highlight the need to consider land use in the deployment of renewable energy sources (Y. H. Wang et al., 2019; Zhou et al., 2012).

In contrast, the feasibility of replacing nuclear power with other energy sources, such as hydropower, wind, and solar power plants, emphasizes the potential of renewable energy sources in electricity generation (Klemeš et al., 2020). Furthermore, the Fukushima Dai-ichi nuclear power plant accident serves as a reminder of the importance of safety buffer zones and the potential environmental impact of nuclear power generation (Hirose, 2012). On the other hand, a climate model shows that large-scale wind and solar farms in the Sahara can increase rain and vegetation, indicating the potential environmental benefits of renewable energy deployment (Y. Li et al., 2018).

Figure 36: The Fukushima I Nuclear Power Plant after the 2011 Tōhoku earthquake and tsunami (fixed aspect ratio and brightened from original image). Digital Globe, CC BY-SA 3.0, via Wikimedia Commons.

Moreover, the massive penetration of renewable energy sources, which are variable and not "dispatchable," may impact the power system supply-demand balance, highlighting the need for careful land use planning in the integration of renewable energy sources (Mazauric et al., 2022). Additionally, combining economic and fluid dynamic models to determine the optimal spacing in very large wind farms reveals the significance of land use optimization for wind energy generation (Stevens et al., 2016). Furthermore, the evaluation of power system reliability when replacing a nuclear power plant with wind power plants emphasizes the need to consider the impact of land use on the reliability of energy generation (Čepin, 2019).

To understand the cost considerations associated with Battery Storage Renewable Energy and Nuclear Energy, it is essential to analyse the initial costs, operating costs, and other relevant economic factors. Battery Storage Renewable Energy is known for its high initial costs, which may decrease over time, coupled with generally lower operating costs (Fouquet, 2018). This is supported by the fact that the integration of renewables and energy storage batteries considerably improves the quality of life, cuts costs by reducing organic fuel consumption, and improves the environmental record (Karamov et al., 2021). Additionally, a study on the techno-economic role of nuclear power in the transition to the net-zero energy system of the Netherlands found that nuclear power investments can reduce demand for variable renewable energy sources in the short term and lead to higher energy independence in the long term (Fattahi et al., 2022).

In contrast, Nuclear Energy involves high initial capital costs for plant construction, coupled with ongoing expenses for safety measures, regulatory compliance, and decommissioning (S. H. Kim et al., 2014). Furthermore, the economic, safety, engineering, and climatology impacts of nuclear energy must be fully understood prior to becoming a sustainable energy source (Laureto & Pearce, 2016). Additionally, a critical review of the premises underlying Korea's nuclear energy policy suggests that nuclear energy may be edged out by renewables in the long term, and a nuclear renaissance may never materialize (Yun, 2015).

The integration of Battery Storage Renewable Energy and Nuclear Energy into the power system also involves considerations of energy storage systems. A study on the usable capacity efficiency and lifespan of hybrid energy storage systems under office building load patterns emphasizes the importance of considering the ratio of different types of batteries to maintain maximum battery capacity and reduce the initial cost of the system (Pikultong et al., 2022). Moreover, the coordinated control of energy storage systems and diesel generators in isolated power systems highlights the role of battery energy storage systems in suppressing deviations of frequency and voltage, indicating their potential economic benefits in ensuring system stability (Goya et al., 2011).

In terms of scalability, Battery Storage Renewable Energy can be easily expanded by adding more installations and increasing storage capacity. This is due to the gaining popularity of renewable energy systems, especially microgrid renewable energy systems, where battery storage charging and discharging are crucial for both islanded and non-islanded modes (Zainurin et al., 2021). Battery storage is considered one of several technological options that can increase the flexibility of a power system and allow integration with renewable energy sources, thus supporting its scalability (Ramelan et al., 2021). Additionally, battery energy storage systems (BESSs) connected to the grid at the same point as renewable energy sources can mitigate the effects of high variability and enhance the stability of grids with high penetration levels of renewable power plants, further supporting the scalability of battery storage renewable energy (Saez-de-Ibarra et al., 2016).

On the other hand, nuclear energy faces challenges in scalability due to the time and regulatory hurdles involved in building new plants. This is in contrast to battery storage renewable energy, which has shown improvement in both technology and cost, allowing for constant power supply and the storage of electricity generated from alternative sources/renewables in large amounts (Farrag et al., 2019). Furthermore, battery energy storage systems are employed in microgrids to efficiently utilize renewable energy sources to meet dynamic load requirements, highlighting their scalability in microgrid applications (Garmabdari et al., 2019).

Public perception and social acceptance significantly influence attitudes towards different forms of energy sources. Battery storage renewable energy is generally viewed positively due to its association with clean and sustainable energy, while nuclear energy faces challenges related to safety concerns, nuclear accidents, and radioactive waste (Genys & Krikštolaitis, 2020; Groot et al., 2012; Ho et al., 2018; McBeth et al., 2022; Sposato & Hampl, 2020; Vavoulioti et al., 2023; Woo et al., 2018; Yıldız & Arı, 2019). The positive public perception of battery storage renewable energy is influenced by its potential to increase the flexibility of power systems and facilitate the integration of renewable energy sources (Hayajneh & Zhang, 2020; Kalair et al., 2020; Mutz et al., 2022; Ramelan et al., 2021; Tariq & Mahmood, 2021). On the other hand, the challenges faced by nuclear energy in terms of public acceptance are attributed to factors such as proximity to nuclear plants, social trust, knowledge about nuclear energy, environmental sensitivity, and sense of security (Vavoulioti et al., 2023). Additionally, the social acceptance of nuclear power has become a decisive factor in framing sustainable energy policies (Woo et al., 2018).

The public's perception of nuclear energy is influenced by various social, cultural, economic, and political factors, and the proximity to nuclear plants has been found to have a negative effect on its acceptance (Genys & Krikštolaitis, 2020). Furthermore, studies have shown that public acceptance of nuclear energy is influenced by factors such as political ideology, trust, and perceived risks (Genys & Krikštolaitis, 2020; McBeth et al., 2022; Shinoda et al., 2021; Vavoulioti et al., 2023). It has been noted that the social acceptance of nuclear energy is necessary for the development of energy technology (Genys & Krikštolaitis, 2020; McBeth et al., 2022; Shinoda et al., 2021; Vavoulioti et al., 2023; Yıldız & Arı, 2019). Moreover, the social and contextual aspects of energy, such as energy security and affordability, can influence the acceptance of specific energy sources (Kim et al., 2019).

In contrast, battery storage renewable energy technologies have been positively associated with egoistic values and perceived benefits, leading to their higher acceptability (Groot et al., 2012). Additionally, the integration of renewable energy sources with battery energy storage systems has been highlighted as a means to improve power quality and facilitate the better integration of renewable energy sources with microgrids (Faisal et al., 2019; Mutz et al., 2022; Ramelan et al., 2021; Tariq & Mahmood, 2021). Furthermore, the use of battery energy storage systems in the generation, transmission, and distribution segments of the electric grid has been identified as an opportunity for large-scale stationary electrical energy storage (Doughty et al., 2010).

Waste management considerations in the context of battery storage for renewable energy and nuclear energy generation are crucial due to the potential environmental challenges posed by the production and disposal of batteries, as well as the generation of radioactive waste. The life cycle assessment of lithium-ion power batteries (LIBs) for electric vehicles has been evaluated to identify key stages contributing to the overall environmental burden and to find effective ways to reduce this burden (Sun et al., 2020). Additionally, the efficient recovery and reutilization of spent lithium-ion batteries are necessary from the viewpoint of cleaner production and green chemistry, emphasizing the importance of sustainable strategies for battery waste management (Liang et al., 2021). Furthermore, a literature review on recycling lithium batteries highlights the significance of limiting the impact of toxic and non-ferrous metals associated with battery production, emphasizing the environmental aspects of battery waste management (Filomeno & Feraco, 2020).

In the context of nuclear energy, the management of radioactive waste is a critical concern. The issues relating to the management of radioactive wastes are well-formulated internationally, and guidelines for radioactive waste disposal are well-documented, emphasizing the need for careful and secure long-term storage solutions for radioactive waste (Ravichandran et al., 2011). Moreover, the overview of the nuclear fuel cycle strategies and spent nuclear fuel management technologies emphasizes the importance of technologies aimed at reducing nuclear waste generation and the burden on dry storage and final disposal in radioactive waste management (Tsai et al., 2020). Additionally, a review on bioremediation as an emerging technology for the treatment of radionuclide waste highlights the need for sustainable and efficient strategies for radioactive waste management, particularly in developing countries (Cheema, 2023).

The environmental challenges associated with battery storage for renewable energy and the generation of radioactive waste from nuclear energy necessitate comprehensive waste management strategies. These strategies should encompass the entire life cycle of batteries, including production, usage, and disposal, to minimize environmental impacts. Additionally, the long-term storage and disposal of radioactive waste from nuclear energy generation require secure and sustainable solutions to mitigate potential environmental and public health risks.

To determine the most suitable energy source between Battery Storage Renewable Energy and nuclear energy, various factors such as geographical considerations, energy demand, environmental priorities, and public acceptance need to be taken into account. Both technologies have their strengths and weaknesses, and many regions are adopting a mix of both approaches to balance these aspects. Nuclear energy has been considered as a potential solution for energy problems due to the limitations of non-renewable sources and the limited energy potential of some renewable sources (Guk & Kalkan, 2015). However, the global trend towards energy transition and environmental sustainability prioritizes renewable energy sources, indicating a growing emphasis on renewable energy (T. Lee, 2020). Moreover, the need for environmental protection and increasing demands for natural resources are forcing companies to reconsider their business models and restructure their supply chain operations, reflecting a global trend towards sustainability and environmental priorities (Wu & Pagell, 2010).

Nuclear energy offers high energy density and convenient transportation and storage of nuclear fuel, making it a reliable backup energy source for renewable energy power plants (Liu & Dai, 2019). However, public acceptance of nuclear energy varies based on geographical and social factors, with country of sample and time period of data collection moderating public perceptions of benefits, risks, and acceptance of nuclear energy (Ho et al., 2018). Additionally, the displacement of high-carbon footprint technology would find favour with environmentalists, who tend to be opposed to nuclear energy, highlighting the environmental concerns associated with nuclear energy (Rao & Gould, 2022).

Geographical considerations play a crucial role in the energy sector, as the geographic smoothing of solar photovoltaic electric power production in the Western USA demonstrates the impact of geographic factors on renewable energy generation (Klima et al., 2018). Furthermore, the development of a competitive nuclear industry requires a strong supply capacity of nuclear energy mineral resources, emphasizing the geographical aspect of resource availability (J. Zhang et al., 2019). The integration of green growth in economic strategies underscores the importance of curbing emissions and pollution for sustainable growth and well-being, aligning with environmental priorities.

To determine the more economically feasible option between battery storage renewable energy and nuclear energy, it is essential to consider various factors such as cost, sustainability, and environmental impact. Renewable energy sources, including battery storage, have gained attention due to their potential economic feasibility and environmental benefits (Hariri et al., 2020; Kim, 2014; Tanoto, 2011). The analysis of a stand-alone hybrid distributed generation system in eastern Indonesia revealed that a PV/battery/wind-hybrid system is the most economically feasible, with a cost of energy at US$ 0.817/kWh, while maintaining system reliability (Tanoto, 2011). Additionally, renewable energy has been highlighted as an alternative to traditional fossil fuels, being more environmentally friendly, abundant, and economically feasible (Hariri et al., 2020). Furthermore, the advancement in renewable energy technologies has made energy generation from onsite solar, wind, or geothermal systems technically and economically feasible (Kim, 2014).

On the other hand, nuclear energy has been a subject of extensive research regarding its relationship with economic growth and its potential as an economically feasible energy source (Apergis & Payne, 2010; Wolde-Rufael & Menyah, 2010; Yoo & Ku, 2009). Studies have shown a positive correlation between nuclear energy consumption and economic growth in various countries, indicating the potential economic benefits of nuclear energy (Kırıkkaleli et al., 2020; Wolde-Rufael & Menyah, 2010; Yoo & Ku, 2009). However, the economic feasibility of nuclear energy is also influenced by factors such as capital costs, safety concerns, and waste management (Chang et al., 2014; Durán et al., 2023; Ferrari, 2023).

In terms of economic feasibility, it is crucial to consider the cost-effectiveness and sustainability of both options. While renewable energy technologies have shown promising economic feasibility and environmental benefits, nuclear energy has also demonstrated potential economic advantages, particularly in its contribution to economic growth. However, the economic feasibility of nuclear energy is also influenced by safety and waste management concerns. Therefore, a comprehensive assessment considering cost, sustainability, and long-term economic implications is necessary to determine the most economically feasible option between battery storage renewable energy and nuclear energy.

That being said, Matthews (2022) argues that despite the undeniable significance of renewable energy, it is crucial to confront the reality that wind and solar power alone have not succeeded in slowing, let alone halting, emissions from fossil fuels. In 2019, only 16 percent of global primary energy was derived from low-carbon sources, with renewables contributing 11.4% and nuclear energy 4.3%. Even the most optimistic projections indicate that transitioning to renewable energy could take more than 30 years, a luxury of time we simply do not possess (Matthews, 2022).

There has been minimal progress in altering the composition of the energy mix. Over the last decade, the share of renewable energy increased modestly from 10.6% to 11.7%, while fossil fuels, including coal and gas, marginally decreased from 80.1% to 79.6% (Matthews, 2022). This stagnation is concerning, especially as energy demand rises, resulting in an overall increase in the use of fossil fuels.

Undeniably, emissions-free renewables play a critical role in our energy landscape. However, the remarkable growth of wind and solar energy, while significant, falls short of the necessary scale to effectively curb the dominance of fossil fuels. To meet our emissions reduction targets, renewables must undergo a massive expansion (Matthews, 2022).

The investment in renewables has not matched the required scale. UN Secretary-General Guterres has called for a substantial increase in spending on renewables, emphasizing the missed opportunities in the past that could have mitigated the current reliance on volatile fossil fuel markets. Matthews (2022) suggests that we have not made the essential investments in renewables, nor are we on the brink of doing so. While renewables have exhibited substantial growth, their pace is far from sufficient to curb the dominance of fossil fuels. The prospect of building adequate solar and wind infrastructure within the available timeframes remains elusive.

Confronting this challenging reality, it appears unlikely that we will deploy enough renewable energy to transition away from fossil fuels within the necessary timeframe (Matthews, 2022). Germany's situation, despite its fervent pursuit of wind and solar energy, illustrates the persistent reliance on coal. The International Energy Agency's World Energy Outlook has starkly stated that the world is not transitioning to clean energy rapidly enough to eliminate emissions by the middle of the century (Matthews, 2022). While efficiency measures are part of the solution, we also need a substantial increase in clean power to meet the surging demand associated with decarbonization through electrification.

In this context, nuclear power's potential to significantly augment emissions-free electricity production should not be overlooked (Matthews, 2022). Nuclear fuel stands out as one of the longest-lasting and abundant sources of energy on the planet. The longevity of nuclear fuel is comparable even to solar power, with astronomers estimating the sun's remaining lifespan at about 7 billion to 8 billion years, while the half-life of thorium-232, a component of nuclear fuel, is approximately 14 billion years (Matthews, 2022).

Matthews (2022) concludes that to attain carbon neutrality, it is imperative to implement a comprehensive suite of clean technologies. Without the widespread adoption of clean energy across the economy, the goal of reducing emissions by 50% by 2030 and achieving complete eradication by 2050 will remain unattainable. Given the pivotal role of energy in addressing the climate crisis, leveraging all available sources of emissions-free power becomes imperative. The combination of renewables and nuclear energy emerges as a potentially optimal solution in our endeavour to eliminate carbon emissions by the midpoint of the century (Matthews, 2022).

Chapter Eight

Social Impact

L ithium-ion batteries have significantly impacted people's lives by revolutionizing modern society, making it more portable, intelligent, and cleaner (H. Li et al., 2021). The development of safer, durable, higher-power, and lower-cost lithium batteries has enabled their widespread use in portable electronics and electric vehicles (Peled & Menkin, 2017). The increasing demand for lithium-ion batteries with higher capacity and longer life has been driven by the need to achieve a decarbonized society (Minami et al., 2022). These batteries have become an essential part of modern daily life, powering mobile phones and laptops, and have exerted a great influence on daily convenience (Lü & Chen, 2022). Furthermore, the improvement in technology has led to longer battery life, enhancing the usability of various devices (G. Kang et al., 2019).

The widespread adoption of lithium-ion batteries has been driven by their impact on modern society, powering devices such as laptops, cellular phones, and MP3 players (Yu et al., 2005). The understanding of the intercalation of lithium in different materials has given birth to rechargeable lithium-ion batteries, further solidifying their impact on society (Reddy et al., 2020). Today, lithium-ion batteries are an indispensable part of modern daily life, powering mobile phones and laptops, and revolutionizing modern society (Ye, 2022). The application of lithium-ion batteries in various industries, including daily life, is increasing, emphasizing the essential need to evaluate their safety and reliability (Zou et al., 2022).

The increasing adoption of electric vehicles has necessitated lithium-ion batteries with high-performance characteristics, such as higher energy density, faster charging time, and prolonged lifetime (Shi et al., 2020). However, despite their significant impact, lithium-ion batteries still face obstacles in the field of large-scale energy storage due to a paucity of lithium supplies, relatively high prices, and low energy density (Liu et al., 2023). To ensure the safe and reliable operation of lithium-ion battery-powered devices and systems, it is crucial to predict their life and performance accurately (Fang et al., 2015). The complex physical and chemical changes in the use process can lead to performance degradation or failure, resulting in serious safety issues and major economic losses (Qu et al., 2019).

The detailed study of lithium-ion batteries, including their electrochemical performance and mechanical safety, has made them the most widely used power battery for pure electric vehicles (Xu et al., 2021). However, factors such as temperature can influence the capacity and internal resistance of lithium-ion batteries, affecting their performance (Lv et al., 2021). Lithium-ion batteries have become the most popular electric energy storage systems for various configurations of hybrid and electric vehicles due to their high energy-to-weight ratio, high energy-to-volume ratio, and excellent cycle life (Liu et al., 2019). Early monitoring and warning of thermal runaway in lithium-ion power batteries are crucial for ensuring safety and preventing safety accidents (A. Gao et al., 2022).

Positive Adaptation

The adoption of electric vehicles (EVs) has been facilitated by the development of lithium-ion batteries, which have played a crucial role in the shift towards cleaner and more sustainable transportation. This transition has led to a reduction in air pollution and greenhouse gas emissions, thereby positively impacting air quality and public health (Notter et al., 2010). Notter et al. (2010) highlighted that the environmental impact of E-mobility caused by the battery is measured to be 15% of the total impact, indicating the significance of lithium-ion batteries in the environmental benefits of EVs (Notter et al., 2010). Furthermore, Peters et al. (2020) found that the electrification of a portion of light-duty vehicles could lead to widespread reductions in ozone and particulate matter, thus contributing to improved air quality. Additionally, (Liou & Wu, 2021)employed the benefit transfer method to estimate the social cost of carbon and the health co-benefits via impact pathway analysis, emphasizing the monetary health co-benefits and greenhouse gas emissions reduction benefits associated with the adoption of EVs.

Moreover, the positive social impacts of EV adoption are influenced by various factors. (Mashayekhi, 2021)) identified seven factors that influence the rate of adoption of battery electric vehicles, including social pressure, social prestige, environmental usefulness, price, perceived risk, and knowledge and information about EVs, highlighting the multifaceted nature of the adoption process. Additionally, (Jayasingh et al., 2021) found that environmental concern, perceived economic benefit, charging infrastructure, and social influence significantly impact consumers' attitudes towards electric two-wheelers, further emphasizing the role of social and economic factors in EV adoption). (Lai et al., 2015) also highlighted the importance of instrumental attributes and individuals' pro-environmental self-identity in influencing the perception of electric vehicles, indicating the complex interplay of psychological and behavioural factors in EV adoption.

The impact of EV adoption on air quality is also influenced by the source of electricity. (Vidhi & Shrivastava, 2018) emphasized that the air quality impact of EVs depends on the source of electricity, indicating the need to consider the environmental implications of power generation in assessing the overall benefits of EV adoption. Additionally, (Ke et al., 2016) demonstrated that the electrification of a significant portion of the vehicle fleet can lead to reductions in average PM concentrations, highlighting the potential for EV adoption to positively impact regional and urban air quality.

Renewable energy storage is crucial for integrating renewable sources like solar and wind into power grids, reducing dependence on fossil fuels, and enhancing grid reliability. Lithium-ion batteries are widely used for this purpose (Li, 2013). The integration of renewable energy into the grid necessitates large-scale energy storage technologies (Li, 2013). Lithium metal batteries are considered promising for high-energy-density energy storage (P. Y. Chen et al., 2021). Additionally, the next-generation "beyond-lithium" battery chemistry is seen as a feasible solution for achieving renewable energy integration goals (Xu et al., 2015).

Temperature significantly influences the capacity of lithium-ion batteries, with discharge capacity being approximated by a cubic polynomial of temperature (Lv et al., 2021). Thermal analysis of lithium-ion batteries is crucial, and various methods, including calorimetry, have been used for this purpose (Madani et al., 2021b). Furthermore, scholars have established relevant thermal models of lithium-ion batteries using simulation software to analyse their thermal properties under different conditions (A. Gao et al., 2022).

In the context of renewable energy integration, distributed charge/discharge control of energy storages, such as battery energy storage systems (BESS), is essential for grid stability and operation in both grid-connected and islanded modes (Eghtedarpour &

Farjah, 2014). The stable operation of microgrids during grid-connected and islanded modes is critical, and the integration of renewable energy sources, such as distributed generation (DG), offers several benefits for the utility grid (Mehrasa et al., 2015). Moreover, distributed energy storage systems (DESS) play a crucial role in stabilizing power grids during the large-scale integration of renewable energy units (Pouresmaeil et al., 2017).

The reliability of the power grid is a key consideration in the integration of renewable energy sources. Smart grid systems are being developed to address the challenges and issues associated with renewable energy integration, and investments and upgrades to the power sector are fundamental in this regard (Dahunsi et al., 2022). Additionally, power system reliability is an important index for evaluating the ability of power supply, and the role of power-grid and communication-system interdependencies on cascading failures is a critical area of study (Rahnamay-Naeini & Hayat, 2013; Zheng et al., 2013).

Portable electronics, such as smartphones and laptops, have been revolutionized by the advent of lithium-ion batteries (LIBs) (Dirican et al., 2014). These batteries have significantly enhanced productivity and connectivity by providing high energy density, long cycle life, and affordable cost (Tinambunan et al., 2022). The demand for high capacity electrode materials for LIBs has been driven by the need to power future portable electronics and electric vehicles (Wu et al., 2012). LIBs have not only dominated the portable electronics market but are also expanding to large-scale electric energy storage applications (Fang et al., 2018). They are considered the most preferred rechargeable battery technology due to their superior properties (Dirican et al., 2014)

Research on sodium-based batteries, as an alternative to LIBs, was largely abandoned due to the success and overall quality of LIBs (Adelhelm et al., 2015). However, for sodium-ion batteries to achieve comparable performance to current LIBs, significant improvements are still required in cathode, anode, and electrolyte materials (Bai et al., 2018). Additionally, rechargeable alkaline-ion batteries, including LIBs, are widely used in portable consumer electronics and electric vehicles (Zeng et al., 2022).

The safety, performance, and cycle lifetime of LIBs are significantly impacted by their operating temperature (Liu et al., 2019). Furthermore, the mechanical properties and thermal runaway of automotive LIBs have been studied extensively, providing valuable insights for the safety of electric vehicle LIBs (Liu et al., 2019; Xu et al., 2021). Moreover, the thermal characteristics of LIBs have been investigated, showing a significant dependence of solid electrolyte interface resistance and electrolyte resistance on battery temperature (Madani et al., 2021b).

The growth of the lithium-ion battery industry has significantly contributed to job creation across various sectors such as manufacturing, research and development, and maintenance, particularly in regions where battery production facilities are established. As highlighted by Zackrisson et al. (2010), the increasing use of lithium-ion batteries in plug-in hybrid electric vehicles has raised critical issues regarding their life cycle assessment. Furthermore, Suryoatmojo and Pratama (2021) emphasized the evolution of lithium-ion batteries from their original use in electronic equipment to larger scale applications in automotive and industrial technology, indicating the expanding job opportunities in these sectors. Additionally, the attention garnered by lithium in recent years due to its application in the battery industry further underscores its role in job creation (Chen et al., 2018).

Moreover, the economic, technical, and environmental aspects of recycling lithium batteries have been extensively reviewed by Filomeno and Feraco (2020), shedding light on the opportunities and processes of recycling treatments for lithium-ion batteries, which in turn can contribute to job creation in the recycling and waste management sector. Additionally, Tao et al. (2020) conducted an economic feasibility analysis of new lithium-ion batteries in photovoltaic energy storage systems, demonstrating the employment of cost-benefit models to analyse the economy of lithium-ion batteries, further emphasizing their potential for job creation in the renewable energy sector.

The increasing importance of lithium-ion batteries in various industrial applications and significant areas, as highlighted by Liu, Luo, et al. (2013), further supports the notion of job creation in sectors utilizing these batteries. Furthermore, the development of a program for comparing the charge and discharge performance of lithium-ion batteries in different temperature ranges, as demonstrated by Ali et al. (2019), indicates the growing research and development activities in the field, potentially leading to job creation in the research and testing domains. Additionally, Liao et al. (2019) conducted an economic analysis of an industrial photovoltaic system coupled with battery storage, revealing insights into the economic aspects of different battery energy storage systems, which can contribute to job creation in the renewable energy sector.

To address the issue of energy access in remote areas, the combination of lithium-ion batteries with renewable energy sources has been identified as a viable solution. Lithium-ion batteries have been widely recognized for their dominance in portable electronics, electric vehicles, and grid-energy storage (Manthiram, 2017). They have proven to be a reliable source of energy for wearables and electric vehicles, making them suitable for providing electricity in remote and off-grid areas (Gowda & Channegowda, 2022). However, it is important to note that there are limited lithium reserves, with more than 60% of accessible lithium resources located in remote or politically sensitive areas (Xia & Xin, 2022). Despite this limitation, lithium-ion batteries remain a popular choice for various applications, including providing electricity in remote areas.

In the context of remote energy access, the use of lithium-ion batteries in renewable energy systems, such as solar-and wind-energy-based power generation, has been highlighted as a cost-effective choice for supplying power to remote communities compared to grid supply (Rezk et al., 2020). Additionally, off-grid electricity generation with renewable energy technologies, such as solar and wind, has been studied extensively, emphasizing the application of these technologies in providing electricity to off-grid areas (Sen & Bhattacharyya, 2014). Furthermore, the integration of lithium-ion batteries with renewable energy sources in microgrids has been explored as a method for energy saving and stabilization of the grid (Vallati et al., 2015).

In the specific case of lithium-ion batteries, their application in providing electricity to remote areas requires careful consideration of factors such as thermal management and energy storage. Research has focused on studying the thermal management of lithium-ion batteries, which is crucial for ensuring their safe and efficient operation in remote and off-grid settings (Madani et al., 2021a). Moreover, the design and sensitivity analysis of hybrid photovoltaic-fuel-cell-battery systems have been investigated to supply power to small communities, highlighting the relevance of integrating different energy sources for remote energy access (Rezk et al., 2020).

Negative Adaptation

On the other hand, the manufacturing of lithium-ion batteries has raised significant social concerns that demand attention due to various issues associated with their production and supply chain. One primary concern is the use of cobalt, which has been linked to human rights challenges in its extraction process (Olivetti et al., 2017). The surge in demand for lithium batteries has led to increased mining activities, often in countries with lax labour regulations, exposing workers to unsafe conditions and inadequate wages (Alonso et al., 2022). Additionally, the extraction of lithium raises concerns for indigenous communities, as mining operations may encroach upon their ancestral lands and disrupt traditional livelihoods, giving rise to concerns about the infringement of indigenous rights and the loss of cultural heritage (Hanna & Vanclay, 2013).

Furthermore, the extraction of materials for lithium-ion batteries often involves mining activities with negative environmental and social consequences, including worries about unethical practices such as child labour and unsafe working conditions (O'Faircheallaigh & Babidge, 2023). Improper disposal of lithium-ion batteries can result in environmental pollution, posing risks to ecosystems and human health (Schlosberg & Carruthers, 2010). The growing demand for lithium-ion batteries also raises concerns about resource depletion and environmental degradation (Mo et al., 2022). Additionally, the widespread adoption of electric vehicles and renewable energy technologies may result in job displacement in traditional industries, such as fossil fuel extraction and refining, necessitating the management of this transition to minimize negative social impacts (Prasetyo et al., 2022).

To address these social concerns, it is recommended to completely eliminate the use of cobalt in lithium-ion batteries and promote battery recycling and land recovery as mitigation strategies (Sehil et al., 2021). Ensuring transparency and traceability in the lithium battery supply chain is crucial to avoid sourcing materials from regions associated with unethical practices, such as child labour or environmental exploitation (Tong et al., 2020). Moreover, promoting the use of alternative materials and advancing battery recycling technologies can contribute to addressing the social problems associated with lithium-ion battery manufacturing (Zhang et al., 2018).

Social Structuring and Interaction

The advent of lithium-ion-powered smartphones has revolutionized global communication and connectivity, transcending conventional boundaries and fostering a more interconnected, informed, and dynamic global society. These devices have played a pivotal role in transforming communication from a localized to a global phenomenon (Simon & Gogotsi, 2008). With the ability to connect to the internet, lithium-ion-powered smartphones facilitate instant communication with individuals worldwide, breaking down geographical barriers and fostering a sense of global interconnectedness (Simon & Gogotsi, 2008). Social media platforms, messaging apps, and email services empower users to engage in real-time conversations, contributing to the rapid dissemination of news and events, enabling users to stay informed about global occurrences in real time (Xu, 2014).

Moreover, these devices have redefined social interactions by providing multiple channels for communication, such as voice and video calls, text messaging, and multimedia sharing, creating diverse avenues for expressing thoughts and emotions (Manthiram, 2017). This diversity enriches communication, allowing individuals to tailor their interactions based on the content and context, fostering deeper connections (Manthiram, 2017). Additionally, social media platforms, facilitated by lithium-ion-powered devices, have become virtual spaces for community building, enabling individuals with shared interests to form communities, exchange ideas, and support one another, contributing to a more diversified social landscape (Chaolong Zhang et al., 2022).

Figure 37: Mobile phone use. Photo by Camilo Jimenez on Unsplash.

During emergencies or natural disasters, lithium-ion-powered devices serve as crucial communication tools, allowing users to send and receive real-time updates, seek help, and coordinate relief efforts, which has proven instrumental in disaster management (Z. Chen et al., 2021). Furthermore, the widespread adoption of lithium-ion-powered devices has contributed to digital inclusion, enabling individuals in remote or economically disadvantaged areas to access communication tools, educational resources, and employment opportunities through affordable smartphones, potentially bridging the digital divide and empowering marginalized communities (Puente et al., 2023).

Beyond personal communication, these devices facilitate professional networking and business interactions, reshaping traditional workplace dynamics, promoting remote work and global business relationships (Chun et al., 2020). They also play a role in cultural exchange by enabling individuals to share and appreciate diverse cultural content, contributing to a more interconnected world where people can explore and understand different cultures (Z. Chen et al., 2020).

The democratization of knowledge through the widespread use of lithium-ion-powered portable devices has significantly transformed various aspects of society. These devices have facilitated access to a wealth of information, educational resources, news updates, entertainment, language and skill development tools, healthcare information, financial services, job opportunities, and environmental awareness (Grainger et al., 2020; Markopoulos et al., 2021; Pearson et al., 2022; Rasche et al., 2018; Rubanovich et al., 2017; Tang & Hai, 2021). The ability to access the internet from virtually anywhere has allowed individuals to explore a vast repository of information, regardless of their geographical location or socioeconomic background (Grainger et al., 2020; Markopoulos et al., 2021; Pearson et al., 2022; Rasche et al., 2018; Rubanovich et al., 2017; Tang & Hai, 2021). This has been particularly impactful in regions with limited access to traditional educational infrastructure, as individuals can pursue self-directed learning journeys and enhance their knowledge on diverse subjects through online courses, educational apps, and digital libraries (Grainger et al., 2020; Krebs & Duncan, 2015; Rubanovich et al., 2017).

Furthermore, the immediate access to news updates and current affairs has contributed to an informed citizenry, fostering civic engagement and awareness (Grainger et al., 2020; Krebs & Duncan, 2015; Rubanovich et al., 2017). The integration of lithium-ion batteries in portable devices has revolutionized entertainment consumption, allowing users to stream movies, music, and multimedia content on-the-go, transforming commuting or travel time into opportunities for leisure and relaxation (Gonka & Kim, 2015; Pearson et al., 2022; Tang & Hai, 2021).

Additionally, language learning apps, skill development platforms, and online tutorials have leveraged the ubiquity of these devices to facilitate continuous learning, contributing to a culture of lifelong learning (Pearson et al., 2022; Gonka & Kim, 2015; Tang & Hai, 2021). Moreover, these devices have enabled users to access healthcare information and resources, empowering individuals to take control of their health and well-being, particularly in remote areas with limited access to healthcare facilities (Tang & Hai, 2021). Furthermore, they have contributed to financial inclusion by providing access to online banking, payment systems, and financial literacy resources, even in regions with limited traditional banking infrastructure (Markopoulos et al., 2021; Rasche et al., 2018).

The ability to access online job portals, freelance platforms, and e-commerce websites has expanded economic opportunities for many, reducing geographical constraints and fostering entrepreneurship (Krebs & Duncan, 2015; Markopoulos et al., 2021). Additionally, information related to environmental issues and sustainability is easily accessible through these devices, contributing to raising awareness about ecological challenges and promoting eco-friendly practices (Markopoulos et al., 2021; Rasche et al., 2018). The democratization of knowledge has also been emphasized in the context of intrapreneurship, corporate entrepreneurship, and innovation strategy formulation and execution (Choi, 2019; Markopoulos et al., 2021).

The emergence of educational apps has made learning more convenient and flexible, meeting the learning needs of contemporary people to a great extent (Tang & Hai, 2021). The observance of democratic processes in classrooms has been highlighted as a means to contribute to the development of knowledge, skills, and attitudes necessary for democratic citizenship (Subramanian, 2016).

The integration of lithium-ion batteries into laptops and tablets has brought about a significant transformation in the way professionals work, leading to a remote work revolution (L. Zhang et al., 2017). This revolution has particularly impacted knowledge-based industries, where work is centred around digital tasks and collaboration. The portability and flexibility provided by lithium-ion-powered devices have enabled professionals to work from various locations, contributing to a better work-life balance and increased productivity (Wu et al., 2016). This flexibility has been especially beneficial for working parents, caregivers, and those seeking a more adaptable work routine.

The rise of digital nomadism is closely linked to the portability and reliability of lithium-ion-powered devices, allowing professionals to travel while staying connected and productive (Lee et al., 2011). This has given rise to a global community of individuals who can work from diverse locations, experiencing new cultures and environments without compromising their professional commitments. Additionally, collaborative tools and platforms accessible through these devices have transformed how teams work together, leading to more inclusive and diverse work environments by breaking down geographical barriers (Lin et al., 2013).

The ability to carry powerful computing devices wherever one goes has significantly increased individual and organizational productivity (F. Zhang et al., 2022). Tasks that traditionally required a fixed workstation can now be accomplished on-the-go, contributing to a more efficient workflow. Moreover, lithium-ion-powered devices have played a crucial role in the rise of entrepreneurship and startups, lowering the barriers to entry for new ventures and fostering innovation and economic dynamism (Meng, 2021).

The impact of lithium-ion-powered devices extends across diverse professions, from creative fields relying on graphic design software to scientific research requiring data analysis tools (P. T. Chen et al., 2020). Furthermore, in times of unforeseen events or emergencies, having access to a reliable computing device becomes crucial, with lithium-ion-powered laptops and tablets serving as essential tools for communication, coordination, and information access during emergencies, contributing to preparedness and resilience (Suryoatmojo & Pratama, 2021).

Portable devices powered by lithium-ion batteries are instrumental in facilitating continuous learning and skill development, allowing professionals to engage in online courses, attend virtual workshops, and acquire new skills relevant to their careers, enhancing their employability and adaptability in a rapidly changing job market (Haizhou, 2017).

The rise of lithium-ion-powered smartphones has significantly contributed to the growth of social media platforms, enabling people to easily connect, share updates, and engage with others on platforms like Facebook, Instagram, Twitter, and more. This has reshaped social relationships and the way individuals build and maintain connections. The impact of social media on various aspects of human life has been extensively studied and documented in academic literature. Research has shown that social media use has both positive and negative effects on individuals. For instance, it has been found that social media can have positive effects such as increased social support, self-expression, and the ability to connect with others (Hussain & Khatoon, 2023). However, it has also been associated with negative outcomes, including negative impacts on mental health, such as depression and social media fatigue (Guo & Cai, 2022; C.-C. Lee et al., 2014; Viola, 2021). Furthermore, the use of social media has been linked to changes in social relationships, with studies highlighting the influence of social media on intimate relationships and the beginning of relationships (Aronowicz, 2020; Whiteside et al., 2018).

The influence of social media on adolescents has been a topic of particular interest, with research exploring its impact on adolescent mental health and behaviour (Bekalu et al., 2019; Guo et al., 2022; Jabbar et al., 2022). Additionally, the addictive nature of social media, particularly among teenagers, has been investigated, with studies focusing on factors such as gadget addiction and the influence of social media on adolescent behaviour (Guo et al., 2022; Santoso et al., 2021).

Moreover, the role of social media in influencing consumer behaviour and marketing strategies has been extensively studied. Research has examined the effects of content like-ability, credibility, and social media engagement on users' acceptance of product place-ment in mobile social networks (Lai & Liu, 2020). Additionally, the use of influencers in digital marketing and the impact of social comparison on adolescents' flourishing through short videos have been explored (Guo et al., 2022; Khan et al., 2022).

The impact of social media on various aspects of society, including family dynamics, education, and healthcare, has also been a subject of investigation. Studies have delved into the role of social media in family relationships, the influence of social media on students' academic performance, and best practices in social media at public, nonprofit, education, and healthcare organizations (Bedua et al., 2021; Khan et al., 2022; Procentese et al., 2019).

Furthermore, the privacy and ethical implications of social media use have been ex-amined, with research focusing on issues such as networked privacy and the impact of social media behaviour on secrecy and privacy (Çavuş et al., 2018; Marwick & boyd, 2014). Additionally, the influence of social media on language learning and its potential for innovative foreign language learning have been explored (Jamshidian, 2020).

The impact of lithium-ion batteries on entertainment and leisure has been profound, revolutionizing the way individuals engage with cultural content and seek leisure activities in the digital age. The portability and versatility of these batteries have catalysed a trans-formation in the entertainment landscape, enabling on-the-go entertainment, streaming services, gaming accessibility, augmented reality (AR) and virtual reality (VR) experi-ences, personalized content consumption, digital reading, social entertainment, content creation, and a shift in traditional entertainment models (Ponce et al., 2017; Sato et al., 2020).

Lithium-ion-powered portable devices, such as smartphones and tablets, have facil-itated on-the-go entertainment, allowing users to access streaming services for movies, TV shows, and music virtually anywhere, transforming daily commutes, travel, and downtime into opportunities for personalized entertainment (Ponce et al., 2017; Sato et al., 2020). This has shifted the paradigm from traditional scheduled TV viewing to individualized, on-demand content consumption. The rise of lithium-ion-powered devices has fuelled the growth of streaming platforms like Netflix, Hulu, and Spotify, offering a vast library of content that users can enjoy at their convenience, thereby transforming the way people engage with entertainment (Ponce et al., 2017; Sato et al., 2020).

Furthermore, the proliferation of mobile gaming has been significantly influenced by lithium-ion batteries, as smartphones and tablets powered by these batteries serve as portable gaming consoles, providing access to a diverse range of games and contributing to the gamification of leisure (Ponce et al., 2017; Sato et al., 2020). Additionally, the integration of AR and VR technologies in lithium-ion-powered devices has expanded the entertainment landscape, offering immersive gaming, storytelling, and interactive experiences (Ponce et al., 2017; Sato et al., 2020).

The personalization of content consumption has been facilitated by lithium-ion-powered devices, enabling users to curate their entertainment experiences through personalized recommendations, playlists, and content algorithms, leading to a more immersive entertainment landscape where users have greater control over what they watch, listen to, or play. Moreover, the digitalization of reading habits and expanded access to literature have been influenced by e-books and audiobooks accessible through portable devices, transforming the way people engage with literature (Ponce et al., 2017; Sato et al., 2020).

Social entertainment experiences have also been enhanced by lithium-ion-powered devices, allowing users to engage in multiplayer gaming, participate in virtual events, and share their entertainment experiences on social media, thereby enhancing the communal nature of entertainment, even when individuals are physically apart. Additionally, the same devices that enable content consumption also empower users to become content creators, contributing to the democratization of content creation and giving rise to new forms of entertainment and creative expression (Ponce et al., 2017; Sato et al., 2020).

The widespread use of lithium-ion-powered devices for entertainment has also influenced cultural trends and references, contributing to shared cultural experiences through memes, viral videos, and online challenges disseminating rapidly through social media. This intersection of technology and entertainment has become a significant cultural force, shaping how individuals engage with cultural content and express creativity in the digital age. The accessibility of entertainment through lithium-ion devices has disrupted traditional entertainment models, leading to the decline of physical media, such as DVDs and

CDs, and the rise of digital streaming, representing a shift in how consumers access and consume content. This transformation has implications for industries, business models, and the concept of ownership in entertainment, reflecting the multidimensional impact of lithium-ion batteries on entertainment and leisure (Ponce et al., 2017; Sato et al., 2020).

The integration of lithium-ion batteries in wearable health devices has revolutionized the way individuals monitor and manage their health and well-being. These devices, such as smartwatches and fitness bands, leverage the capabilities of lithium-ion batteries to track various health metrics, including steps taken, distance travelled, calories burned, heart rate, sleep patterns, and stress levels (Mercer et al., 2016). The use of wearable fitness trackers has been found to increase physical activity levels and impact motivational constructs such as enjoyment, challenge, affiliation, and positive health motivation, particularly in young adults (Kerner & Goodyear, 2017). Additionally, wearable devices provide continuous health monitoring, contributing to preventive health measures by enabling early detection of irregularities in sleep, activity, or vital signs, prompting individuals to seek medical attention or make lifestyle adjustments before more serious health issues arise (Henriksen et al., 2018).

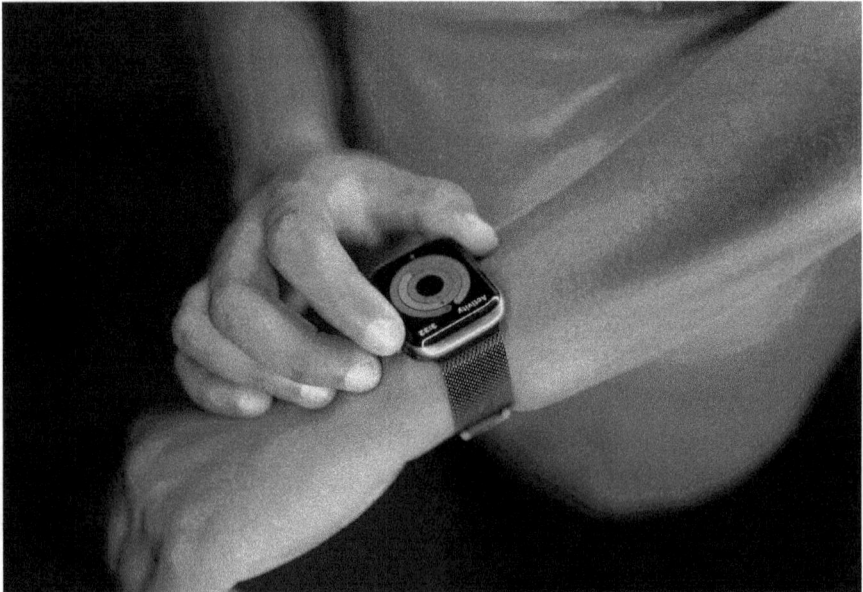

Figure 38: Using an Apple Watch. Photo by Solen Feyissa on Unsplash.

Moreover, wearable devices equipped with lithium-ion batteries offer features such as heart rate monitoring, stress tracking, calorie management, activity reminders, and goal setting, which contribute to a holistic understanding of an individual's well-being and promote a conscious approach to nutrition and physical activity (Y. Wang et al., 2017). These devices also seamlessly integrate with health and fitness applications on smartphones, creating a comprehensive health ecosystem and enhancing the overall user experience and the effectiveness of health-tracking initiatives (Henriksen et al., 2018). Furthermore, the use of wearable activity trackers has been shown to be beneficial for older adults, as they are the fastest-growing segment of the population regarding Internet use and smartphone ownership, facilitating the potential adoption of wearable activity trackers in their daily lives (Holko et al., 2022).

While wearable devices have shown promise in promoting health and well-being, it is essential for future studies to use larger study samples, longer intervention and follow-up periods, and integrative and personalized innovative mobile technologies to provide comprehensive and sustainable support for patients and health service providers (Tedesco et al., 2017). Additionally, the impact of wearable devices on life and health insurance companies, as well as the potential benefits for the industry and the wearables market, has been analysed, highlighting the potential emerging market drivers for such technology in the future (Bianchi et al., 2022).

Accessibility and inclusivity are crucial aspects of modern society, and the integration of lithium-ion batteries in portable devices has significantly contributed to enhancing accessibility for individuals with disabilities. The use of assistive technologies, such as screen reader applications, speech-to-text and text-to-speech functionalities, and educational apps tailored to different learning styles, has revolutionized the way individuals with disabilities interact with digital content and engage in educational activities (Bouck et al., 2011; Mankoff et al., 2010; Mortenson et al., 2012). Furthermore, the availability of lithium-ion-powered devices has opened up new employment opportunities, with workplace accommodations and remote work options supported by these devices, thereby increasing access to employment for individuals with disabilities (Bryant & Seay, 2020; Mortenson et al., 2012). Additionally, GPS-enabled portable devices have facilitated accessible navigation for individuals with visual impairments, contributing to enhanced mobility and independence in navigating public spaces (Crudden et al., 2017; Mortenson et al., 2012).

Moreover, the flexibility of lithium-ion-powered devices allows for customization to accommodate individual preferences and accessibility needs, promoting a more user-friendly experience (Gherardini et al., 2020; Mortenson et al., 2012). These devices also play a critical role in emergency assistance, as individuals with disabilities can utilize communication apps to seek help and access information tailored to specific accessibility needs, thereby contributing to their safety and well-being in various situations (Day & Huefner, 2003; Mortenson et al., 2012). The widespread use of lithium-ion-powered devices has also prompted a shift towards inclusive design practices, with developers and manufacturers increasingly considering accessibility features during the design and development phases, aiming to create products that are usable by the widest possible audience (Gherardini et al., 2020; Mortenson et al., 2012). Furthermore, legislation and advocacy efforts complement the impact of lithium-ion-powered devices on accessibility, with many countries enacting regulations that mandate accessibility standards for digital products and services, and advocacy organizations working to promote the development of technologies that prioritize inclusivity (Bryant & Seay, 2020; Mortenson et al., 2012).

Digital literacy and skill development have been significantly influenced by the ubiquity of portable devices powered by lithium-ion batteries. These devices, including tablets, laptops, and smartphones, have revolutionized education, skill acquisition, career advancement, entrepreneurship, social and civic engagement, multimedia creation, adaptability, global connectivity, cybersecurity awareness, and efforts to close the digital divide (Carter et al., 2016). The impact of portable devices on digital literacy extends beyond individual skill development, influencing education, career paths, entrepreneurship, and societal engagement. The use of portable devices has led to the development of proficiencies that are increasingly relevant in the digital age, ranging from basic computer literacy to advanced technical skills. Educational apps and e-learning platforms, often utilized on these devices, contribute to a more interactive and dynamic learning experience (Carter et al., 2016). Furthermore, the ability to navigate online platforms, manage digital content, and utilize e-commerce solutions has become integral to entrepreneurial success (Carter et al., 2016).

Portable devices have also facilitated social and civic engagement, as individuals use them to stay informed about current events, engage with social issues through online platforms, and participate in digital advocacy, contributing to informed citizenship and the shaping of public discourse (Carter et al., 2016). Additionally, the democratization of multimedia creation through portable devices has empowered individuals to express themselves creatively and share their perspectives with a global audience (Carter et al., 2016). The use of portable devices encourages an adaptive mindset and a willingness to engage in continuous learning, as users become accustomed to updating software, exploring new apps, and adapting to changing digital landscapes (Carter et al., 2016). Moreover, portable devices enhance global connectivity, allowing individuals to collaborate and communicate with people worldwide, fostering cross-cultural understanding and collaboration. Digital literacy also includes an understanding of online security practices, such as creating strong passwords, recognizing phishing attempts, and safeguarding personal information, which is essential for protecting oneself in the digital realm (Carter et al., 2016). Efforts are needed to address the digital divide and ensure equitable access to digital skills, especially in underserved communities, to create a more inclusive digital society (Carter et al., 2016).

The widespread use of lithium-ion-powered smartphones has significantly transformed cultural and social expression, providing individuals with powerful tools to document, share, and celebrate their cultural heritage. Smartphones have democratized storytelling, allowing diverse voices to contribute to the broader cultural tapestry (Harmon & Mazmanian, 2013). Through social media platforms, individuals can share aspects of their daily lives, celebrations, and traditions, fostering cultural expression and connection with others who share similar backgrounds. Furthermore, smartphones facilitate global cultural exchange by allowing users to access and engage with diverse cultural content, promoting a richer understanding of global cultures and fostering a sense of shared humanity (Harmon & Mazmanian, 2013). Additionally, smartphones play a pivotal role in social activism and advocacy, enabling users to document and share instances of social injustice, human rights violations, and environmental issues in real-time, thereby contributing to a more socially conscious global community (Harmon & Mazmanian, 2013).

Moreover, smartphones enable individuals to actively participate in online movements that align with their cultural or social values, providing avenues for collective expression and solidarity (Harmon & Mazmanian, 2013). They also contribute to the preservation of indigenous cultures by allowing community members to document and share their languages, traditions, and rituals, thereby helping in the transmission of cultural knowledge across generations and raising awareness about the importance of preserving indigenous heritage. In regions facing language endangerment, smartphones serve as tools for language revitalization, enabling users to engage with and revitalize their native languages through language-learning apps, digital storytelling, and online language communities (Harmon & Mazmanian, 2013).

The accessibility of smartphones has democratized artistic expression, allowing users to create digital art, photography, and multimedia content that reflect their cultural influences (Harmon & Mazmanian, 2013). Social media platforms provide a global stage for sharing this digital artwork, fostering cross-cultural appreciation and dialogue (Harmon & Mazmanian, 2013). Additionally, smartphones have become integral to the documentation and celebration of cultural festivals and events, allowing users to capture and share moments of cultural significance, thereby contributing to the promotion and understanding of cultural diversity (Harmon & Mazmanian, 2013). Furthermore, smartphone-enabled cultural expression plays a role in cultural diplomacy, as individuals, organizations, and governments use digital platforms to share aspects of their culture with a global audience, fostering cross-cultural understanding and contributing to soft power diplomacy (Harmon & Mazmanian, 2013)

The increasing prevalence of lithium-ion batteries in electronic devices has raised concerns about electronic waste (e-waste) due to the potential environmental risks associated with their disposal. These batteries contain toxic components such as lithium, cobalt, and other heavy metals, which can leach into the environment if not managed responsibly, posing a threat to ecosystems, wildlife, and human health (Albuquerque et al., 2021; Tischner & Charter, 2017). In response to these concerns, there has been a growing awareness of the environmental impact of lithium-ion batteries, leading to initiatives promoting responsible disposal practices and the recycling of valuable materials like lithium and cobalt (Kousar et al., 2022). Additionally, educational campaigns have been launched to inform consumers about the importance of responsible e-waste management, encouraging them to participate in recycling programs and utilize manufacturer-sponsored take-back programs (Danish et al., 2019). Furthermore, the concept of a circular economy, which emphasizes recycling, reusing, and reducing waste, aligns with the goal of minimizing the environmental impact of lithium-ion batteries, promoting the design of products with recycling in mind (Ziadat, 2009).

Regulatory measures have also been introduced to address the environmental impact of e-waste, including guidelines for responsible disposal of electronic devices and battery recycling (S. Wang et al., 2022). Moreover, there has been innovation in sustainable product design, with manufacturers exploring ways to create electronic devices with eco-friendly materials and modular components to facilitate the recycling process, thereby reducing the overall environmental impact (Zarei et al., 2022). Life cycle assessments have become increasingly important in evaluating the environmental impact of lithium-ion batteries, guiding decision-makers in identifying opportunities to minimize environmental harm at each stage, from raw material extraction to manufacturing, use, and eventual disposal or recycling (Chen & Tsai, 2015).

Global collaboration has emerged as a crucial aspect of addressing the environmental impact of lithium-ion batteries, with the sharing of best practices, technologies, and research findings on responsible disposal and recycling contributing to a more coordinated global effort. In summary, the widespread use of lithium-ion batteries has spurred environmental awareness, prompting initiatives, regulations, and innovations aimed at mitigating the environmental impact of electronic waste. Responsible disposal practices, recycling programs, and a shift toward a circular economy are integral components of addressing environmental concerns associated with lithium-ion batteries.

Lithium-Ion Batteries through a Social Theory Lens

Social theories serve as frameworks or conceptual tools for interpreting and analysing social phenomena, offering systematic ways to understand the complexities of human societies and uncover underlying principles that shape social interactions, institutions, and structures (Chernoff, 2019). These theories provide insights into the dynamics of society, aiding researchers and scholars in making sense of social processes and behaviours (Chernoff, 2019). The study of social theories spans various disciplines, including sociology, anthropology, political science, and psychology, reflecting the interdisciplinary nature of social theory (Chernoff, 2019).

One of the fundamental aspects of social theories is the examination of social structures and their impact on human behaviour and interactions. Social structures are conceptualized as patterns of social arrangements and relationships that influence individuals within a society (Musolf, 2003). These structures are seen as integral to understanding social life, with the interplay between structure and agency providing a comprehensive explanation of social reality (Musolf, 2003). Moreover, the significance of social structures is highlighted in the context of gender, race, and social inequality, positioning them as central elements in shaping the dynamics of society (Bonilla-Silva, 1997; Risman, 2004).

Furthermore, social theories encompass diverse perspectives, such as symbolic interactionism and structural functionalism, which offer distinct lenses for understanding social phenomena (Barbano, 1968; Musolf, 1992). Symbolic interactionism emphasizes the role of symbols and interactions in shaping social behaviour, while structural functionalism focuses on the interplay between social institutions and their functions within society (Barbano, 1968; Musolf, 1992). These theoretical perspectives contribute to the multifaceted nature of social theories, providing researchers with varied analytical tools to explore and interpret social phenomena.

In addition to social structures, social theories also delve into the dynamics of social interactions and networks. The concept of social interaction is central to understanding the mechanisms through which individuals engage with one another, influencing the formation of social ties and relationships (Cohen & Felson, 1979; Marwick, 2012). Moreover, the emergence of online social networking has expanded the scope of social interactions, leading to discussions on the impact of virtual interactions on individuals' well-being and sense of community (Marwick, 2012; Oh et al., 2014).

The relevance of social theories extends beyond academic discourse, as they have practical implications for fields such as social entrepreneurship and social policy. Social entrepreneurship theory draws on cognitive frameworks, such as the theory of planned behaviour, to understand and promote sustainable social impact (Ebrashi, 2013). Similarly, social policy is informed by social theories, as they provide insights into social structures, human agency, and the ontological state of social life, influencing the design and implementation of policies (Musolf, 2003)

Structural Functionalism, as proposed by Emile Durkheim and Talcott Parsons, focuses on the functions of social institutions and their contribution to societal stability and cohesion (Longhofer & Winchester, 2016). This theory emphasizes the interdependence of social structures and their role in maintaining order within society. Conflict Theory, associated with Karl Marx and Max Weber, highlights the influence of power, coercion, and competition in social structures, exploring how conflicts between different groups contribute to social change. It emphasizes the role of struggle and competition in shaping social dynamics.

Symbolic Interactionism, advocated by George Herbert Mead and Erving Goffman, examines micro-level interactions between individuals and the role of symbols and language in shaping social reality. This theory focuses on the subjective meanings attached to symbols and the significance of human interpretation in shaping social interactions. Feminist Theory, with key thinkers such as Simone de Beauvoir and bell hooks, centres on gender-based inequalities and advocates for the rights and perspectives of women in various social contexts. It critically analyses power dynamics and societal structures that perpetuate gender disparities.

Critical Theory, associated with Theodor Adorno and Herbert Marcuse, aims to critique and transform society, focusing on issues like power, domination, and the role of culture in maintaining social structures. This theory seeks to uncover and challenge oppressive systems within society. Postmodernism, as proposed by Jean-François Lyotard and Michel Foucault, questions established narratives and emphasizes the diversity of individual experiences and perspectives, challenging the notion of objective truth. It deconstructs traditional norms and questions established truths.

Structural Functionalism

Structural functionalism is a sociological theory that focuses on the structures and functions of society. According to Levy and Coser (1957), this theory emphasizes the interrelated parts of society and how they contribute to the stability and functioning of the whole. It views society as a complex system with interconnected parts, each serving a specific function to maintain the overall stability and equilibrium. Rome and Coser (1956) further elaborate on the functions of social conflicts within this framework, highlighting how conflicts can serve specific functions in maintaining the stability of social structures. This perspective aligns with the idea that even conflicts within society can have functional aspects that contribute to the overall balance and functioning of the social system.

Structural Functionalism is a sociological perspective that perceives society as a complex system with interconnected parts working collaboratively to preserve stability and order. This theoretical framework places a strong emphasis on understanding the functions of diverse social institutions, structures, and norms in maintaining social equilibrium. At its core, Structural Functionalism posits that each element within society serves a specific function, contributing to the overall stability and functioning of the entire social system.

The social structure, according to Structural Functionalism, is envisioned as a composition of various institutions, such as family, education, government, and religion. These institutions are attributed specific roles, playing crucial parts in upholding social order. Each component of society, encompassing both institutions and norms, is believed to possess a designated function or purpose. These functions collectively contribute to the overall stability and cohesion of the social system.

Interconnectedness is a fundamental concept within Structural Functionalism, underscoring the interdependence of different societal components. Changes occurring in one part are seen to have ripple effects throughout the entire social structure, highlighting the intricate relationships that bind various elements together.

The theory introduces the notions of manifest and latent functions. Manifest functions are the recognized and intended consequences of social structures, while latent functions represent unintended or less obvious consequences. Both types of functions are considered contributors to the stability of the social system.

Structural Functionalism directs its primary focus toward understanding how social institutions and structures work together to maintain social order and stability. It posits the idea that societies naturally tend to move toward a state of equilibrium. However, the theory also acknowledges that not all aspects of society contribute positively to stability. Dysfunction refers to elements that may disrupt the equilibrium and have negative consequences for society.

In its historical context, Structural Functionalism wielded significant influence in sociology during the mid-20th century. Over time, criticisms emerged, particularly regarding its inclination to prioritize stability and order over the examination of social conflict and change. Despite these criticisms, certain elements of Structural Functionalism persist in influencing sociological thought, maintaining its status as an essential perspective for understanding specific aspects of social systems.

Esteve-Calvo and Lloret-Climent (2006) discuss the utility of structural input-output functions, emphasizing the application of graph theory to study the relationships between variables. This approach aligns with the structural functionalist perspective, as it underscores the interconnectedness and interdependence of various elements within a system, reflecting the core principles of structural functionalism. In addition, Yoho et al. (2019) highlight the significance of understanding structure and function as a key learning concept across multiple disciplines, particularly in STEM fields. This underscores the broad applicability and relevance of the concept of structure and function, emphasizing its importance in understanding various systems and phenomena.

From a Structural Functionalism perspective, the impact of lithium-ion-powered smartphones on global society's stability and cohesion can be analysed through key principles of Structural Functionalism.

Global Communication and Connectivity: Lithium-ion-powered smartphones serve as tools that transcend geographical boundaries, fostering global interconnectedness. This aligns with the functionalist view of technology contributing to the stability of society. Instant communication enables the rapid dissemination of information, contributing to a more informed and connected global citizenry. Beyond individual communication, this interconnectedness contributes to a sense of community and shared global identity, reinforcing social cohesion.

Social Media and Community Building: Social media platforms, facilitated by lithium-ion-powered devices, provide spaces for community building and interaction. Users can connect with like-minded individuals, share ideas, and provide support, enhancing social cohesion by fostering virtual communities. These online communities contribute to a more diversified social landscape, allowing individuals with shared interests to form connections, aligning with the functionalist emphasis on social stability.

Emergency Communication: During emergencies or natural disasters, lithium-ion-powered devices play a crucial role in communication, coordination, and information dissemination. This functionality supports disaster management efforts, ensuring a quick response and potentially saving lives. The use of these devices in emergencies reinforces the importance of a shared communication infrastructure, promoting societal resilience and cooperation.

Digital Inclusion and Access: The widespread adoption of lithium-ion-powered devices contributes to digital inclusion, enabling access to communication tools, educational resources, and job opportunities. Individuals in remote or economically disadvantaged areas gain access to resources, potentially bridging the digital divide and promoting societal equity. Increased accessibility fosters a sense of inclusion and equality, aligning with functionalist ideals of a harmonious and stable society.

Professional Networking and Global Business: Lithium-ion-powered devices facilitate professional networking and reshape traditional workplace dynamics. Remote work and global business relationships become more prevalent, contributing to a flexible and interconnected professional landscape. The breaking down of geographical barriers promotes a more inclusive and diverse global workforce, aligning with the functionalist notion of societal stability through cooperation.

Knowledge Democratization: The democratization of knowledge is facilitated by portable devices with lithium-ion batteries, providing access to information, education, and various resources. Individuals gain access to educational resources, news updates, entertainment, and skill development tools, fostering a more informed and educated society. The widespread availability of knowledge contributes to social awareness, civic engagement, and a shared understanding of societal issues, reinforcing stability.

Environmental Awareness and Responsibility: The awareness of the environmental impact of lithium-ion batteries leads to initiatives promoting responsible disposal and recycling. Recycling programs and responsible disposal practices contribute to environmental sustainability, aligning with the functionalist view of societal stability. Environmental awareness fosters a sense of responsibility and collective action, promoting cooperation for the common good.

In a structural functionalist analysis, these functions collectively contribute to the stability, cohesion, and functionality of global society through the integration of lithium-ion-powered devices. The emphasis is on the positive functions that contribute to the well-being and interconnectedness of individuals within the broader societal framework.

Conflict Theory

Conflict theory is a sociological perspective that examines the role of conflict, power, and social inequality in shaping societies. Originating from the works of Karl Marx and later expanded by other sociologists, this theory asserts that society is characterized by inherent conflicts and struggles over resources, power, and social status. In contrast to structural functionalism, which accentuates stability and consensus, conflict theory directs attention to the tensions and inequalities inherent in social structures.

This theory posits that society is characterized by competition and struggle over limited resources, and that these conflicts arise from the unequal distribution of power and resources among different groups within society (Ferrare & Phillippo, 2021). According to Ferrare and Phillippo (2021), conflict theory lacks a comprehensive definition, but it is generally understood as a framework for understanding contemporary struggles over various societal issues, including education policy. This perspective is rooted in the idea that social order is maintained through the dominance of certain groups over others, leading to tensions and conflicts between them (Overholt, 1977).

Social inequality is a central theme in conflict theory, emphasizing disparities in economics, class divisions, and the uneven distribution of resources as sources of conflict in society. The theory also underscores the pivotal role of power in social relationships, where those in power can control resources, influence decision-making processes, and shape societal norms and institutions.

Class struggle, a key concept in conflict theory, denotes the ongoing conflict and competition between different social classes. Marx, a prominent figure in conflict theory, focused on the struggle between the bourgeoisie (owners of the means of production) and the proletariat (working class). Conflict theory views social change as a consequence of conflicts and struggles between groups with opposing interests, suggesting that change occurs through social revolutions or transformative movements challenging existing power structures.

Ideology and hegemony are critical aspects highlighted by conflict theorists, contending that dominant groups maintain power by promoting ideologies that legitimize their position. Hegemony refers to the cultural and ideological dominance of one group over others. Symbolic violence, a concept introduced by sociologist Pierre Bourdieu, explores how dominant groups impose values, norms, and beliefs on others through subtle and symbolic means.

Conflict theorists critically examine social institutions, including education, government, and the legal system, highlighting how these structures may perpetuate inequality and serve the interests of dominant groups. Moreover, conflict theory often incorporates intersectionality, recognizing that social inequalities are interconnected and influenced by factors such as race, gender, ethnicity, and other identity markers.

The application of conflict theory extends beyond sociology and can be found in various disciplines. For instance, in the field of psychology, conflict theory has been used to understand the dynamics of individual and group conflicts. It has been suggested that a modern conflict theory should incorporate the concept of the ego, emphasizing the role of internal psychological conflicts in shaping human behaviour (Frank, 2007; Lettieri, 2005). Additionally, the role of affect and reward in conflict-triggered cognitive control has been explored from the perspective of conflict monitoring theory, highlighting the interplay between emotions and cognitive processes in conflict situations (Dreisbach & Fischer, 2012).

In organizational management, conflict theory has been employed to manage cross-cultural conflicts within organizations, emphasizing the importance of understanding and addressing conflicts arising from diverse cultural backgrounds (Mayer & Louw, 2012). Furthermore, conflict theory has been utilized to model complex societal interactions, providing insights into the dynamics of conflicts within societies and the factors influencing their resolution (Medler et al., 2008).

In analysing social phenomena, conflict theory provides researchers and sociologists with a lens that emphasizes the role of conflict and power dynamics in understanding societal structures and processes. It complements other sociological perspectives by shedding light on aspects of society that might be overlooked when focusing solely on stability and consensus.

From a Conflict Theory perspective, the advent of lithium-ion-powered smartphones can be analysed through the lens of power dynamics, social inequality, and conflicts over resources. The transformative impact of these devices on global communication and connectivity is not uniformly distributed across all segments of society. Instead, it reflects and exacerbates existing power differentials and social inequalities. Conflict theorists argue that the benefits of this technological revolution are not distributed equitably, influenced by economic disparities, creating a digital divide (Manthiram, 2020). The ability to access and utilize these devices is influenced by economic disparities, creating a digital divide between those who can afford and seamlessly integrate these technologies into their lives and those who face barriers due to economic limitations (B. S. Lee, 2020).

Moreover, the emphasis on the positive aspects of smartphones may obscure the potential negative consequences. Conflict theorists would scrutinize how these devices might contribute to social inequalities, reinforcing existing power structures. For example, the role of social media platforms, enabled by lithium-ion-powered devices, in community building is acknowledged, but the potential for these platforms to amplify existing social hierarchies or create new forms of inequality is not explored (Chitrakar et al., 2014). In times of emergencies or natural disasters, the conflict perspective would prompt an analysis of how the use of lithium-ion-powered devices for communication during such events might disproportionately benefit certain social groups, potentially leaving marginalized communities at a disadvantage due to limited access or resources (Cai et al., 2021).

The discussion about the democratization of knowledge and the transformative impact on various aspects of society is also viewed critically through the conflict lens. Conflict theorists might argue that the knowledge accessible through these devices is not neutral but is influenced by the ideologies and interests of dominant groups. The potential for symbolic violence, where certain values and norms are imposed on others, could be explored in the context of the information disseminated through these devices

(S. Dong et al., 2021). Furthermore, the analysis of the impact on entertainment, cultural exchange, and the workplace is likely to be examined through the conflict perspective. Conflict theorists would scrutinize how these devices might reinforce or challenge existing power dynamics, including issues related to cultural representation, labour exploitation in global business relationships, and the potential for these technologies to perpetuate or disrupt traditional workplace hierarchies (Ingried et al., 2022).

In terms of health and well-being, conflict theorists might question how the benefits of wearable health devices are distributed across different socioeconomic groups. Are these devices equally accessible, and do they contribute to reinforcing health disparities? Additionally, the potential commodification of health data and its implications for privacy and surveillance would be areas of concern (P. Xu et al., 2020). Concerns about electronic waste and environmental impact are also seen through the conflict perspective, with questions about who bears the brunt of environmental harm. Conflict theorists might explore whether regulations and initiatives to address e-waste disproportionately burden certain communities or whether the burden is distributed more equitably (Jianchuan Liu et al., 2021).

Symbolic Interactionism

Symbolic Interactionism is a sociological perspective that focuses on the micro-level interactions and symbolic meanings in everyday life. This theoretical framework emphasizes the role of symbols and shared meanings in shaping individual behaviour and social interactions (Carter & Fuller, 2016). Developed primarily by George Herbert Mead and later expanded by Herbert Blumer, symbolic interactionism provides insights into how people create, interpret, and modify symbols to construct their social reality (Carter & Fuller, 2016).

One of the key concepts of symbolic interactionism is symbols, which are objects, words, gestures, or other elements that carry shared meanings in a particular social context. These symbols serve as a basis for communication and interaction between individuals (Carter & Fuller, 2016). The core emphasis of symbolic interactionism is on the subjective meanings that individuals attach to symbols. People act based on the meaning they give to things, and these meanings are not inherent but socially constructed through interaction (Carter & Fuller, 2016).

The development of self is crucial in symbolic interactionism. According to Mead, individuals go through stages of development, including the "I" (the subjective and impulsive self) and the "Me" (the objective self shaped by social interactions). The self is formed through the process of taking the perspective of the other (Burbank & Martins, 2010). Symbolic interactionists argue that individuals learn to understand and interpret the world by taking on the roles of others. Through role-taking, people anticipate how others will respond to their actions, influencing their behaviour (Bett et al., 2020).

The theory emphasizes face-to-face interactions and the importance of communication in shaping social reality. Everyday interactions contribute to the creation and reinforcement of shared meanings within a society. Within symbolic interactionism, labelling theory explores how individuals and groups are labelled by society, and how these labels can affect their behaviour and social outcomes. Labels can create self-fulfilling prophecies, influencing individuals to conform to societal expectations.

The looking glass self, coined by Charles Horton Cooley, is the concept that individuals develop their self-concept by imagining how others perceive them. People use others as a mirror to form their own self-image. Symbolic interactionism also suggests that social change occurs through the modification and reinterpretation of symbols. As meanings evolve through interaction, societal norms and values can also change over time.

Symbolic Interactionism is often applied to various areas of sociology, including education, family dynamics, deviance, and the understanding of social institutions. Researchers using this perspective often engage in qualitative research methods, such as participant observation and in-depth interviews, to explore the intricate details of social interactions and meanings in specific contexts.

From the perspective of Symbolic Interactionism, the advent of lithium-ion-powered smartphones can be analysed by focusing on the symbolic meanings and social interactions associated with these devices and their impact on various aspects of society. Symbolic Interactionism emphasizes the role of symbols, meanings, and social interactions in shaping individual and societal behaviour (Ramadesigan et al., 2010).

Lithium-ion-powered smartphones symbolize a shift from localized to global communication, breaking down geographical barriers and fostering global interconnectedness. They represent a transformative tool for shaping social interactions, communication channels, and the way individuals express thoughts and emotions (Ramadesigan et al., 2010). The smartphones facilitate instant communication, creating a sense of global interconnectedness by enabling real-time conversations and breaking down geographical barriers. Social media platforms, messaging apps, and email services provided by these devices become virtual spaces for community building, symbolizing a shift in how people form connections (Ramadesigan et al., 2010).

Individuals adapt their communication styles and interactions based on the diverse channels offered by smartphones, contributing to a richer and more varied social landscape. The use of smartphones during emergencies demonstrates the role-taking process, where individuals use these devices to seek help and coordinate relief efforts, symbolizing their instrumental role in disaster management (Ramadesigan et al., 2010). The smartphones contribute to digital inclusion by providing access to communication tools, educational resources, and employment opportunities, symbolizing a potential bridging of the digital divide. Labelling theory can be applied to how these devices label individuals as connected, informed, and empowered, influencing their self-perception and societal roles (Ramadesigan et al., 2010).

Smartphones symbolize a tool for cultural exchange, enabling individuals to share and appreciate diverse cultural content, contributing to a more interconnected world. The impact of smartphones on language learning, skill development, and the sharing of cultural heritage symbolizes the democratization of knowledge and cultural expression (Ramadesigan et al., 2010). The integration of lithium-ion batteries in portable devices is portrayed as transformative, impacting not only individual behaviours but also shaping societal structures, including education, entertainment, and healthcare. The adoption of smartphones is linked to identity formation, as individuals engage in continuous learning, skill development, and cultural expression through these devices (Ramadesigan et al., 2010).

Social media, facilitated by smartphones, plays a significant role in shaping the looking glass self, where individuals observe and present themselves based on perceived societal expectations. The positive and negative effects of social media highlight the impact of symbolic interactions on self-perception and social relationships (Ramadesigan et al., 2010). The availability of lithium-ion-powered devices contributes to inclusive design practices, symbolizing a societal shift towards considering accessibility features during product development. Legislation and advocacy efforts symbolize a recognition of the importance of inclusivity and accessibility in the design and use of digital technologies (Ramadesigan et al., 2010).

The environmental impact of lithium-ion batteries symbolizes a growing awareness of electronic waste and the need for responsible disposal practices. Initiatives, regulations, and innovations in sustainable product design represent symbolic responses to environmental concerns associated with electronic waste (Ramadesigan et al., 2010).

Feminist Theory

To understand and address the social, political, economic, and cultural inequalities between men and women, feminist theory has emerged as a diverse and interdisciplinary body of thought (Jackson, 1985). It aims to analyse, critique, and challenge the structures and norms that contribute to gender-based disparities (Kandiyoti, 1988). Key concepts within feminist theory include patriarchy, gender roles and expectations, intersectionality, the social construction of gender, empowerment and agency, and feminist activism (Hunnicutt, 2009). These concepts are crucial in understanding the complexities of gender inequality and the various factors that contribute to it.

Patriarchy is often identified as a central social structure that perpetuates gender inequality (Kandiyoti, 1988). It refers to a system where men hold primary power and women are systematically disadvantaged. Feminist theorists also examine how societal expectations and norms around femininity and masculinity shape individuals' behaviours and opportunities (Hunnicutt, 2009). They critique rigid gender roles that limit individuals based on their assigned gender. Furthermore, feminist theory recognizes that gender intersects with other social categories such as race, class, sexual orientation, and ethnicity (Srivastava & Willoughby, 2022). This intersectionality emphasizes the interconnectedness of various social identities and how they intersect to create unique experiences of oppression or privilege.

Moreover, feminist theorists argue that gender is not solely a biological or natural phenomenon but is socially constructed (Narayanan, 2019). This challenges essentialist views that link specific characteristics or roles exclusively to biological sex. Additionally, feminism promotes the empowerment of women and the recognition of their agency (Aldoory, 2005). It advocates for women's rights, opportunities, and the ability to make choices without facing discrimination or coercion. Many feminist theorists are actively engaged in feminist activism, advocating for policy changes, legal reforms, and social transformations to address gender inequality (Fawcett, 2022). Activism is seen as a crucial component of feminist theory's commitment to social change.

Feminist theory encompasses various branches and perspectives, each offering unique insights into gender issues. Some major schools of thought within feminist theory include liberal feminism, radical feminism, Marxist feminism, Black feminism, and postmodern feminism (Garlick, 2017). Each of these perspectives offers distinct approaches to addressing gender inequality and challenging existing social structures. It's important to note that feminist theory is not a monolithic or homogeneous perspective; it encompasses a wide range of ideas and approaches (Williams, 2010). Scholars and activists within feminist theory may hold different views on specific issues, strategies for change, and the root causes of gender inequality.

From a feminist theory perspective, the analysis of the impact of lithium-ion-powered devices on society can be approached through a lens that considers the intersection of technology, gender, and power dynamics. This approach is crucial to critically examine how these developments may intersect with gender relations and contribute to or challenge existing inequalities. The feminist analysis encourages us to assess the implications of technological advancements on gender relations, ensuring that the benefits are equitably distributed and that potential risks and disparities are acknowledged and addressed (Bellini et al., 2022). It also calls for an intersectional approach, considering how gender intersects with other social categories such as race, class, and geography in shaping individuals' experiences with technology (Floegel & Costello, 2021).

Access and Representation: The design, development, and representation within the technology industry can include gender biases in the creation of apps, devices, or online platforms (Shaw & Gant, 2002). Women may be excluded from benefiting from these technological advancements due to gender-based disparities in access and utilization of lithium-ion-powered devices, particularly in certain regions or socio-economic groups (Costello & Floegel, 2021).

Gendered Impacts of Connectivity: Changes in communication and connectivity can impact women differently, leading to positive outcomes or reinforcing existing gender stereotypes or risks (Barbala, 2022). Increased connectivity may not equally benefit marginalized groups of women, including those in rural areas or with limited economic resources (Schurr et al., 2023).

Representation in Digital Spaces: In the context of social media and community building, women's representation and equal opportunities for participation and leadership in online communities are important considerations (Nichols et al., 2021). Social media platforms, powered by lithium-ion devices, may facilitate or perpetuate gender-based harassment or discrimination (Reyes-Sosa et al., 2022).

Impact on Work and Entrepreneurship: Lithium-ion-powered devices influence gender dynamics in the workplace, especially in industries where remote work and entrepreneurship are prominent. Increased connectivity may contribute to more inclusive work environments for women, but new challenges may also be introduced (Alemann et al., 2021). Gender-specific impacts on entrepreneurship, including access to funding, mentorship, and networking facilitated by these devices, are crucial to consider (L. Wang et al., 2019).

Digital Literacy and Education: The gendered aspects of digital literacy and access to educational resources through these devices, as well as disparities in how men and women engage with technology for educational purposes, should be addressed (Ly-Le, 2022). These devices have the potential to empower women through online learning, skill development, and access to information (Floegel & Costello, 2021).

Health and Well-being: Wearable health devices may impact women's health and contribute to gendered expectations related to body image and fitness (Qian & Ji, 2022). Concerns about data privacy and the potential misuse of health-related information, particularly for women, should be acknowledged and addressed (Radhakrishnan, 2021).

Environmental Impact and Gender Justice: The discussion of electronic waste and responsible disposal should account for any gendered implications, including gender-specific consequences of exposure to toxic materials from discarded devices (Benschop, 2021). Women may be disproportionately affected by environmental degradation caused by the production and disposal of lithium-ion batteries (Halford et al., 2015).

Critical Theory

Critical Theory, which originated from the Frankfurt School in the mid-20th century, is characterized by its interdisciplinary nature and its aim to challenge established social structures and power relations. This school of thought involves a deep critique of society, focusing on power dynamics, historical context, and the role of intellectuals in contributing to social change. Notable thinkers associated with Critical Theory include Max Horkheimer, Theodor Adorno, Herbert Marcuse, and Walter Benjamin, and it has evolved to encompass various forms such as critical race theory and feminist theory (Bronner, 2017; Carrington & Selva, 2010; Datta, 2009; Rexhepi & Torres, 2011).

Critical Theory emphasizes the examination of power dynamics and how they operate in various social, political, economic, and cultural contexts. It often explores how certain groups maintain dominance over others and how this dominance is perpetuated (Friedman & Landsberg, 2013; Kellner, 1990). Additionally, it considers historical context crucial for understanding contemporary social issues and examines how historical events and structures have shaped the current state of society, with a focus on social change over time (Bronner, 2017; Datta, 2009).

The theory also explores how dominant groups establish and maintain cultural hegemony, shaping norms, values, and ideologies to maintain their social and political control. Cultural products, such as media, literature, and art, are analysed for their role in this process (Kong, 2023). Furthermore, Critical Theory is not only about critique; it is also concerned with the possibilities for social transformation and emancipation. It seeks to identify avenues for challenging oppressive structures and fostering a more just and equitable society (Wilner et al., 2012).

Critical Theory is inherently interdisciplinary, borrowing ideas and methods from various fields. Scholars associated with Critical Theory engage with philosophy, sociology, psychology, literature, and other disciplines to analyse and critique societal structures (Miller et al., 2008; Silver, 2022). The role of the intellectual is also central in the tradition of Critical Theory, as intellectuals are expected to engage with and challenge the status quo (Carrington & Selva, 2010; Rexhepi & Torres, 2011).

From a Critical Theory perspective, the implications for societal structures, power relations, and broader social dynamics can be analysed through various lenses. Critical Theory prompts an examination of who benefits the most from these technological advancements and the potential consequences for societal structures and power dynamics. This analysis aligns with the core tenets of Critical Theory, which encourages questioning and examining the underlying assumptions, power structures, and potential consequences of technological advancements (Selwyn, 2004).

Global Connectivity and Inequality: Critical Theory prompts an examination of whether the global connectivity facilitated by smartphones contributes to a more equitable distribution of resources, opportunities, and information or if it exacerbates existing inequalities. Access to such devices and the benefits they bring may not be uniform across different socio-economic groups and regions, leading to potential disparities in resource access and opportunities (Korupp & Szydlik, 2005).

Cultural Hegemony and Commodification: The widespread adoption of lithium-ion-powered devices, especially smartphones, raises questions about cultural homogenization or cultural imperialism. Critical Theory encourages scrutiny of whether certain cultural expressions and content are privileged over others, potentially leading to the dominance of particular narratives, values, or ideologies. Additionally, the role of corporations in shaping and commodifying culture through these devices should be critically evaluated (Bonfils & Askheim, 2014).

Environmental Concerns and Capitalism: Critical Theory draws attention to the environmental impact of lithium-ion batteries, emphasizing the potential contradictions between technological advancements, consumerism, and ecological sustainability. The production and disposal of electronic devices, including smartphones, contribute to electronic waste and environmental degradation, prompting questions about the sustainability of the current technological paradigm and the extent to which capitalist structures prioritize profit over environmental well-being (L. Wang et al., 2019).

Digital Divide and Access Inequities: Critical Theory encourages an examination of the digital divide, considering who has access to lithium-ion-powered devices and the internet, and who is excluded. The root causes of these disparities, including economic inequalities, structural barriers, and the role of corporate interests, should be critically analysed to understand the implications for societal structures and power relations (Buchinskaia & Stremousova, 2021).

Surveillance and Privacy Concerns: While the text highlights the transformative impact of smartphones on various aspects of life, including entertainment, education, and healthcare, Critical Theory prompts an exploration of the potential downsides, such as increased surveillance, erosion of privacy, and the commodification of personal data by tech companies. This analysis aligns with Critical Theory's focus on questioning power dynamics and potential consequences of technological advancements (Schneider-Kamp & Askegaard, 2019).

Labour and Capital in the Technology Industry: Critical Theory encourages an analysis of the labour conditions and power relations within the technology industry, including the working conditions of those involved in the production of lithium-ion batteries and devices. This analysis aligns with Critical Theory's emphasis on examining power structures and labour-capital dynamics within technological advancements (Schneider-Kamp & Askegaard, 2019).

Resistance and Empowerment: Critical Theory acknowledges the potential for resistance and empowerment and would be interested in how individuals and communities use these devices not just as consumers but also as agents of change. The possibilities and limitations of these tools in fostering social and political transformation should be critically explored, aligning with Critical Theory's focus on transformative change and social justice (Salazar et al., 2022).

Postmodernism

Postmodernism, a philosophical and cultural movement that emerged in the mid-20th century, represents a departure from modernism and is characterized by scepticism towards grand theories and ideologies, a rejection of absolute truths, and a general suspicion of overarching metanarratives (Stock, 2020). Postmodernism challenges the idea of universal narratives and emphasizes the existence of multiple, localized, and often conflicting narratives (Stock, 2020). It involves breaking down traditional structures, embracing fragmentation and deconstruction, and challenging the idea of a singular, stable reality (Stock, 2020). Postmodernism also questions traditional sources of authority and expertise, highlighting the role of power dynamics and subjectivity in shaping what is considered valid knowledge (Stock, 2020). Additionally, it is associated with epistemic relativism, suggesting that truth and meaning are contingent on individual perspectives, cultural contexts, and historical circumstances (Stock, 2020).

Furthermore, postmodern works often display a sense of playfulness, irony, and self-awareness, using parody and pastiche to subvert traditional forms and challenge established norms (Stock, 2020). The concept of hyperreality is introduced, blurring the line between reality and representation, where simulacra are given significance, making it difficult to discern what is real and what is simulated (Stock, 2020). Postmodernism also acknowledges and celebrates cultural diversity and hybridity, recognizing that cultures are constantly evolving through interactions and exchanges (Stock, 2020). Moreover, it critiques consumer culture and the commodification of art and culture, exploring how consumerism influences identity, representation, and social relationships (Stock, 2020).

The movement has faced criticism for its perceived relativism and scepticism, and the term "postmodernism" itself is sometimes contested, with scholars using alternative terms such as "poststructuralism" or "late capitalism" to capture specific aspects of the movement. Postmodernism has had a profound impact on various disciplines, including philosophy, literature, art, architecture, and cultural theory.

The transformative impact of lithium-ion-powered devices on various aspects of society, including communication, education, entertainment, and healthcare, aligns with several postmodern themes and considerations. The use of these devices has led to a fragmentation of communication channels and cultural expressions, contributing to a multiplicity of voices and perspectives, which resonates with postmodern ideas of fragmentation and rejection of grand narratives in favour of diverse, localized narratives (Zuo et al., 2017). Additionally, lithium-ion-powered smartphones have played a role in breaking down geographical barriers, enabling global communication and interconnectedness, aligning with postmodern notions of deconstructing traditional boundaries and fostering a more fluid, interconnected world (Zuo et al., 2017). The impact of lithium-ion devices on cultural exchange and the preservation of diverse cultural practices also aligns with postmodern ideas of cultural pluralism, contributing to a more inclusive representation of cultural diversity (Zuo et al., 2017).

Furthermore, the text emphasizes the role of lithium-ion devices in transforming entertainment, creating a shift from traditional media consumption to on-the-go, personalized experiences, reflecting aspects of hyperreality, where the boundaries between the real and the simulated become blurred (Zuo et al., 2017). While the text emphasizes the positive aspects of lithium-ion-powered devices, a postmodern perspective might encourage a critical examination of the potential negative effects, such as questioning the impact on mental health, privacy concerns, and the environmental implications of electronic waste (Zuo et al., 2017). This critical examination aligns with postmodern scepticism and critique.

The global impact of lithium-ion devices on communication, education, and cultural exchange, as well as their role in addressing local challenges, resonates with postmodern considerations of navigating multiple scales simultaneously (Zuo et al., 2017). Additionally, the section discussing the environmental impact of lithium-ion batteries introduces a critical dimension to the narrative, emphasizing the importance of responsible disposal and recycling in a more sustainable, circular economy, which aligns with postmodern concerns about consumerism and environmental sustainability (Zuo et al., 2017).

Structuration Theory

Structuration Theory, developed by sociologist Anthony Giddens, is a social theory that aims to elucidate the relationship between social structure and individual agency (Giddens, 1979). The theory, introduced in Giddens' 1979 book "Central Problems in Social Theory," revolves around the concept of the duality of structure, which posits that structure and agency are not separate entities but are intertwined (Giddens, 1979). According to , structures are not external forces that act upon individuals but are produced and reproduced through the actions of individuals, emphasizing that structures and agency are two sides of the same coin (Giddens, 1979).

In the context of Structuration Theory, structures are the enduring patterns of social relations and practices that shape social life, encompassing institutions, norms, rules, and cultural frameworks that provide the context for social interactions (Giddens, 1979). These structures provide the "rules of the game" in society, influencing and constraining the actions of individuals (Giddens, 1979). On the other hand, agency refers to the capacity of individuals to act, make choices, and influence social outcomes (Giddens, 1979). Giddens argues that individuals are not simply constrained by social structures; they actively contribute to the creation and reproduction of these structures through their everyday actions (Giddens, 1979).

The relationship between structure and agency is recursive, signifying that each influences and shapes the other in an ongoing and dynamic process (Giddens, 1979). Individuals draw upon existing structures as they act, but their actions also contribute to the reproduction or transformation of those structures over time (Giddens, 1979). Giddens also emphasizes the importance of time and space in social life, highlighting that the duality of structure operates across both time and space, and social practices are influenced by historical structures as well as the ongoing reproduction of structures in the present (Giddens, 1979).

The structuration process involves the continuous interplay between rules and resources (structure) and the actions of individuals (agency) (Giddens, 1979). The reproduction of social structures relies on the routine practices of individuals, and any change in those structures is contingent upon changes in those practices (Giddens, 1979). Moreover, Structuration Theory recognizes that social actions can have unintended consequences, and the reproduction of structures may lead to outcomes that were not originally intended or anticipated by individuals (Giddens, 1979).

From a Structuration Theory perspective, the advent of lithium-ion-powered smartphones has significantly transformed global communication and societal dynamics. Structuration Theory underscores the interdependent relationship between individuals and the social structures in which they are embedded, highlighting the co-constitutive nature of human agency and social systems (Egid et al., 2021).

Lithium-ion-powered smartphones have not only revolutionized communication but have also actively influenced and been influenced by the social structures within which they operate. These devices have contributed to the emergence of new social norms, communication patterns, and interconnected networks, showcasing the duality of structure as they shape and are shaped by societal dynamics (Yoon & Yun, 2021). The global impact of lithium-ion-powered smartphones is rooted in their ability to transcend geographical boundaries, enabling instant communication and fostering a sense of global interconnectedness. Individuals, through their agency, utilize these devices to engage in real-time conversations, disseminate news, and reshape social interactions, thereby exemplifying the structuration processes inherent in the reciprocal relationship between human agency and social structures (Mandić & Praničević, 2019).

Furthermore, these smartphones have facilitated the structuration of disaster management by serving as crucial communication tools during emergencies. The reciprocal relationship between individuals and the structure of disaster response is evident as users leverage their agency to coordinate relief efforts and seek help, thereby demonstrating the active role of these devices in shaping the structuration of disaster management (Sobianowska-Turek et al., 2021). In addition, the widespread use of lithium-ion-powered devices has led to the structuration of knowledge democratization, as these devices have become structures that facilitate access to information, educational resources, and various services. Individuals, through their agency, pursue self-directed learning, stay informed, and access opportunities that were previously limited, thereby contributing to the reciprocal influence between human agency and the structuration of knowledge dissemination (Ishimaru et al., 2022).

Moreover, within the realm of entertainment, the structuration of content consumption has undergone a transformative process due to the portability and accessibility of smartphones powered by lithium-ion batteries. These devices have become structures that enable on-the-go entertainment, streaming services, and gaming accessibility, with users actively engaging with these structures and shaping trends, preferences, and the nature of entertainment (Gorman, 2016). Furthermore, wearable devices powered by lithium-ion batteries have acted as structures that individuals use to monitor and manage their health, thereby contributing to the structuration of health and well-being. Users actively engage with these devices, promoting preventive health measures and a conscious approach to well-being, thus exemplifying the reciprocal relationship between human agency and the structuration of health management (Avolio et al., 2022).

The structuration of accessibility for individuals with disabilities has also been influenced by lithium-ion-powered devices, as these devices have become structures that enable assistive technologies, opening up new avenues for employment, navigation, and customization based on individual needs. Legislative measures and advocacy efforts further contribute to the reciprocal shaping of these structures, highlighting the active role of these devices in shaping societal structures. Additionally, the ubiquity of lithium-ion-powered devices has influenced the structuration of digital literacy, encompassing education, skill development, and societal engagement. These devices have become structures that individuals engage with, fostering an adaptive mindset, continuous learning, and global connectivity, thereby reflecting the reciprocal nature of the structuration of digital literacy (Wilson, 2021).

Cultural and social expression have also undergone a structuration process as smartphones have become structures that democratize storytelling, promote cultural exchange, and contribute to social activism. The reciprocal relationship is evident as users actively participate in shaping the cultural narrative and expressing their identities, showcasing the active role of these devices in the structuration of cultural and social dynamics (Richter, 2023). Furthermore, the structuration of societal responses to the potential environmental risks associated with lithium-ion batteries has highlighted the collective effort to address environmental concerns. Initiatives, regulations, and innovations have become structures that individuals engage with, reflecting the reciprocal nature of societal responses to environmental challenges, thereby emphasizing the active role of these devices in shaping environmental awareness and responsible disposal practices (Leonard et al., 2023).

World-Systems Theory

World-Systems Theory, developed by Immanuel Wallerstein in the 1970s, offers a comprehensive framework for understanding the interconnected nature of the world, shaped by economic, political, and cultural factors (Lawson & Wallerstein, 1978). This theory categorizes nations into core, semi-peripheral, and peripheral regions, each playing distinct roles in the global economy (Lawson & Wallerstein, 1978). Core nations are economically dominant and technologically advanced, exerting control over global economic systems, while peripheral nations, being less economically developed, often provide raw materials and cheap labour, making them vulnerable to exploitation (Lawson & Wallerstein, 1978). Semi-peripheral nations occupy an intermediate position, sharing characteristics of both core and peripheral nations, and may exploit peripheral nations while being exploited by core nations (Lawson & Wallerstein, 1978).

World-Systems Theory emphasizes historical development and change over time, tracing the evolution of the modern world-system from the 16th century, with the rise of capitalism and the development of a global division of labour (Lawson & Wallerstein, 1978). It argues that there is a systematic and structured inequality in the global exchange of goods and services, with core nations benefiting more from trade than peripheral nations (Lawson & Wallerstein, 1978). This unequal exchange perpetuates dependency and exploitation, as peripheral nations rely on core nations for technology, capital, and market access (Lawson & Wallerstein, 1978). The theory also highlights the cyclical nature of world-systems, influenced by economic, political, and social factors, and underscores that global inequality is deeply embedded in the structure of the world-system, rather than being a result of cultural differences or individual decisions (Lawson & Wallerstein, 1978).

Furthermore, World-Systems Theory places a strong emphasis on the capitalist system as a driving force behind the world-system, suggesting that capitalism inherently leads to the creation of a global economic structure with core, semi-peripheral, and peripheral components (Lawson & Wallerstein, 1978). This perspective conceptualizes the world as a single interacting unit, a world-economy, where economic relationships between nations are fundamental in shaping global dynamics (Lawson & Wallerstein, 1978).

From a World-Systems Theory perspective, the proliferation of lithium-ion-powered devices, particularly smartphones, has catalysed a global transformation, transcending conventional boundaries and shaping interconnected dynamics within the world-system. These devices, driven by capitalist structures, have become central to the global division of labour, contributing to a complex web of economic, social, and cultural interactions.

The advent of lithium-ion-powered smartphones has reinforced the global hierarchy, with technologically advanced regions (core nations) at the forefront of innovation and production, while less economically developed areas (peripheral nations) often serve as suppliers of raw materials or labour for the production processes (Lawson & Wallerstein, 1978). This reinforces the core-periphery dynamics, where core nations benefit dispro-portionately from the global trade in these devices, perpetuating an unequal exchange that favours the core at the expense of the periphery (Lawson & Wallerstein, 1978).

World-Systems Theory underscores the cyclical nature of global systems, and the evo-lution of lithium-ion-powered devices fits into a historical trajectory, marking phases of expansion and contraction within the world-system (Lawson & Wallerstein, 1978). The inequality in the global distribution and use of lithium-ion-powered devices is deeply rooted in the structural framework of the world-system, emphasizing that it is not merely a result of individual choices or cultural differences (Lawson & Wallerstein, 1978).

Lithium-ion-powered devices, driven by capitalist forces, have become integral com-ponents of the world-economy, shaping the global structure (Lawson & Wallerstein, 1978). While these devices contribute to digital inclusion and cultural exchange, their impact is not evenly distributed, potentially perpetuating existing global inequalities (Lawson & Wallerstein, 1978). Core nations often benefit more, potentially overshad-owing the cultural expressions of peripheral nations (Lawson & Wallerstein, 1978).

The environmental impact of lithium-ion batteries becomes a global issue, necessi-tating coordinated efforts and global collaboration (Nurqomariah & Fajaryanto, 2018a). Regulations, recycling programs, and innovations in sustainable design emerge as re-sponses to the ecological challenges posed by electronic waste (Nurqomariah & Fajaryan-to, 2018a).

Network Society Theory

The Network Society Theory, developed by sociologist Manuel Castells, offers a com-prehensive framework for understanding the societal transformations resulting from the widespread integration of information and communication technologies (ICTs) (Robot-ham, 2005). This theory emphasizes the central role of ICTs, particularly the internet, in shaping social, economic, and cultural processes (Robotham, 2005). It highlights the shift from hierarchical and centralized organizational structures to decentralized and flexible forms, where networks play a crucial role in shaping social structures (Robotham, 2005). Castells introduces the concept of "informational capitalism," emphasizing the centrality of information in economic activities and the prominence of knowledge-based industries (Robotham, 2005). Moreover, the theory acknowledges the influence of ICTs on cultural production, consumption, and identity formation, as well as the coexistence of global and local dynamics in the networked world (Robotham, 2005).

The theory also addresses power shifts, resistance, and activism facilitated by networked communication, recognizing the emergence of new forms of power and the potential for resistance and social movements using digital platforms (Robotham, 2005). Furthermore, it explores the fluid and fragmented nature of identity in the Network Society, where individuals can construct multiple identities in various online spaces (Robotham, 2005). Additionally, the Network Society is characterized by spatial and temporal flexibility, enabling individuals to connect and communicate across geographical distances with less constraint by traditional schedules (Robotham, 2005).

The integration of lithium-ion-powered smartphones has significantly transformed global communication and connectivity, aligning with the principles of Network Society Theory proposed by Castells. These devices have transcended conventional boundaries, fostering a more interconnected, informed, and dynamic global society (Simon & Gogotsi, 2008). Simon and Gogotsi (2008) emphasize the pivotal role of lithium-ion-powered smartphones in reshaping communication dynamics, facilitating instant global connections, and breaking down geographical barriers (Simon & Gogotsi, 2008). Furthermore, Xu (2014) highlights the empowerment of users through internet connectivity, enabling real-time conversations and rapid dissemination of news and events through various communication channels (Simon & Gogotsi, 2008). This aligns with the Network Society's emphasis on decentralized and flexible organization, breaking traditional barriers, and potentially bridging the digital divide (Simon & Gogotsi, 2008).

In addition to individual communication, the widespread adoption of lithium-ion-powered devices contributes to digital inclusion, providing access to communication tools, educational resources, and employment opportunities, particularly in remote or economically disadvantaged areas (Simon & Gogotsi, 2008). This reflects the Network Society's broader socio-economic impacts and emphasis on breaking down traditional barriers (Simon & Gogotsi, 2008). Moreover, these devices reshape traditional workplace dynamics, promoting remote work and global business relationships, illustrating the spatial and temporal flexibility characteristic of the Network Society (Simon &

Gogotsi, 2008). The rise of digital nomadism exemplifies the transformative impact on work practices and the emergence of a global community with diverse work environments (Simon & Gogotsi, 2008).

The democratization of knowledge through the proliferation of lithium-ion-powered devices significantly transforms various aspects of society, facilitating access to information, educational resources, news updates, entertainment, language and skill development tools, healthcare information, financial services, job opportunities, and environmental awareness (Simon & Gogotsi, 2008). This aligns with the Network Society's emphasis on the central role of information in economic activities (Simon & Gogotsi, 2008). Furthermore, the integration of lithium-ion batteries into laptops and tablets enhances the mobility of professionals, contributing to a remote work revolution and transforming the dynamics of knowledge-based industries (Simon & Gogotsi, 2008).

In the realm of entertainment and leisure, lithium-ion-powered devices revolutionize the landscape, enabling on-the-go entertainment, streaming services, gaming accessibility, augmented reality (AR), virtual reality (VR), and personalized content consumption (Simon & Gogotsi, 2008). The integration of AR and VR technologies illustrates the transformative impact of these devices on cultural trends and shared experiences in the Network Society (Simon & Gogotsi, 2008). Additionally, the impact of lithium-ion batteries extends to wearable health devices, contributing to the Network Society's focus on individual well-being and self-monitoring, aligning with the role of these devices in fostering a holistic understanding of health and well-being (Simon & Gogotsi, 2008).

Moreover, the integration of lithium-ion batteries in assistive technologies promotes accessibility for individuals with disabilities, aligning with the Network Society's commitment to inclusivity (Simon & Gogotsi, 2008). These devices enable individuals with disabilities to interact with digital content, access educational apps, and participate in the workforce, contributing to increased accessibility and employment opportunities. Digital literacy and skill development, essential components of the Network Society, are significantly influenced by the ubiquity of portable devices powered by lithium-ion batteries, fostering an adaptive mindset, enabling interactive and dynamic learning experiences, and contributing to the development of proficiencies relevant in the digital age (Simon & Gogotsi, 2008).

In the context of cultural and social expression, lithium-ion-powered smartphones empower individuals to document, share, and celebrate their cultural heritage, contributing to a more interconnected global community (Simon & Gogotsi, 2008). The Network

Society's emphasis on diverse voices, cultural exchange, and digital activism aligns with the transformative impact of these devices on storytelling, cultural preservation, and social advocacy. However, the Network Society perspective also acknowledges challenges, such as the environmental impact of lithium-ion batteries, giving rise to concerns about electronic waste (e-waste). Initiatives, regulations, and innovations aimed at responsible disposal and recycling reflect the Network Society's commitment to addressing environmental concerns (Simon & Gogotsi, 2008).

Actor-Network Theory

Actor-Network Theory (ANT) is a theoretical framework that challenges traditional notions of agency, structure, and the separation between human and non-human entities (Bruni & Teli, 2007). Developed by French sociologists Michel Callon, Bruno Latour, and John Law in the late 20th century, ANT offers a distinctive perspective on how social order and networks are formed and maintained (Denis et al., 2007). It views the social world as a network of interconnected actors, regardless of whether they are human or non-human, and emphasizes the symmetry of actors, treating both human and non-human entities as equal in terms of their ability to shape the network (Xu et al., 2022). The theory introduces the concept of "translation," which refers to the process through which actors bring others into their networks, involving negotiation, persuasion, and coercion as actors seek to align the interests of different entities (Montenegro & Bulgacov, 2014). Additionally, ANT challenges the anthropocentric view by attributing agency not only to humans but also to non-human elements, such as technologies, objects, and ideas (Walsham, 1997).

The application of ANT has extended to various fields, including science and technology studies, organizational studies, urban studies, and environmental studies (Gareau, 2012). Its strength lies in providing a framework for understanding the complex interplay between human and non-human elements in the formation of socio-technical networks. However, ANT has also faced criticism for its challenging and abstract language, as well as its rejection of broader social structures and historical contexts in favour of localized networks and interactions (Pouloudi et al., 2004).

The Actor-Network Theory (ANT) provides a comprehensive framework for understanding the intricate dynamics and interdependencies involving lithium-ion-powered devices, particularly smartphones, in shaping global communication networks and influencing various aspects of contemporary society. This theory emphasizes the agency of both human and non-human entities, treating these devices as pivotal actors in socio-technical interactions (Simon & Gogotsi, 2008).

The proliferation of lithium-ion-powered devices, such as smartphones, has led to the formation of transformative networks that transcend conventional boundaries, fostering interconnectedness and reshaping global communication (Simon & Gogotsi, 2008). These devices, as active actors within the network, play a crucial role in breaking down geographical barriers and enabling instant communication, thereby contributing to the richness of interactions among users (Simon & Gogotsi, 2008). Social media platforms, messaging apps, and email services act as intermediaries in this network, facilitating real-time conversations and the rapid dissemination of information (Simon & Gogotsi, 2008).

During emergencies or natural disasters, these devices serve as vital intermediaries, allowing users to send and receive real-time updates, seek help, and coordinate relief efforts, highlighting their versatility and importance in critical situations (Simon & Gogotsi, 2008). Furthermore, the ANT perspective acknowledges the role of these devices in digital inclusion, as they enable access to communication tools, educational resources, and job opportunities, potentially bridging the digital divide and reshaping traditional workplace dynamics (Simon & Gogotsi, 2008).

In the realm of education, healthcare, and entertainment, these devices act as intermediaries in accessing information, educational resources, news updates, entertainment, language and skill development tools, healthcare information, financial services, and job opportunities, contributing to a more inclusive digital society and fostering civic engagement (Simon & Gogotsi, 2008). Additionally, they revolutionize content consumption, transform commuting or travel time into opportunities for leisure and relaxation, and enable professionals to work from diverse locations while staying connected and productive (Simon & Gogotsi, 2008).

The integration of these devices into wearable health technologies creates a network that revolutionizes health monitoring and management, leveraging lithium-ion batteries to track various health metrics and promote preventive health measures, thereby contributing to a holistic approach to well-being (Simon & Gogotsi, 2008). Moreover, initiatives promoting responsible disposal practices, recycling programs, and a circular economy are recognized as efforts within the socio-technical network to mitigate the environmental impact of lithium-ion batteries (Simon & Gogotsi, 2008).

References

Abe, T., Miyatake, K., Shimoida, Y., & Horie, H. (2007). Research and Development Work on Lithium-Ion Batteries for Environmental Vehicles. *World Electric Vehicle Journal*.

Acosta, L. A., Magcale-Macandog, D. B., Kumar, K. S. K., Cui, X., Eugenio, E. A., Macandog, P. B. M., Salvacion, A. R., & Eugenio, J. M. A. (2016). The Role of Bioenergy in Enhancing Energy, Food and Ecosystem Sustainability Based on Societal Perceptions and Preferences in Asia. *Agriculture*.

Adamovic, D., Ishiyama, D., Djordjievski, S., Ogawa, Y., Stevanović, Z., Kawaraya, H., Sato, H., Obradovic, L., Marinkovic, V., Petrović, J., & Gardic, V. (2021). Estimation and Comparison of the Environmental Impacts of Acid Mine Drainage-bearing River Water in the Bor and Majdanpek Porphyry Copper Mining Areas in Eastern Serbia. *Resource Geology*.

Adelhelm, P., Hartmann, P., Bender, C. L., Busche, M. R., Eufinger, C., & Janek, J. (2015). From Lithium to Sodium: Cell Chemistry of Room Temperature Sodium–air and Sodium–sulfur Batteries. *Beilstein Journal of Nanotechnology*.

Agusdinata, D. B., Liu, W., Eakin, H., & Romero, H. (2018). Socio-Environmental Impacts of Lithium Mineral Extraction: Towards a Research Agenda. *Environmental Research Letters*.

Agyarko, K., Dartey, E., Kuffour, R. A., & Sarkodie, P. A. (2014). Assessment of Trace Elements Levels in Sediment and Water in Some Artisanal and Small-Scale Mining (ASM) Localities in Ghana. *Current World Environment*.

Ahmed, S., Nelson, P. A., Gallagher, K. G., & Dees, D. W. (2016). Energy Impact of Cathode Drying and Solvent Recovery During Lithium-Ion Battery Manufacturing. *Journal of Power Sources*.

Aji, I. P., & Afianti, H. (2021). Comparison of Lithium Ion and Lithium Polymer Performance as Solar Panel Energy Storage. *Jeecs (Journal of Electrical Engineering and Computer Sciences)*.

Akinbile, B. J., Matsinha, L. C., Ambushe, A. A., & Makhubela, B. C. E. (2021). Catalytic Conversion of CO_2 to Formate Promoted by a Biochar-Supported Nickel Catalyst Sourced From Nickel Phytoextraction Using Cyanogen-Rich Cassava. *Acs Earth and Space Chemistry*.

Al-Ammar, E. A., Malik, N. H., & Usman, M. (2011). Application of Using Hybrid Renewable Energy in Saudi Arabia. *Engineering Technology & Applied Science Research*.

Alblawi, A., Elkholy, M. H., & Talaat, M. (2019). ANN for Assessment of Energy Consumption of 4 kW PV Modules Over a Year Considering the Impacts of Temperature and Irradiance. *Sustainability*.

Albrecht-Schönzart, T. E., Zhang, H., Li, A., Li, K., Wang, Z., Xu, X., Wang, Y., Sheridan, M. V., Hu, H.-S., Xu, C., Alekseev, E. V., Chai, Z., & Wang, S. (2022). Ultrafiltration Separation of Nanoscale Am(VI)-Polyoxometalate Clusters From Lanthanides.

Albuquerque, C. A. d., Mello, C. H. P., Gomes, J. H. d. F., & Santos, V. C. d. (2021). Bibliometric Analysis of Studies Involving E-Waste: A Critical Review. *Environmental Science and Pollution Research*.

Aldoory, L. (2005). A (Re)Conceived Feminist Paradigm for Public Relations: A Case for Substantial Improvement. *Journal of Communication*.

Alemann, A. v., Gruhlich, J., Horwath, I., & Weber, L. (2021). A Plea to Reflect on the Entanglements of Gendered Work Patterns and Digital Technologies. *Gender and Research*.

Ali, M. F., Shi, L., Jamil, I., & Aurangzeb, M. (2019). Research on Charging and Discharging of Lithium Ion Battery Based on Temperature Controlled Technique. *International Journal of Engineering Works*.

Alibaba, M., Pourdarbani, R., Manesh, M. H. K., Herrera-Miranda, I., Gallardo-Bernal, I., & Hernández-Hernández, J. L. (2020). Conventional and Advanced Exergy-Based Analysis of Hybrid Geothermal–Solar Power Plant Based on ORC Cycle. *Applied Sciences*.

Aljarrah, S., Alsabbagh, A., & Al-Mahasneh, M. A. (2022). Selective Recovery of Lithium From Dead Sea End Brines Using UBK10 Ion Exchange Resin. *The Canadian Journal of Chemical Engineering*.

Alkhateeb, E., & Abu-Hijleh, B. (2019). Potential for Retrofitting a Federal Building in the UAE to Net Zero Electricity Building (nZEB). *Heliyon*.

Allied Market Research. (2023). *Lithium-ion Battery Market Research, 2032*. Retrieved 13/12/2023 from

Almanaseer, N., Abbassi, B., Dunlop, C., Friesen, K., & Nestico-Semianiw, E. (2020). Multi-Criteria Analysis of Waste-to-Energy Technologies in Developed and Developing Countries. *Environmental Research Engineering and Management*.

Alola, A. A., & Alola, U. V. (2018). Agricultural Land Usage and Tourism Impact on Renewable Energy Consumption Among Coastline Mediterranean Countries. *Energy & Environment*.

Alonso, E., Pineault, D., & Nassar, N. T. (2022). Streamlined Approach for Assessing Embedded Consumption of Lithium and Cobalt in the United States. *Journal of Industrial Ecology*.

Alves, S. A., Grzebielucka, E. C., Borges, C. F. P., Souza, E. C. F., Marques, J. A., Arrua, M. E. P., & Antunes, S. R. M. (2023). The Brazilian Potential in the Development of Clean Energy Sources: Brief Historical Development and Actions to Mitigate Environmental Impacts.

An, Z. (2021). Research on Residual Life Prediction Method of Lithium Ion Battery for Pure Electric Vehicle. *International Journal of Materials and Product Technology*.

André, G., & Godin, M. (2013). Child Labour, Agency and Family Dynamics. *Childhood*.

Angel, H., Stovall, J. P., Williams, H. M., Farrish, K. W., Oswald, B. P., & Young, J. L. (2018). Surface and Subsurface Tillage Effects on Mine Soil Properties and Vegetative Response. *Soil Science Society of America Journal*.

Ansari, S., Ayob, A., Lipu, M. S. H., Hussain, A., & Saad, M. H. M. (2021). Multi-Channel Profile Based Artificial Neural Network Approach for Remaining Useful Life Prediction of Electric Vehicle Lithium-Ion Batteries. *Energies*.

Apergis, N., & Payne, J. E. (2010). A Panel Study of Nuclear Energy Consumption and Economic Growth. *Energy Economics*.

Aristizábal, L. M., Rúa, S., Gaviria, C. E., Osorio, S. P., Zuluaga, C. A., Posada, N. L., & Vásquez, R. E. (2016). Design of an Open Source-Based Control Platform for an Underwater Remotely Operated Vehicle. *Dyna*.

Armand, M., & Tarascon, J.-M. (2008). Building better batteries. *Nature, 451*(7179), 652-657.

Arnold, M., Hyer, K., Small, B. J., Chisolm, T. H., Saunders, G. H., McEvoy, C. L., Lee, D. H., Dhar, S., & Bainbridge, K. E. (2019). Hearing Aid Prevalence and Factors Related to Use Among Older Adults From the Hispanic Community Health Study/Study of Latinos. *Jama Otolaryngology–head & Neck Surgery*.

Aronowicz, S. (2020). How Does Social Media Negatively Affects the Beginning of Relationships?

Asbar, Y., Biby, S., Razif, R., & Siregar, W. V. (2021). Millennial Generation and Smartphone Purchase Intention. *Management Research and Behavior Journal*.

Ashwin, T. R., Chung, Y. M., & Wang, J. (2016). Capacity Fade Modelling of Lithium-Ion Battery Under Cyclic Loading Conditions. *Journal of Power Sources*.

Ashworth, D. C., Fuller, G. W., Toledano, M. B., Font, A., Elliott, P., Hansell, A., & Hoogh, K. d. (2013). Comparative Assessment of Particulate Air Pollution Exposure From Municipal Solid Waste Incinerator Emissions. *Journal of Environmental and Public Health*.

Avolio, M. L., Komatsu, K. J., Koerner, S. E., Grman, E., Isbell, F., Johnson, D. S., Wilcox, K. R., Alatalo, J. M., Baldwin, A. H., Beierkuhnlein, C., Britton, A. J., Foster, B. L., Harmens, H., Kern, C. C., Li, W., McLaren, J. R., Souza, L., Yu, Q., & Zhang, Y. (2022). Making Sense of Multivariate Community Responses in Global Change Experiments. *Ecosphere*.

Awada, H., Kontopoulou, M., & Docoslis, A. (2022). A Note on Using Expanded Graphite for Achieving Energy- and Time-efficient Production of Graphene Nanoplatelets via Liquid Phase Exfoliation. *The Canadian Journal of Chemical Engineering*.

Azahar, A. A., Jamail, N. A. M., Isa, A. H. M., Sani, F. N. M., Razali, D. H. M., Busu, M. B. M., & Kamarudin, Q. E. (2021). Economical Electricity Home System Using Solar. *Journal of Electronic Voltage and Application*.

Babayomi, O. O., Dahoro, D. A., & Zhang, Z. (2022). Affordable clean energy transition in developing countries: Pathways and technologies. *IScience*, *25*(5), 104178.

Bai, Q., Yang, L., Chen, H., & Mo, Y. (2018). Computational Studies of Electrode Materials in Sodium-Ion Batteries. *Advanced Energy Materials*.

Bai, Y., Essehli, R., Jafta, C. J., Livingston, K., & Belharouak, I. (2021). Recovery of Cathode Materials and Aluminum Foil Using a Green Solvent. *Acs Sustainable Chemistry & Engineering*.

Bak, S. M., Hu, E., Zhou, Y. N., Yu, X., Senanayake, S. D., Cho, S. J., Kim, K. B., Chung, K. Y., Yang, X. Q., & Nam, K. W. (2014). Structural Changes and Thermal Stability of Charged $LiNi_xMn_yCo_zO_2$ Cathode Materials Studied by Combined *In Situ* Time-Resolved XRD and Mass Spectroscopy. *Acs Applied Materials & Interfaces*.

Baldo, L., Vedova, M. D. L. D., Querques, I., & Maggiore, P. (2023). Prognostics of Aerospace Electromechanical Actuators: Comparison Between Model-Based Metaheuristic Methods. *Journal of Physics Conference Series*.

Barbala, A. M. (2022). Transcending Instagram: Affective Swedish Hashtags Taking Intimate Feminist Entanglements From Viral to 'IRL'. *Media Culture & Society*.

Barbano, F. (1968). Social Structures and Social Functions: The Emancipation of Structural Analysis in Sociology*. *Inquiry*.

Bardi, U. (2010). Extracting Minerals From Seawater: An Energy Analysis. *Sustainability*.

Barnhart, C. J., Brandt, A. R., & Benson, S. M. (2013). The Energetic Implications of Curtailing Versus Storing Solar- And Wind-Generated Electricity. *Energy & Environmental Science*.

Barré, A., Deguilhem, B., Grolleau, S., Gérard, M., Suard, F., & Riu, D. (2013). A Review on Lithium-Ion Battery Ageing Mechanisms and Estimations for Automotive Applications. *Journal of Power Sources*.

Barzkar, A., & Ghassemi, M. (2020). Electric Power Systems in More and All Electric Aircraft: A Review. *Ieee Access*.

Bates, J. B. (2000). Thin-Film Lithium and Lithium-Ion Batteries. *Solid State Ionics*.

Bedua, A. B. S. V., Bengan, C. V. P., Erich, P., Goleng, D. J. G., Posanso, R. G. D., Pueblo, C. T., & Abusama, H. (2021). Social Media on the Students' Academic Performance. *Indonesian Journal of Educational Research and Technology*.

Bekalu, M. A., McCloud, R. F., & Viswanath, K. (2019). Association of Social Media Use With Social Well-Being, Positive Mental Health, and Self-Rated Health: Disentangling Routine Use From Emotional Connection to Use. *Health Education & Behavior*.

Bellini, R., Meissner, J. L., Finnigan, S. M., & Strohmayer, A. (2022). Feminist Human–computer Interaction: Struggles for Past, Contemporary and Futuristic Feminist Theories in Digital Innovation. *Feminist Theory*.

Benschop, Y. (2021). Grand Challenges, Feminist Answers. *Organization Theory*.

Bett, H. K., Nguru, F., & Kiruhi, T. M. (2020). Construction of Followership Identity Among Kenyan Teachers. *Industrial and Commercial Training*.

Bianchi, C., Tuzovic, S., & Kuppelwieser, V. G. (2022). Investigating the Drivers of Wearable Technology Adoption for Healthcare in South America. *Information Technology and People*.

Bilgili, F., & Öztürk, İ. (2015). Biomass Energy and Economic Growth Nexus in G7 Countries: Evidence From Dynamic Panel Data. *Renewable and Sustainable Energy Reviews*.

Bird, G., Brewer, P., Macklin, M. G., Nikolova, M., Kotsev, T., Mollov, M., & Swain, C. H. (2009). Dispersal of Contaminant Metals in the Mining-Affected Danube and Maritsa Drainage Basins, Bulgaria, Eastern Europe. *Water Air & Soil Pollution*.

Bode, C., & Wagner, S. M. (2015). Structural Drivers of Upstream Supply Chain Complexity and the Frequency of Supply Chain Disruptions. *Journal of Operations Management*.

Boisvert, L., Turgeon, K., Boulanger, J.-F., Bazin, C., & Houlachi, G. (2020). Recovery of Cobalt From the Residues of an Industrial Zinc Refinery. *Metals*.

Boll, C. (2023). *How Long Do Lithium-ion Batteries Last?* . Retrieved 27/12/2023 from

Bond, T. C., Covert, D. S., Kramlich, J. C., Larson, T. V., & Charlson, R. J. (2002). Primary Particle Emissions From Residential Coal Burning: Optical Properties and Size Distributions. *Journal of Geophysical Research Atmospheres*.

Bonfils, I. S., & Askheim, O. P. (2014). Empowerment and Personal Assistance – Resistance, Consumer Choice, Partnership or Discipline? *Scandinavian Journal of Disability Research*.

Bonilla-Silva, E. (1997). Rethinking Racism: Toward a Structural Interpretation. *American Sociological Review*.

Bouck, E. C., Maeda, Y., & Flanagan, S. (2011). Assistive Technology and Students With High-Incidence Disabilities. *Remedial and Special Education*.

Boxall, N. (2023). *Lithium-ion battery recycling*. CSIRO. Retrieved 4/1/2023 from

Bronner, S. E. (2017). *Critical Theory: A Very Short Introduction*.

Bruce, P. G., Scrosati, B., & Tarascon, J. M. (2008). Nanomaterials for Rechargeable Lithium Batteries. *Angewandte Chemie*.

Bruni, A., & Teli, M. (2007). Reassembling the Social—An Introduction to Actor Network Theory. *Management learning*.

Brusselen, D. V., Kayembe-Kitenge, T., Mbuyi-Musanzayi, S., Kasole, T. L., Ngombe, L. K., Obadia, P. M., Mukoma, D. K. w., Herck, K. V., Avonts, D., Devriendt, K., Smolders, E., Nkulu, C. B. L., & Nemery, B. (2020). Metal Mining and Birth Defects: A Case-Control Study in Lubumbashi, Democratic Republic of the Congo. *The Lancet Planetary Health*.

Bryant, B. R., & Seay, P. C. (2020). Republication of the Technology-Related Assistance to Individuals With Disabilities Act: Relevance to Individuals With Learning Disabilities and Their Advocates. *Journal of Learning Disabilities*.

Buchinskaia, O., & Stremousova, E. (2021). Inequality of Digital Access Between Russian Regions. *E3s Web of Conferences*.

Bugga, R. V., Smart, M. C., & Whitcanack, L. D. (2010). Storage Characteristics of Lithium-Ion Cells. *Ecs Transactions*.

Buqa, H., Goers, D., Holzapfel, M., Spahr, M. E., & Novák, P. (2005). High Rate Capability of Graphite Negative Electrodes for Lithium-Ion Batteries. *Journal of the Electrochemical Society*.

Burbank, P. M., & Martins, D. C. (2010). Symbolic Interactionism and Critical Perspective: Divergent or Synergistic? *Nursing Philosophy*.

Burger, J., Gochfeld, M., Clarke, J. H., Powers, C. W., & Kosson, D. S. (2013). An Ecological Multidisciplinary Approach to Protecting Society, Human Health, and the Environment at Nuclear Facilities. *Remediation Journal*.

Burns, J. T., Stevens, D. A., & Dahn, J. R. (2015). In-Situ Detection of Lithium Plating Using High Precision Coulometry. *Journal of the Electrochemical Society*.

Busarac, N., Adamović, D., Grujović, N., & Živić, F. (2022). Lightweight Materials for Automobiles. *Iop Conference Series Materials Science and Engineering*.

Byrne, K., Hawker, W., & Vaughan, J. (2017). Effect of Key Parameters on the Selective Acid Leach of Nickel From Mixed Nickel-Cobalt Hydroxide.

Cai, P., Zou, K., Deng, X., Wang, B., Min, Z., Li, L., Hou, H., Zou, G., & Ji, X. (2021). Comprehensive Understanding of Sodium-Ion Capacitors: Definition, Mechanisms, Configurations, Materials, Key Technologies, and Future Developments. *Advanced Energy Materials*.

Cai, W., Zhang, Y., Li, J., Sun, Y., & Cheng, H. (2014). Single-Ion Polymer Electrolyte Membranes Enable Lithium-Ion Batteries With a Broad Operating Temperature Range. *Chemsuschem*.

Calisaya-Azpilcueta, D., Herrera-León, S., Lucay, F. A., & Cisternas, L. A. (2020). Assessment of the Supply Chain Under Uncertainty: The Case of Lithium. *Minerals*.

Calvin, K., Cowie, A., Berndes, G., Arneth, A., Cherubini, F., Portugal-Pereira, J., Grassi, G., House, J., Johnson, F. X., Popp, A., Rounsevell, M., Slade, R., & Smith, P. (2021). Bioenergy for Climate Change Mitigation: Scale and Sustainability. *GCB Bioenergy*.

Camargos, P. H., Santos, P. H. J. D., Santos, I. R., Ribeiro, G. S., & Caetano, R. E. (2022). Perspectives On<scp>Li-ion</Scp>battery Categories for Electric Vehicle Applications: A Review of State of the Art. *International Journal of Energy Research*.

Can, H., & Korkmaz, Ö. (2019). The Relationship Between Renewable Energy Consumption and Economic Growth. *International Journal of Energy Sector Management*.

Cao, J., Chow, J. C., Lee, S. C., Li, Y., Chen, S. W., An, Z., Fung, K., Watson, J. G., Zhu, C., & Liu, S. X. (2005). Characterization and Source Apportionment of Atmospheric Organic and Elemental Carbon During Fall and Winter of 2003 in Xi'an, China. *Atmospheric Chemistry and Physics*.

Cao, X., Dreisinger, D., Lü, J., & Belanger, F. (2017). Electrorefining of High Purity Manganese. *Hydrometallurgy*.

Carrington, S., & Selva, G. (2010). Critical Social Theory and Transformative Learning: Evidence in Pre-service Teachers' Service-learning Reflection Logs. *Higher Education Research & Development*.

Carter, B., Rees, P., Hale, L., Bhattacharjee, D., & Paradkar, M. (2016). Association Between Portable Screen-Based Media Device Access or Use and Sleep Outcomes. *Jama Pediatrics*.

Carter, M. J., & Fuller, C. (2016). Symbols, Meaning, and Action: The Past, Present, and Future of Symbolic Interactionism. *Current Sociology*.

Catsaros, O. (2023, 1/1/2024). Lithium-Ion Battery Pack Prices Hit Record Low of $139/kWh.

Çavuş, S., Akman, E., & Ayhan, B. (2018). Transformation of Secrecy and Privacy: Social Media Behavior of Turkish and Kyrgyz Students. *Selçuk Üniversitesi Türkiyat Araştırmaları Dergisi*.

Čepin, M. (2019). Evaluation of the Power System Reliability if a Nuclear Power Plant Is Replaced With Wind Power Plants. *Reliability Engineering & System Safety*.

Cerrillo-González, M. d. M., Villén-Guzmán, M., Acedo-Bueno, L. F., Rodríguez-Maroto, J. M., & Paz-Garcia, J. M. (2020). Hydrometallurgical Extraction of Li and Co From LiCoO2 Particles–Experimental and Modeling. *Applied Sciences*.

Chagnes, A., & Pośpiech, B. (2013). A Brief Review on Hydrometallurgical Technologies for Recycling Spent Lithium-ion Batteries. *Journal of Chemical Technology & Biotechnology*.

Chang, J., Huang, Q., Gao, Y., & Zheng, Z. (2021). Pathways of Developing High-Energy-Density Flexible Lithium Batteries. *Advanced Materials*.

Chang, T., Gatwabuyege, F., Gupta, R., Inglesi-Lotz, R., Manjezi, N., & Simo-Kengne, B. D. (2014). Causal Relationship Between Nuclear Energy Consumption and Economic Growth in G6 Countries: Evidence From Panel Granger Causality Tests. *Progress in Nuclear Energy*.

Chaudhury, D. R. (2023). China's lithium battery industry detrimental for its economy. *The Economic Times*.

Cheema, H. A. (2023). A Review on Bioremediation - An Emerging Technology for Treatment of Radionuclide Waste. *Journal of Modern Agriculture and Biotechnology*.

Chen-Glasser, M., & DeCaluwe, S. C. (2022). A Review on the Socio-Environmental Impacts of Lithium Supply for Electric Aircraft. *Frontiers in Aerospace Engineering*.

Chen, C., Jiang, J., He, W., Lei, W., Hao, Q., & Zhang, X. (2020). 3D Printed High-Loading Lithium-Sulfur Battery Toward Wearable Energy Storage. *Advanced Functional Materials*.

Chen, C. L., & Tsai, C.-H. (2015). Marine Environmental Awareness Among University Students in Taiwan: A Potential Signal for Sustainability of the Oceans. *Environmental Education Research*.

Chen, G., Liu, Y., Yang, Y., & Wang, B. (2023). A Temperature and Voltage Coupling Equivalent Electrical Behavior Model of Lithium-Ion Battery Pack for Electric Unmanned Aerial Vehicle Under Variable Load Conditions. *Energy Technology*.

Chen, I. C., Hamano, H., Iwasaki, H., & Yamada, K. (2019). An Economic-Environmental Analysis of Lithium Ion Batteries Based on Process Design and a Manufacturing Equipment Database. *Journal of Chemical Engineering of Japan*.

Chen, J., Gao, F., Li, X., Yang, K., Wang, S., & Yang, R. (2017). The Study of the Toxicity of the Gas Released on Lithium Ion Battery During Combustion.

Chen, K. H., & Ding, Z. (2015). Lithium-Ion Battery Lifespan Estimation for Hybrid Electric Vehicle.

Chen, L., Du, W., He, Y., Wang, Q., Zhao, W., & Cao, Z. (2022). The Influence Characteristics of Open-Pit Coal Mining on Groundwater Level in Baorixile.

Chen, M., Ma, X., Chen, B., Arsenault, R., Karlson, P., Simon, N. L., & Wang, Y. (2019). Recycling End-of-Life Electric Vehicle Lithium-Ion Batteries. *Joule*.

Chen, P. T., Zeng, F.-Y., Zhang, X.-H., Chung, R. J., Yang, C. J., & Huang, K. (2020). Composite Sinusoidal Waveform Generated by Direct Digital Synthesis for Healthy Charging of Lithium-Ion Batteries. *Energies*.

Chen, P. Y., Yan, C., Chen, P., Zhang, R., Yao, Y., Peng, H. J., Yan, L. T., Kaskel, S., & Zhang, Q. (2021). Selective Permeable Lithium-Ion Channels on Lithium Metal for Practical Lithium–Sulfur Pouch Cells. *Angewandte Chemie*.

Chen, T., Jin, Y., Lv, H., Yang, A., Liu, M., Chen, B., Xie, Y., & Chen, Q. (2020). Applications of Lithium-Ion Batteries in Grid-Scale Energy Storage Systems. *Transactions of Tianjin University*, *26*(3), 208-217.

Chen, T., Zhang, X., Wang, J., Li, J., Wu, C., Hu, M., & Bian, H. (2020). A Review on Electric Vehicle Charging Infrastructure Development in the UK. *Journal of Modern Power Systems and Clean Energy*.

Chen, W. S., Lee, C. H., & Ho, H. J. (2018). Purification of Lithium Carbonate From Sulphate Solutions Through Hydrogenation Using the Dowex G26 Resin. *Applied Sciences*.

Chen, X., Huang, H., Pan, L., Liu, T., & Niederberger, M. (2019). Fully Integrated Design of a Stretchable Solid-State Lithium-Ion Full Battery. *Advanced Materials*.

Chen, X. Y., Li, Q., Jiang, S., Zhang, M., Fu, L., & Shen, B. (2022). A Novel Combo-transmission System of Cold Energy and Electricity for Aluminium Profile Production: Using Liquid Nitrogen and Superconductor Technologies. *Energy Science & Engineering*.

Chen, Z., Li, J., Wu, M., & Liao, J. (2021). High Stability Gel Electrolytes for Long Life Lithium Ion Solid State Supercapacitor. *E3s Web of Conferences*.

Chen, Z., Li, L.-F., Cui, W., Yang, S., Wang, Y., & Wang, D. (2023). Remaining Useful Life Prognostics of Lithium-Ion Batteries Based on a Coordinate Reconfiguration of Degradation Trajectory and Multiple Linear Regression. *Frontiers in Energy Research*.

Chen, Z., Li, W., Yang, J., Liao, J., Chen, C., Song, Y., Shah, S. A. A., Xu, Z., & Wu, M. (2020). Excellent Electrochemical Performance of Potassium Ion Capacitor Achieved by a High Nitrogen Doped Activated Carbon. *Journal of the Electrochemical Society*.

Cheng, F., Liang, J., Tao, Z., & Chen, J. (2011). Functional Materials for Rechargeable Batteries. *Advanced Materials*.

Cheng, X. B., Lin, H., Yuan, H., Peng, H. J., Tang, C., Huang, J. Q., & Zhang, Q. (2021). A Perspective on Sustainable Energy Materials for Lithium Batteries. *Susmat*.

Chernoff, C. (2019). The sociological imagination in the era of thought experiments. *The British Journal of Sociology, 70*(5), 2169-2175.

Cheung, M. L., Pires, G. D., & Rosenberger, P. J. (2020). The Influence of Perceived Social Media Marketing Elements on Consumer–brand Engagement and Brand Knowledge. *Asia Pacific Journal of Marketing and Logistics*.

Chien, W. W. (2012). Prevalence of Hearing Aid Use Among Older Adults in the United States. *Archives of Internal Medicine*.

Chitrakar, R., Makita, Y., Ooi, K., & Sonoda, A. (2014). Lithium Recovery From Salt Lake Brine by H_2TiO_3. *Dalton Transactions*.

Choi, A. (2019). Microscopic Discourse on Democratic Knowledge: The Knowledge of Freedom Unto Equality.

Chow, A. T. (2022). Proactive Approach to Minimize Lithium Pollution. *Journal of Environmental Quality*.

Chun, H., Kim, J., Yu, J., & Han, S. (2020). Real-Time Parameter Estimation of an Electrochemical Lithium-Ion Battery Model Using a Long Short-Term Memory Network. *Ieee Access*.

Ciez, R. E., & Whitacre, J. (2016). The Cost of Lithium Is Unlikely to Upend the Price of Li-Ion Storage Systems. *Journal of Power Sources*.

Cohen, L. E., & Felson, M. (1979). Social Change and Crime Rate Trends: A Routine Activity Approach. *American Sociological Review*.

Cooke, S. J., Venturelli, P. A., Twardek, W. M., Lennox, R. J., Brownscombe, J. W., Skov, C., Hyder, K., Suski, C. D., Diggles, B. K., Arlinghaus, R., & Danylchuk, A. J. (2021). Technological Innovations in the Recreational Fishing Sector: Implications for Fisheries Management and Policy. *Reviews in Fish Biology and Fisheries*.

Costello, K. L., & Floegel, D. (2021). The Potential of Feminist Technoscience for Advancing Research in Information Practice. *Journal of Documentation*.

Crawford, I. (2022). *How much CO2 is emitted by manufacturing batteries?* Climate Portal. Retrieved 28/12/2023 from

Crawford, S. E., Kim, K. J., & Baltrus, J. P. (2022). A Portable Fiber Optic Sensor for the Luminescent Sensing of Cobalt Ions Using Carbon Dots. *Journal of Materials Chemistry C*.

Crespo-Cebada, E., Díaz-Caro, C., Nevado Gil, M. T., & Mirón Sanguino, Á. S. (2020). Does water pollution influence willingness to accept the installation of a mine near a city? Case study of an open-pit lithium mine. *Sustainability*, *12*(24), 10377.

Crudden, A., Antonelli, K., & O'Mally, J. (2017). A Customized Transportation Intervention for Persons With Visual Impairments. *Journal of Visual Impairment & Blindness*.

Cui, C., Sun, C., Liu, Y., Jiang, X., & Chen, Q. (2019). Determining Critical Risk Factors Affecting Public-private Partnership Waste-to-energy Incineration Projects in China. *Energy Science & Engineering*.

Dahunsi, F. M., Abdul-Lateef, O. A., Melodi, A. O., Ponnle, A. A., Sarumi, O. A., & Adedeji, K. A. (2022). Smart Grid Systems in Nigeria: Prospects, Issues, Challenges and Way Forward. *Fuoye Journal of Engineering and Technology*.

Dai, Q., Kelly, J. G., Gaines, L., & Wang, M. (2019). Life Cycle Analysis of Lithium-Ion Batteries for Automotive Applications. *Batteries*.

Danish, M., Ali, S., Ahmad, M., & Zahid, H. (2019). The Influencing Factors on Choice Behavior Regarding Green Electronic Products: Based on the Green Perceived Value Model. *Economies*.

Daryanani, S. (2022). Lithium-Ion Batteries for Electric Vehicles. *Power Electronic News*, *November, 2022*.

Datta, R. P. (2009). Critical Theory and Social Justice: Review of Honneth's Pathologies of Reason: On the Legacy of Critical Theory. *Studies in Social Justice*.

Davis, S. J., & Haltiwanger, J. (1992). Gross Job Creation, Gross Job Destruction, and Employment Reallocation. *The Quarterly Journal of Economics*.

Day, J. N., & Huefner, D. S. (2003). Assistive Technology: Legal Issues for Students With Disabilities and Their Schools. *Journal of Special Education Technology*.

Delnavaz, A., & Voix, J. (2014). Energy Harvesting for in-Ear Devices Using Ear Canal Dynamic Motion. *Ieee Transactions on Industrial Electronics*.

Demiralay, S., & Kılınçarslan, E. (2019). The Impact of Geopolitical Risks on Travel and Leisure Stocks. *Tourism Management*.

Demirel, Y. (2018). Feasibility of Power and Methanol Production by an Entrained-Flow Coal Gasification System. *Energy & Fuels*.

Deng, Z., Mo, Y., & Ong, S. P. (2016). Computational Studies of Solid-State Alkali Conduction in Rechargeable Alkali-Ion Batteries. *NPG Asia Materials*.

Denholm, P., Ela, E., Kirby, B., & Milligan, M. (2010). Role of Energy Storage With Renewable Electricity Generation.

Denis, J. L., Langley, A., & Rouleau, L. (2007). Strategizing in Pluralistic Contexts: Rethinking Theoretical Frames. *Human Relations*.

Deutzmann, J. S., Kracke, F., Gu, W., & Spormann, A. M. (2022). Microbial Electrosynthesis of Acetate Powered by Intermittent Electricity. *Environmental Science & Technology*.

Di, Z., Liu, Z.-F., Li, H.-R., Liu, Z., & Li, C. (2023). Enhancing the Stability of Poly(ionic Liquids)@MOFs@COFs *via* Core–shell Protection Strategy for $^{99}TcO_4^-$ Sequestration. *Inorganic Chemistry Frontiers*.

Diab, D., Lefebvre, F., Nassar, G., Smagin, N., Isber, S., Omar, F. E., & Naja, A. (2019). An Autonomous Low-Power Management System for Energy Harvesting From a Miniaturized Spherical Piezoelectric Transducer. *Review of Scientific Instruments*.

Diao, W., Xing, Y., Saxena, S., & Pecht, M. (2018). Evaluation of Present Accelerated Temperature Testing and Modeling of Batteries. *Applied Sciences*.

Diekmann, J., Hanisch, C., Froböse, L., Schälicke, G., Loellhoeffel, T., Fölster, A.-S., & Kwade, A. (2016). Ecological Recycling of Lithium-Ion Batteries From Electric Vehicles With Focus on Mechanical Processes. *Journal of the Electrochemical Society*.

Digernes, M. N., Rudi, L., Andersson, H., Stålhane, M., Wasbø, S. O., & Knudsen, B. R. (2018). Global Optimisation of Multi-Plant Manganese Alloy Production. *Computers & Chemical Engineering*.

Dirican, M., Yanılmaz, M., Fu, K., Yıldız, Ö., Kızıl, H., Hu, Y., & Zhang, X. (2014). Carbon-Confined PVA-Derived Silicon/Silica/Carbon Nanofiber Composites as Anode for Lithium-Ion Batteries. *Journal of the Electrochemical Society*.

Dong, S., Lv, N., Wu, Y., Zhu, G., & Dong, X. (2021). Lithium-Ion and Sodium-Ion Hybrid Capacitors: From Insertion-Type Materials Design to Devices Construction. *Advanced Functional Materials*.

Dong, W., Li, C., Wang, C., Wu, L., Hu, Z. Y., Liu, J., & Su, B. L. (2021). Phase Conversion Accelerating "Zn-Escape" Effect in ZnSe-CFs Heterostructure for High Performance Sodium-Ion Half/Full Batteries. *Small*.

Dreisbach, G., & Fischer, R. (2012). The Role of Affect and Reward in the Conflict-Triggered Adjustment of Cognitive Control. *Frontiers in Human Neuroscience*.

Dreisinger, D., Glück, T., & Lü, J. (2012). The Recovery of Cobalt From the Boleo Deposit Using Leach, SX and EW.

Drobinski, P., & Tantet, A. (2021). Integration of Climate Variability and Climate Change in Renewable Energy Planning. *Physics-Uspekhi*.

Du, W., Chen, L., He, Y., Wang, Q., Gao, P., & Li, Q. (2022). Spatial and Temporal Distribution of Groundwater in Open-Pit Coal Mining: A Case Study From Baorixile Coal Mine, Hailaer Basin, China. *Geofluids*.

Dunn, B., Kamath, H. R., & Tarascon, J. M. (2011). Electrical Energy Storage for the Grid: A Battery of Choices. *Science*.

Dunn, J. B., Gaines, L., Sullivan, J. L., & Wang, M. Q. (2012). Impact of Recycling on Cradle-to-Gate Energy Consumption and Greenhouse Gas Emissions of Automotive Lithium-Ion Batteries. *Environmental Science & Technology*.

Dupraz, C., Marrou, H., Talbot, G., Dufour, L., Nogier, A., & Ferard, Y. (2011). Combining Solar Photovoltaic Panels and Food Crops for Optimising Land Use: Towards New Agrivoltaic Schemes. *Renewable Energy*.

Durán, I. A., Saqib, N., & Mahmood, H. (2023). Assessing the Connection Between Nuclear and Renewable Energy on Ecological Footprint Within the EKC Framework: Implications for Sustainable Policy in Leading Nuclear Energy-Producing Countries. *International Journal of Energy Economics and Policy*.

Ebrashi, R. E. (2013). Social Entrepreneurship Theory and Sustainable Social Impact. *Social Responsibility Journal*.

Ecker, M., Gerschler, J. B., Vogel, J., Käbitz, S., Hust, F., Dechent, P., & Sauer, D. U. (2012). Development of a Lifetime Prediction Model for Lithium-Ion Batteries Based on Extended Accelerated Aging Test Data. *Journal of Power Sources*.

Eghtedarpour, N., & Farjah, E. (2014). Distributed Charge/Discharge Control of Energy Storages in a Renewable-energy-based DC Micro-grid. *Iet Renewable Power Generation*.

Egid, B. R., Roura, M., Aktar, B., Quach, J. A., Chumo, I., Dias, S., Hegel, G., Jones, L. P., Karuga, R., Lar, L. A., Lopez, Y., Pandya, A., Norton, T., Sheikhattari, P., Tancred, T., Wallerstein, N., Zimmerman, E. B., & Ozano, K. (2021). 'You Want to Deal With Power While Riding on Power': Global Perspectives on Power in Participatory Health Research and Co-Production Approaches. *BMJ Global Health*.

Ellingsen, L. A. W., Hung, C. R., & Strømman, A. H. (2017). Identifying Key Assumptions and Differences in Life Cycle Assessment Studies of Lithium-Ion Traction Batteries With Focus on Greenhouse Gas Emissions. *Transportation Research Part D Transport and Environment*.

Energy5. (2023). *The Use of Lithium-Ion Batteries in Medical Devices*. Retrieved 28/12/2023 from

Esteve-Calvo, P. F., & Lloret-Climent, M. (2006). Coverage and invariability by structural functions. *International Journal of General Systems*, *35*(6), 699-706.

Eswarlal, V. K., Vasudevan, G., Dey, P. K., & Vasudevan, P. (2014). Role of Community Acceptance in Sustainable Bioenergy Projects in India. *Energy Policy*.

Etacheri, V., Marom, R., Elazari, R., Salitra, G., & Aurbach, D. (2011). Challenges in the Development of Advanced Li-Ion Batteries: A Review. *Energy & Environmental Science*.

Faisal, M., Hannan, M. A., Ker, P. J., & Uddin, M. N. (2019). Backtracking Search Algorithm Based Fuzzy Charging-Discharging Controller for Battery Storage System in Microgrid Applications. *Ieee Access*.

Fan, W., Zhang, H., Wang, H., Zhao, X., Sun, S., Shi, J., Huang, M., Liu, W., Zheng, Y., & Li, P. (2019). Dual-Doped Hierarchical Porous Carbon Derived From Biomass for Advanced Supercapacitors and Lithium Ion Batteries. *RSC Advances*.

Fang, H., Fan, H., Ma, H., Shen, H., & Dong, Y. (2015). Lithium-Ion Batteries Life Prediction Method Basedon Degenerative Characters and Improved Particle Filter.

Farrag, M. E., Hepburn, D. M., & García, B. (2019). Quantification of Efficiency Improvements From Integration of Battery Energy Storage Systems and Renewable Energy Sources Into Domestic Distribution Networks. *Energies*.

Fattahi, A., Sijm, J. P. M., Broek, M. v. d., Gordon, R. M., Diéguez, M. S., & Faaij, A. (2022). Analyzing the Techno-Economic Role of Nuclear Power in the Transition to the Net-Zero Energy System of the Netherlands Using the IESA-Opt-N Model.

Fawcett, B. (2022). Feminisms in Social Work and Social Care: Backwards, Forwards or Something in Between. *International Social Work*.

Fei, Z., Zhang, Y., Meng, Q., Dong, P., Fei, J., Zhou, S., & Kwon, K. (2021). Dual-Function Regeneration of Waste Lithium Cobalt Oxide for Stable High Voltage Cycle Performance. *Acs Sustainable Chemistry & Engineering*.

Fell, C. R., Sun, L., Hallac, P. B., Metz, B., & Sisk, B. (2015). Investigation of the Gas Generation in Lithium Titanate Anode Based Lithium Ion Batteries. *Journal of the Electrochemical Society*.

Feng, L., Xuan, Z., Zhao, H., Bai, Y., Guo, J., Su, C., & Chen, X. (2014). MnO2 Prepared by Hydrothermal Method and Electrochemical Performance as Anode for Lithium-Ion Battery. *Nanoscale Research Letters*.

Feng, X., Ren, D., He, X., & Ouyang, M. (2020). Mitigating Thermal Runaway of Lithium-Ion Batteries. *Joule*.

Ferrare, J. J., & Phillippo, K. (2021). Conflict Theory, Extended: A Framework for Understanding Contemporary Struggles Over Education Policy. *Educational Policy*.

Ferrari, L. A. (2023). Must Nuclear Energy Be Increased on Brazilian Energy Mix in a Post-Covid-19 World? *Brazilian Journal of Radiation Sciences*.

Ferré-Huguet, N., Nadal, M., Schuhmacher, M., & Domingo, J. L. (2005). Environmental Impact and Human Health Risks of Polychlorinated Dibenzo-P-Dioxins and Dibenzofurans in the Vicinity of a New Hazardous Waste Incinerator: a Case Study. *Environmental Science & Technology*. +

Filomeno, G., & Feraco, S. (2020). Economic, Technical and Environmental Aspects of Recycling Lithium Batteries: A Literature Review. *Global Journal of Researches in Engineering*.

Finegan, D. P., Darcy, E., Keyser, M., Tjaden, B., Heenan, T. M. M., Jervis, R., Bailey, J. J., Malik, R., Vo, N. T., Magdysyuk, O. V., Atwood, R. C., Drakopoulos, M., DiMichiel, M., Rack, A., Hinds, G., & Brett, D. J. L. (2017). Characterising Thermal Runaway Within Lithium-Ion Cells by Inducing and Monitoring Internal Short Circuits. *Energy & Environmental Science*.

Finegan, D. P., Darcy, E., Keyser, M., Tjaden, B., Heenan, T. M. M., Jervis, R., Bailey, J. S., Vo, N. T., Magdysyuk, O. V., Drakopoulos, M., Michiel, M. D., Rack, A., Hinds, G., Brett, D. J. L., & Shearing, P. R. (2017). Identifying the Cause of Rupture of Li-Ion Batteries During Thermal Runaway. *Advanced Science*.

Finley, J. W. (2004). Does Environmental Exposure to Manganese Pose a Health Risk to Healthy Adults? *Nutrition Reviews*.

Finocchi, E. (2021). Standardizing a Unique Renewable Energy Supply Chain: The SURESC Model. *F1000research*.

Fleischmann, J., Hanicke, M., Horetsky, E., Ibrahim, D., Jautelat, S., Linder, M., Schaufuss, P., Torscht, L., & Rijt, A. v. d. (2023). Battery 2030: Resilient, sustainable, and circular. *McKinsey Insights, January 2023.*

Floegel, D., & Costello, K. L. (2021). Methods for a Feminist Technoscience of Information Practice: Design Justice and Speculative Futurities. *Journal of the Association for Information Science and Technology.*

Fouquet, R. (2018). The Economics of Renewable Energy.

Fowler, M., & Fowler, M. (2020). A Review of Lithium-Ion Battery Fault Diagnostic Algorithms: Current Progress and Future Challenges. *Algorithms.*

Frank, G. (2007). A Response to Richard Lettieri's "The Ego Revisited.". *Psychoanalytic Psychology.*

Freeman, M., Robo, E., Birsner, C., MacNair, D., & Renard, Y. (2023). *The Macroeconomic Impact of Increased U.S. Electric Vehicle Battery Production.* E. G. Company.

Friedman, E., & Landsberg, A. S. (2013). Hierarchical Networks, Power Laws, and Neuronal Avalanches. *Chaos an Interdisciplinary Journal of Nonlinear Science.*

Fu, K., Gong, Y., Dai, J., Gong, A., Han, X., Yao, Y., Wang, C., Wang, Y., Chen, Y., Yan, C., Li, Y., Wachsman, E. D., & Hu, L. (2016). Flexible, Solid-State, Ion-Conducting Membrane With 3D Garnet Nanofiber Networks for Lithium Batteries. *Proceedings of the National Academy of Sciences.*

Furberg, A., Arvidsson, R., & Molander, S. (2018). Live and Let Die? Life Cycle Human Health Impacts From the Use of Tire Studs. *International journal of environmental research and public health.*

Gaikwad, A. M., Khau, B. V., Davies, G., Hertzberg, B., Steingart, D. A., & Arias, A. C. (2014). A High Areal Capacity Flexible Lithium-Ion Battery With a Strain-Compliant Design. *Advanced Energy Materials.*

Gailani, A., Mokidm, R., El-Dalahmeh, M. a., El-Dalahmeh, M. d., & Al-Greer, M. (2020). Analysis of Lithium-Ion Battery Cells Degradation Based on Different Manufacturers.

Gaines, L., & Singh, M. K. (1995). Energy and Environmental Impacts of Electric Vehicle Battery Production and Recycling.

Galeotti, M., Salini, S., & Verdolini, E. (2020). Measuring Environmental Policy Stringency: Approaches, Validity, and Impact on Environmental Innovation and Energy Efficiency. *Energy Policy.*

Gallagher, K. G., Goebel, S. G., Greszler, T. A., Mathias, M. F., Oelerich, W., Eroğlu, D., & Srinivasan, V. (2014). Quantifying the Promise of Lithium–air Batteries for Electric Vehicles. *Energy & Environmental Science*.

Galos, J., Pattarakunnan, K., Best, A. S., Kyratzis, I. L., Wang, C., & Mouritz, A. P. (2021). Energy Storage Structural Composites With Integrated Lithium-Ion Batteries: A Review. *Advanced Materials Technologies*.

Gandoman, F. H., Ahmed, E. M., Ali, Z. M., Berecibar, M., Zobaa, A. F., & Aleem, S. H. E. A. (2021). Reliability Evaluation of Lithium-Ion Batteries for E-Mobility Applications From Practical and Technical Perspectives: A Case Study. *Sustainability*.

Gangaja, B., Nair, S. V., & Santhanagopalan, D. (2021). Reuse, Recycle, and Regeneration of LiFePO$_4$ Cathode From Spent Lithium-Ion Batteries for Rechargeable Lithium- And Sodium-Ion Batteries. *Acs Sustainable Chemistry & Engineering*.

Gao, A., Xu, F., & Dong, W. (2022). The Concept of Early Monitoring and Warning of Thermal Runaway of Lithium-Ion Power Battery Using Parameter Analysis. *Journal of Physics Conference Series*.

Gao, D., Yong, Z., Wang, T., & Wang, Y. (2020). A Method for Predicting the Remaining Useful Life of Lithium-Ion Batteries Based on Particle Filter Using Kendall Rank Correlation Coefficient. *Energies*.

Gao, J., Jia, Q., & Yao, Y. (2022). The Impact of Lithium Price on Electric Vehicle Supply Chain Based on Multi-Factors Fama-French Models.

Gao, W., Zhang, X., Zheng, X., Lin, X., Cao, H., Zhang, Y., & Sun, Z. (2017). Lithium Carbonate Recovery From Cathode Scrap of Spent Lithium-Ion Battery: A Closed-Loop Process. *Environmental Science & Technology*.

Gao, Y., Zhang, X., Cheng, Q., Guo, B., & Yang, J. (2019). Classification and Review of the Charging Strategies for Commercial Lithium-Ion Batteries. *Ieee Access*.

Gareau, B. J. (2012). Worlds Apart: A Social Theoretical Exploration of Local Networks, Natural Actors, and Practitioners of Rural Development in Southern Honduras. *Sustainability*.

Garlick, S. (2017). The Return of Nature: Feminism, Hegemonic Masculinities, and New Materialisms. *Men and Masculinities*.

Garmabdari, R., Moghimi, M., Yang, F., Gray, E., & Lu, J. (2019). Optimal Power Flow Scheduling of Distributed Microgrid Systems Considering Backup Generators.

Gauthier, M., Carney, T. J., Grimaud, A., Giordano, L., Pour, N., Chang, H.-J., Fenning, D. P., Lux, S. F., Paschos, O., Bauer, C., Maglia, F., Lupart, S., Lamp, P., & Shao-Horn, Y. (2015). Electrode–Electrolyte Interface in Li-Ion Batteries: Current Understanding and New Insights. *The Journal of Physical Chemistry Letters*.

Genys, D., & Krikštolaitis, R. (2020). Clusterization of Public Perception of Nuclear Energy in Relation to Changing Political Priorities. *Insights Into Regional Development*.)

Ghazali, E., Johari, M. A. M., & Fauzi, M. A. (2022). An Overview of Characterisation, Utilisation, and Leachate Analysis of Clinical Waste Incineration Ash. *International Journal of Environmental Research*.

Gherardini, F., Petruccioli, A., Dalpadulo, E., Bettelli, V., Mascia, M. T., & Leali, F. (2020). A Methodological Approach for the Design of Inclusive Assistive Devices by Integrating Co-Design and Additive Manufacturing Technologies.

Ghorpade, Y. V. (2019). To Study the Effect of Cryogenic Treatment on Cathode Electrode of Lithium Cobalt Oxide (C-LiCoO2) Material. *International Journal for Research in Applied Science and Engineering Technology*.

Gibson, C., Eshraghi, N., Kemalyan, N., & Mueller, C. F. (2018). Electronic Cigarette Burns: A Case Series. *Trauma*.

Giddens, A. (1979). *Central Problems in Social Theory*.

Giegerich, M., Akdere, M., Freund, C., Fühner, T., Grosch, J., Koffel, S., Schwarz, R., Waldhör, S., Wenger, M., Lorentz, V. R. H., & März, M. (2016). Open, Flexible and Extensible Battery Management System for Lithium-Ion Batteries in Mobile and Stationary Applications.

Gissey, G. C., Dodds, P. E., & Radcliffe, J. (2018). Market and Regulatory Barriers to Electrical Energy Storage Innovation. *Renewable and Sustainable Energy Reviews*.

Golroudbary, S. R., Krasławski, A., Wilson, B. P., & Lundström, M. (2023). Assessment of Environmental Sustainability of Nickel Required for Mobility Transition. *Frontiers in Chemical Engineering*.

Gómez-Urbano, J. L., Enterría, M., Monterrubio, I., Larramendi, I. R. d., Carriazo, D., Ortiz-Vitoriano, N., & Rojo, T. (2020). An Overview of Engineered Graphene-Based Cathodes: Boosting Oxygen Reduction and Evolution Reactions in Lithium– And Sodium–Oxygen Batteries. *Chemsuschem*.

Gonka, J., & Kim, J. (2015). The Role of Information Technology (Apps) in FPMRS. *Current Urology Reports*.

Gonzales-Calienes, G., Kannangara, M., & Bensebaa, F. (2023). Economic and Environmental Viability of Lithium-Ion Battery Recycling—Case Study in Two Canadian Regions with Different Energy Mixes. *Batteries, 9*(7), 375.

Goodenough, J. B., & Kim, Y. S. (2009). Challenges for Rechargeable Li Batteries. *Chemistry of Materials*.

Gorman, R. (2016). Changing Ethnographic Mediums: The Place-based Contingency of Smartphones and Scratchnotes. *Area*.

Gorman, S. F., Pathan, T. S., & Kendrick, E. (2019). The 'Use-by Date' for Lithium-Ion Battery Components. *Philosophical Transactions of the Royal Society a Mathematical Physical and Engineering Sciences*.

Gowda, H., & Channegowda, J. (2022). Contrastive Learning for Practical Battery Synthetic Data Generation Using Seasonal and Trend Representations. *International Journal of Energy Research*.

Gowda, S. R., Gallagher, K. G., Croy, J. R., Bettge, M., Thackeray, M. M., & Balasubramanian, M. (2014). Oxidation State of Cross-Over Manganese Species on the Graphite Electrode of Lithium-Ion Cells. *Physical Chemistry Chemical Physics*.

Goya, T., Uchida, K., Kinjyo, Y., Senjyu, T., Yona, A., & Funabashi, T. (2011). Coordinated Control of Energy Storage System and Diesel Generator in Isolated Power System. *International Journal of Emerging Electric Power Systems*.

Grabmann, S., Kriegler, J., Harst, F., Günter, F. J., & Zaeh, M. F. (2021). Laser Welding of Current Collector Foil Stacks in Battery Production–mechanical Properties of Joints Welded With a Green High-Power Disk Laser. *The International Journal of Advanced Manufacturing Technology*.

Graham, J. D., Rupp, J. A., & Brungard, E. (2021). Lithium in the Green Energy Transition: The Quest for Both Sustainability and Security. *Sustainability*.

Grainger, R., Devan, H., Sangelaji, B., & Hay-Smith, J. (2020). Issues in Reporting of Systematic Review Methods in Health App-Focused Reviews: A Scoping Review. *Health Informatics Journal*.

Greatbatch, W., Holmes, C. F., Takeuchi, E. S., & Ebel, S. J. (1996). Lithium/Carbon Monofluoride (Li/CFx): A New Pacemaker Battery. *Pacing and Clinical Electrophysiology*.

Grispos, G., Flynn, T., Glisson, W. B., & Choo, K. K. R. (2021). Investigating Protected Health Information Leakage From Android Medical Applications.

Groot, J. I. M. d., Steg, L., & Poortinga, W. (2012). Values, Perceived Risks and Benefits, and Acceptability of Nuclear Energy. *Risk Analysis*.

Grosspietsch, D., Thömmes, P., Girod, B., & Hoffmann, V. H. (2018). How, When, and Where? Assessing Renewable Energy Self-Sufficiency at the Neighborhood Level. *Environmental Science & Technology*.

Gu, M., Belharouak, I., Genç, A., Wang, Z., Wang, D., Amine, K., Gao, F., Zhou, G., Thevuthasan, S., Baer, D. R., Zhang, J. G., Browning, N. D., Li, J., & Wang, C. (2012). Conflicting Roles of Nickel in Controlling Cathode Performance in Lithium Ion Batteries. *Nano Letters*.

Gu, X., Qin, L., & Zhang, M. (2023). The Impact of Green Finance on the Transformation of Energy Consumption Structure: Evidence Based on China. *Frontiers in Earth Science*.

Guedes, R. S., Ramos, S. J., Gastauer, M., Júnior, C. F. C., Martins, G. C., Júnior, W. d. R. N., Souza-Filho, P. W. M., & Siqueira, J. O. (2021). Challenges and Potential Approaches for Soil Recovery in Iron Open Pit Mines and Waste Piles.

Guk, E., & Kalkan, N. (2015). The Importance of Nuclear Energy for the Expansion of World's Energy Demand. *Advances in Energy Research*.

Guo, S., Bi, K., Zhang, L., & Jiang, H. (2022). How Does Social Comparison Influence Chinese Adolescents' Flourishing Through Short Videos? *International journal of environmental research and public health*.

Guo, Y., & Cai, Y. (2022). The Impact of Social Media Usage on Depression Cognition and Help-Seeking Behavior: A Study Based on Grounded Theory.

Guo, Y. G., & Hu, J. S. (2008). Nanostructured Materials for Electrochemical Energy Conversion and Storage Devices. *Advanced Materials*.

Gupta, N. (2021). Li-ion cell manufacturing: A look at processes and equipment. *Emerging Technology News, May-June 2021*.

Gutiérrez, J. S., Moore, J. N., Donnelly, J., Dorador, C., Navedo, J. G., & Senner, N. R. (2022). Climate Change and Lithium Mining Influence Flamingo Abundance in the Lithium Triangle. *Proceedings of the Royal Society B Biological Sciences*.

Habib, S., & Arefin, P. (2022). Adoption of Hydrogen Fuel Cell Vehicles and Its Prospects for the Future (A Review). *Oriental Journal of Chemistry*.

Hafner, S., Jones, A., Anger-Kraavi, A., & Monasterolo, I. (2021). Modelling the Macroeconomics of a 'Closing the Green Finance Gap' Scenario for an Energy Transition. *Environmental Innovation and Societal Transitions*.

Hagemeister, J., Stock, S., Linke, M., Fischer, M., Drees, R., Kurrat, M., & Daub, R. (2022). Lean Cell Finalization in Lithium-Ion Battery Production: Determining the Required Electrolyte Wetting Degree to Begin the Formation. *Energy Technology*.

Haizhou, Z. (2017). Modeling of Lithium-Ion Battery for Charging/Discharging Characteristics Based on Circuit Model. *International Journal of Online Engineering (Ijoe)*.

Halford, S., Kukarenko, N., Lotherington, A. T., & Obstfelder, A. (2015). Technical Change and the Un/Troubling of Gendered Ageing in Healthcare Work. *Gender Work and Organization*.

Haltiwanger, J., Jarmin, R. S., & Miranda, J. (2013). Who Creates Jobs? Small Versus Large Versus Young. *The Review of Economics and Statistics*.

Han, B., Bøckman, O., Wilson, B. P., Lundström, M., & Louhi-Kultanen, M. (2019). Purification of Nickel Sulfate by Batch Cooling Crystallization. *Chemical Engineering & Technology*.

Hanna, P., & Vanclay, F. (2013). Human Rights, Indigenous Peoples and the Concept of Free, Prior and Informed Consent. *Impact Assessment and Project Appraisal*.

Hanson, E. D., Mayekar, S., & Dravid, V. P. (2017). Applying Insights From the Pharma Innovation Model to Battery Commercialization—pros, Cons, and Pitfalls. *Mrs Energy & Sustainability*.

Hao, M., Li, J., Park, S., Moura, S. J., & Dames, C. (2018). Efficient Thermal Management of Li-Ion Batteries With a Passive Interfacial Thermal Regulator Based on a Shape Memory Alloy. *Nature Energy*.

Haraldsson, J., & Johansson, M. (2018). Review of Measures for Improved Energy Efficiency in Production-Related Processes in the Aluminium Industry – From Electrolysis to Recycling. *Renewable and Sustainable Energy Reviews*.

Hariri, M. H. M., Desa, M. K. M., Masri, S., & Zainuri, M. A. A. M. (2020). Grid-Connected PV Generation System—Components and Challenges: A Review. *Energies*.

Harmon, E., & Mazmanian, M. (2013). Stories of the Smartphone in Everyday Discourse.

Harper, E. M., Kavlak, G., & Graedel, T. E. (2011). Tracking the Metal of the Goblins: Cobalt's Cycle of Use. *Environmental Science & Technology*.

Harpprecht, C., Oers, L. v., Northey, S., Yang, Y., & Steubing, B. (2021). Environmental Impacts of Key Metals' Supply and Low-carbon Technologies Are Likely to Decrease in the Future. *Journal of Industrial Ecology*.

Hartanto, R., Paranoan, R. R., Uzhary, S. A., & Rahmi, A. (2023). Erosion Rate in Post-Coal Mining Reclamation Area in Kutai Kartanegara District, Indonesia. *Journal of Agriculture and Ecology Research International, 24*(4), 13-21.

Haruna, H., Takahashi, S., & Tanaka, Y. (2016). Accurate Consumption Analysis of Vinylene Carbonate as an Electrolyte Additive in an 18650 Lithium-Ion Battery at the First Charge-Discharge Cycle. *Journal of the Electrochemical Society.*

Hasan, A. A. M. (2014). Integrated Model for Sustainable Development (IMSD): Impact of Technology Integration.

Hashemi, S. R., Baghbadorani, A. B., Esmaeeli, R., Mahajan, A., & Farhad, S. (2020). Machine Learning-based Model for Lithium-ion Batteries In <scp>BMS</Scp> of Electric/Hybrid Electric Aircraft. *International Journal of Energy Research.*

Hawari, M. T. (2022). Premium on a Budget: Second-Hand iPhones in Indonesia. *International Journal of Current Science Research and Review.*

Hayajneh, H. S., & Zhang, X. (2020). Logistics Design for Mobile Battery Energy Storage Systems. *Energies.*

Hayyat, M. U., Nawaz, R., Siddiq, Z., Shakoor, M. B., Mushtaq, M., Ahmad, S. R., Ali, S., Hussain, A., Irshad, M., Alsahli, A. A., & Alyemeni, M. N. (2021). Investigation of Lithium Application and Effect of Organic Matter on Soil Health. *Sustainability.*

He, K., Lin, F., Stach, E. A., Mo, Y., Xin, H. L., & Su, D. (2015). Contrasting Reaction Modality Between Electrochemical Sodiation and Lithiation in NiO Conversion Electrode Materials. *Microscopy and Microanalysis.*

He, L., Sun, Q., Chen, C., Oh, J. A. S., Sun, J., Li, M., Tu, W., Zhou, H., Zeng, K., & Li, L. (2019). Failure Mechanism and Interface Engineering for NASICON-Structured All-Solid-State Lithium Metal Batteries. *Acs Applied Materials & Interfaces.*

Heath, G. A., Ravikumar, D., Hansen, B., & Kupets, E. (2022). A critical review of the circular economy for lithium-ion batteries and photovoltaic modules–status, challenges, and opportunities. *Journal of the Air & Waste Management Association, 72*(6), 478-539.

Heidrich, O., Ford, A., Dawson, R. J., Manning, D., Mohareb, E., Raugei, M., Baars, J., & Rajaeifar, M. A. (2022). LAYERS: A Decision-Support Tool to Illustrate and Assess the Supply and Value Chain for the Energy Transition. *Sustainability.*

Hein, S., Danner, T., & Latz, A. (2020). An Electrochemical Model of Lithium Plating and Stripping in Lithium Ion Batteries. *Acs Applied Energy Materials.*

Heine, C., Browning, C. J., & Gong, C. H. (2019). Sensory Loss in China: Prevalence, Use of Aids, and Impacts on Social Participation. *Frontiers in Public Health*.

Hejazi, G. (2016). Integrated Energy Solution Towards Sustainable Isolated Communities. *International Journal of Environmental Science and Development*.

Hendrickx, M., Paulus, A., Kirsanova, M. A., Bael, M. K. V., Abakumov, A. M., Hardy, A., & Hadermann, J. (2022). The Influence of Synthesis Method on the Local Structure and Electrochemical Properties of Li-Rich/Mn-Rich NMC Cathode Materials for Li-Ion Batteries. *Nanomaterials*.

Henriksen, A., Mikalsen, M. H., Woldaregay, A. Z., Mužný, M., Hartvigsen, G., Hopstock, L. A., & Grimsgaard, S. (2018). Using Fitness Trackers and Smartwatches to Measure Physical Activity in Research: Analysis of Consumer Wrist-Worn Wearables. *Journal of Medical Internet Research*.

Hernandez, R. R., Easter, S. B., Murphy-Mariscal, M. L., Maestre, F. T., Tavassoli, M., Allen, E. B., Barrows, C. W., Belnap, J., Ochoa-Hueso, R., Ravi, S., & Allen, M. F. (2014). Environmental Impacts of Utility-Scale Solar Energy. *Renewable and Sustainable Energy Reviews*.

Hernández, Y., Nicolosi, V., Lotya, M., Blighe, F. M., Sun, Z., De, S. J., McGovern, I. T., Holland, B. N., Byrne, M. T., Gun'ko, Y. K., Boland, J. J., Niraj, P., Duesberg, G. S., Krishnamurthy, S., Goodhue, R., Hutchison, J. L., Scardaci, V., Ferrari, A. C., & Coleman, J. N. (2008). High-Yield Production of Graphene by Liquid-Phase Exfoliation of Graphite. *Nature Physics*.

Herrmann, C., Dewulf, W., Hauschild, M. Z., Kaluza, A., Kara, S., & Skerlos, S. (2018). Life Cycle Engineering of Lightweight Structures. *Cirp Annals*.

Hesse, H. C., Martins, R., Musilek, P., Naumann, M., Truong, C. N., & Jossen, A. (2017). Economic Optimization of Component Sizing for Residential Battery Storage Systems. *Energies*.

Hesse, H. C., Schimpe, M., Kucevic, D., & Jossen, A. (2017). Lithium-Ion Battery Storage for the Grid—A Review of Stationary Battery Storage System Design Tailored for Applications in Modern Power Grids. *Energies*.

Hewathilake, H. P. T. S., Balasooriya, N. W. B., Nakamura, Y., Pitawala, H. M. T. G. A., Wijayasinghe, H. W. M. A. C., & Satish-Kumar, M. (2018). Geochemical, Structural and Morphological Characterization of Vein Graphite Deposits of Sri Lanka: Witness to Carbon Rich Fluid Activity. *Journal of Mineralogical and Petrological Sciences*.

Hibino, M., Harimoto, R., Ogasawara, Y., Kido, R., Sugahara, A., Kudo, T., Tochigi, E., Shibata, N., Ikuhara, Y., & Mizuno, N. (2013). A New Rechargeable Sodium Battery Utilizing Reversible Topotactic Oxygen Extraction/Insertion of $CaFeO_z$ ($2.5 \leq z \leq 3$) in an Organic Electrolyte. *Journal of the American Chemical Society*.

Hidayat, S., Susanty, S., Riveli, N., Suroto, B. J., & Rahayu, I. (2018). Synthesis and characterization of CMC from water hyacinth for lithium-ion battery applications. AIP Conference Proceedings,

Hirose, K. (2012). 2011 Fukushima Dai-Ichi Nuclear Power Plant Accident: Summary of Regional Radioactive Deposition Monitoring Results. *Journal of Environmental Radioactivity*.

Hiroshi Kawamura, Marcelo LaFleur, Kenneth Iversen, & Cheng, H. W. J. (2021). *Frontier Technology Issues: Lithium-ion batteries: a pillar for a fossil fuel-free economy?* United Nations.

Ho, S. S., Leong, A. D., Looi, J., Chen, L., Pang, N., & Tandoc, E. C. (2018). Science Literacy or Value Predisposition? A Meta-Analysis of Factors Predicting Public Perceptions of Benefits, Risks, and Acceptance of Nuclear Energy. *Environmental Communication*.

Holko, M., Litwin, T. R., Muñoz, F., Theisz, K. I., Salgin, L., Jenks, N. P., Holmes, B., Watson-McGee, P., Winford, E., & Sharma, Y. (2022). Wearable Fitness Tracker Use in Federally Qualified Health Center Patients: Strategies to Improve the Health of All of Us Using Digital Health Devices. *NPJ Digital Medicine*.

Hosaka, T., Shimamura, T., Kubota, K., & Komaba, S. (2018). Polyanionic Compounds for Potassium-Ion Batteries. *The Chemical Record*.

Höschele, P., Heindl, S. F., Schneider, B., Sinz, W., & Ellersdorfer, C. (2022). Method for in-Operando Contamination of Lithium Ion Batteries for Prediction of Impurity-Induced Non-Obvious Cell Damage. *Batteries*.

Hosoi, K., & Nakabaru, M. (2009). Status of National Project for SOFC Development in Japan. *Ecs Transactions*.

Hsiao, C. Y. L., Lin, W., Wei, X., Yan, G., Li, S., & Sheng, N. (2019). The Impact of International Oil Prices on the Stock Price Fluctuations of China's Renewable Energy Enterprises. *Energies*.

Hu, H., Xi, X., Sun, X., Li, R., Zhang, Y., & Fu, J. (2020). Numerical Study on the Inhibition Control of Lithium-Ion Battery Thermal Runaway. *Acs Omega*.

Huang, C. J., Hu, K. W., & Cheng, H.-W. (2023). An Electric Vehicle Assisted Charging Mechanism for Unmanned Aerial Vehicles. *Electronics*.

Huang, J. (2023, 27/12/2023). Leading Power Tool Brands and the Lithium-ion Battery Technology Innovation: A Global Overview.

Huang, T.-Y., Pérez-Cardona, J. R., Zhao, F., Sutherland, J. W., & Paranthaman, M. (2021). Life Cycle Assessment and Techno-Economic Assessment of Lithium Recovery From Geothermal Brine. *Acs Sustainable Chemistry & Engineering*.

Huang, W., Huang, B., Bi, X., Lin, Q., Liu, M., Ren, Z., Zhang, G., Wang, X., Sheng, G., & Fu, J. (2014). Emission of PAHs, NPAHs and OPAHs From Residential Honeycomb Coal Briquette Combustion. *Energy & Fuels*.

Hunnicutt, G. (2009). Varieties of Patriarchy and Violence Against Women. *Violence Against Women*.

Hussain, S., & Khatoon, N. (2023). The Impact of Social Media on the Mental Health of Adolescents in Kargil, Ladakh: A Qualitative Study. *Cross-Currents an International Peer-Reviewed Journal on Humanities and Social Sciences*.

Imanishi, N., Takeda, Y., & Yamamoto, O. (2012). Aqueous Lithium-Air Rechargeable Batteries. *Electrochemistry*.

Inglesi-Lotz, R. (2016). The Impact of Renewable Energy Consumption to Economic Growth: A Panel Data Application. *Energy Economics*.

Ingried, V. F., Haryati, S., & Syarif, N. (2022). Hydrothermal LiTiO2 Cathode and Polyurethane Binder of High Current Lithium Ion Batteries. *International Journal on Advanced Science Engineering and Information Technology*.

International Energy Agency. (2023). *Trends in batteries* Retrieved 13/12/2023 from

Iorember, P. T., Gbaka, S., Jelilov, G., Alymkulova, N., & Usman, O. (2022). Impact of International Trade, Energy Consumption and Income on Environmental Degradation in Africa's OPEC Member Countries. *African Development Review*.

Ishimaru, A. M., Barajas-López, F., Min, S., Scarlett, K., & Anderson, E. M. (2022). Transforming the Role of RPPs in Remaking Educational Systems. *Educational Researcher*.

Ismail, A., & Mulyaman, D. (2021). Hyundai Investment on Electric Vehicles in Indonesia. *Intermestic Journal of International Studies*.

Iturrondobeitia, M., Vallejo, C., Berroci, M., Akizu-Gardoki, O., Mínguez, R., & Lizundia, E. (2022). Environmental Impact Assessment of $LiNi_{1/3}Mn_{1/3}Co_{1/3}O_2$

Hydrometallurgical Cathode Recycling From Spent Lithium-Ion Batteries. *Acs Sustainable Chemistry & Engineering*.

Jabbar, J., Dharmarajan, S., Raveendranathan, R. P., Syamkumar, D., & Jasseer, A. (2022). Influence of Social Media on Adolescent Mental Health. *International Journal of English Literature and Social Sciences*.

Jackson, S. (1985). Feminist Theory: From Margin to Center. *Ufahamu a Journal of African Studies*.

Jamshidian, H. S. F. (2020). Innovative Social Media for Foreign Language Learning: A Review of Social Media Types and Their Effects. *Journal of Critical Studies in Language and Literature*.

Jardim, E. (2017). *What 10 years of smartphone use means for the planet*. Greenpeace. Retrieved 27/12/2023 from

Jaworski, A. J., Lejda, K., Lubas, J., & Mądziel, M. (2019). Comparison of Exhaust Emission From Euro 3 and Euro 6 Motor Vehicles Fueled With Petrol and LPG Based on Real Driving Conditions. *Combustion Engines*.

Jayaprakash, N., Das, S., & Archer, L. A. (2011). The Rechargeable Aluminum-Ion Battery. *Chemical Communications*.

Jayasingh, S., Girija, T., & Sivakumar, A. (2021). Factors Influencing Consumers' Purchase Intention Towards Electric Two-Wheelers. *Sustainability*.

Jayawardana, A., Agalgaonkar, A. P., Robinson, D., & Fiorentini, M. (2019). Optimisation Framework for the Operation of Battery Storage Within Solar-rich Microgrids. *Iet Smart Grid*.

Jebli, M. B., & Youssef, S. B. (2018). Investigating the Interdependence Between Non-Hydroelectric Renewable Energy, Agricultural Value Added, and Arable Land Use in Argentina. *Environmental Modeling & Assessment*.

Ji, X., Lee, K. T., & Nazar, L. F. (2009). A Highly Ordered Nanostructured Carbon–sulphur Cathode for Lithium–sulphur Batteries. *Nature Materials*.

Jia, L., Wang, D., Yin, T., Li, X., Li, L., Dai, Z., & Zheng, L. (2022). Experimental Study on Thermal-Induced Runaway in High Nickel Ternary Batteries. *Acs Omega*.

Jiang, S., & Song, Z. (2021). Estimating the State of Health of Lithium-Ion Batteries With a High Discharge Rate Through Impedance. *Energies*.

Jiang, Y., Chen, X., Yan, S., Ou, Y., & Zhou, T. (2022). Mechanochemistry-Induced Recycling of Spent Lithium-Ion Batteries for Synergistic Treatment of Mixed Cathode Powders. *Green Chemistry*.

Jin, S., Jiang, Y., Ji, H., & Yu, Y. (2018). Advanced 3D Current Collectors for Lithium-Based Batteries. *Advanced Materials*.

Jin, Y., Lee, M. E., Kim, G., Seong, H., Nam, W., Kim, S. K., Moon, J. H., & Choi, J. (2023). Hybrid Nano Flake-Like Vanadium Diselenide Combined on Multi-Walled Carbon Nanotube as a Binder-Free Electrode for Sodium-Ion Batteries. *Materials*.

Johnstone, N., Haščič, I., & Popp, D. (2008). Renewable Energy Policies and Technological Innovation: Evidence Based on Patent Counts.

Jolly, M., & Dai, X. J. (2011). Energy Efficiency Improvement by Implementation of the Novel CRIMSON Aluminium Casting Process.

Joseph, T. E., & Charles, A. C. (2021). Renewable Energy Consumption, Environmental Sustainability, and Economic Growth in Developing Countries. *Asian Bulletin of Energy Economics and Technology*.

Joshi, T., Eom, K., Yushin, G., & Fuller, T. F. (2014). Effects of Dissolved Transition Metals on the Electrochemical Performance and SEI Growth in Lithium-Ion Batteries. *Journal of the Electrochemical Society*.

Jung, S., & Jeong, H. S. (2017). Extended Kalman Filter-Based State of Charge and State of Power Estimation Algorithm for Unmanned Aerial Vehicle Li-Po Battery Packs. *Energies*.

Kader, Z. A., Marshall, A., & Kennedy, J. (2021). A Review on Sustainable Recycling Technologies for Lithium-Ion Batteries. *Emergent Materials*.

Kadiyala, A., Kommalapati, R. R., & Huque, Z. (2016). Quantification of the Lifecycle Greenhouse Gas Emissions From Nuclear Power Generation Systems. *Energies*.

Kalair, A. R., Abas, N., Saleem, M., & Khan, N. (2020). Role of Energy Storage Systems in Energy Transition From Fossil Fuels to Renewables. *Energy Storage*.

Kamath, D., Arsenault, R., Kim, H. C., & Anctil, A. (2020). Economic and Environmental Feasibility of Second-Life Lithium-Ion Batteries as Fast-Charging Energy Storage. *Environmental Science & Technology*.

Kandiyoti, D. (1988). Bargaining With Patriarchy. *Gender & Society*.

Kang, B., & Ceder, G. (2009). Battery Materials for Ultrafast Charging and Discharging. *Nature*.

Kang, D. H., Chen, M., & Ogunseitan, O. A. (2013). Potential Environmental and Human Health Impacts of Rechargeable Lithium Batteries in Electronic Waste. *Environmental Science & Technology*.

Kang, G., Wu, L., Guan, Y., & Peng, Z. (2019). A Virtual Sample Generation Method Based on Differential Evolution Algorithm for Overall Trend of Small Sample Data: Used for Lithium-Ion Battery Capacity Degradation Data. *Ieee Access*.

Kang, M., Park, Y.-K., & Kim, K. T. (2019). Economic Feasibility Through the Optimal Capacity Calculation Model of an Energy Storage System Connected to Solar Power Generator. *Energy & Environment*.

Karamov, D., Muftahov, I., & Жуков, А. В. (2021). Increasing Storage Battery Lifetime in Autonomous Photovoltaic Systems With Power Generation Structure Varying Throughout the Year. *E3s Web of Conferences*.

Karasoy, A., & Akçay, S. (2019). Effects of Renewable Energy Consumption and Trade on Environmental Pollution. *Management of Environmental Quality an International Journal*.

Karimi, S., & Kwon, S. (2021). Comparative Analysis of the Impact of Energy-aware Scheduling, Renewable Energy Generation, and Battery Energy Storage on Production Scheduling. *International Journal of Energy Research*.

Karlsson, J., & Byström, P. (2005). Littoral Energy Mobilization Dominates Energy Supply for Top Consumers in Subarctic Lakes. *Limnology and Oceanography*.

Kaya, Ş., Dittrich, C., Stopić, S., & Friedrich, B. (2017). Concentration and Separation of Scandium From Ni Laterite Ore Processing Streams. *Metals*.

Ke, W., Zhang, S., Wu, Y., Zhao, B., Wang, Y., & Hao, J. (2016). Assessing the Future Vehicle Fleet Electrification: The Impacts on Regional and Urban Air Quality. *Environmental Science & Technology*.

Keil, P., Schuster, S. F., Wilhelm, J., Travi, J., Hauser, A., Karl, R. C., & Jossen, A. (2016). Calendar Aging of Lithium-Ion Batteries. *Journal of the Electrochemical Society*.

Kellner, D. (1990). Critical Theory and the Crisis of Social Theory. *Sociological Perspectives*.

Kerner, C., & Goodyear, V. A. (2017). The Motivational Impact of Wearable Healthy Lifestyle Technologies: A Self-Determination Perspective on Fitbits With Adolescents. *American Journal of Health Education*.

Khaliq, A., Rhamdhani, M. A., Brooks, G., & Masood, S. H. (2014). Metal Extraction Processes for Electronic Waste and Existing Industrial Routes: A Review and Australian Perspective. *Resources*.

Khan, A. (2020). An Charging Interface for Lithium Ion Batteries Compatible With Current in-Use UPS System. *International Journal of Engineering Works*.

Khan, H. u. R., Awan, U., Zaman, K., Nassani, A. A., Haffar, M., & Abro, M. M. Q. (2021). Assessing Hybrid Solar-Wind Potential for Industrial Decarbonization Strategies: Global Shift to Green Development. *Energies*.

Khan, S., Zaman, I., Khan, M. I., & Musleha, Z. (2022). Role of Influencers in Digital Marketing: The Moderating Impact of Follower's Interaction. *Gmjacs*.

Khemir, M., Rojas, M., Popova, R., Feizi, T., Heinekamp, J. F., & Strunz, K. (2022). Real-World Application of Sustainable Mobility in Urban Microgrids. *Ieee Transactions on Industry Applications*.

Kilgore, G. (2023). *Carbon Footprint of Lithium-Ion Battery Production (vs Gasoline, Lead-Acid)*. 8 Billion Trees. Retrieved 3/1/2024 from

Kim, H., Park, G. O., Kim, Y., Muhammad, S., Yoo, J., Balasubramanian, M., Cho, Y. H., Kim, M., Lee, B., Kang, K., Kim, H., Kim, J. M., & Yoon, W. S. (2014). New Insight Into the Reaction Mechanism for Exceptional Capacity of Ordered Mesoporous SnO_2 Electrodes via Synchrotron-Based X-Ray Analysis. *Chemistry of Materials*.

Kim, H. C., Wallington, T. J., Arsenault, R., Bae, C., Ahn, S., & Lee, J. (2016). Cradle-to-Gate Emissions From a Commercial Electric Vehicle Li-Ion Battery: A Comparative Analysis. *Environmental Science & Technology*.

Kim, J.-J. (2014). Energy Self-Sufficiency of Office Buildings in Four Asian Cities. *Advances in Energy Research*.

Kim, S.-K., Kim, S.-Y., & Kang, H.-B. (2016). An Analysis of the Effects of Smartphone Push Notifications on Task Performance With Regard to Smartphone Overuse Using ERP. *Computational Intelligence and Neuroscience*.

Kim, S., Lee, J. H., Lee, M., & Lee, c.-k. (2023). A Study of the Manufacturing Characteristics of New Materials for an Electric Vehicle Drive Motor.

Kim, S. H., Wada, K., Kurosawa, A., & Roberts, M. D. (2014). Nuclear Energy Response in the EMF27 Study. *Climatic Change*.

Kim, S. W., Tanim, T. R., Dufek, E. J., Scoffield, D., Pennington, T. D., Gering, K. L., Colclasure, A. M., Mai, W., Meintz, A., & Bennett, J. (2022). Projecting Recent Advancements in Battery Technology to Next-Generation Electric Vehicles. *Energy Technology*.

Kim, Y. (2013). Encapsulation of LiNi0.5Co0.2Mn0.3O2 With a Thin Inorganic Electrolyte Film to Reduce Gas Evolution in the Application of Lithium Ion Batteries. *Physical Chemistry Chemical Physics*.

Kimura, S., Zhu, Q., Zhu, W., & Xiaoying, Y. (2022). Influence of Aerogel Felt With Different Thickness on Thermal Runaway Propagation of 18650 Lithium-Ion Battery. *Electrochemistry*.

Kırıkkaleli, D., Adedoyin, F. F., & Bekun, F. V. (2020). Nuclear Energy Consumption and Economic Growth in the <scp>UK</Scp>: Evidence From Wavelet Coherence Approach. *Journal of Public Affairs*.

Kittner, N., Lill, F., & Kammen, D. M. (2017). Energy Storage Deployment and Innovation for the Clean Energy Transition. *Nature Energy*.

Klein, F., Pieber, S. M., Ni, H., Stefenelli, G., Bertrand, A., Kılıç, D., Pospíšilová, V., Temime-Roussel, B., Marchand, N., Haddad, I. E., Slowik, J. G., Baltensperger, U., Cao, J., Huang, R. J., & Prévôt, A. S. H. (2018). Characterization of Gas-Phase Organics Using Proton Transfer Reaction Time-of-Flight Mass Spectrometry: Residential Coal Combustion. *Environmental Science & Technology*.

Klein, S., Bärmann, P., Beuse, T., Borzutzki, K., Frerichs, J. E., Kasnatscheew, J., Winter, M., & Placke, T. (2020). Exploiting the Degradation Mechanism of NCM523Graphite Lithium-Ion Full Cells Operated at High Voltage. *Chemsuschem*.

Klemeš, J. J., Klimeš, L., Charvát, P., & Pospíšil, J. (2020). Feasibility of Replacement of Nuclear Power With Other Energy Sources in the Czech Republic. *Thermal Science*.

Klima, K., Apt, J., Bandi, M., Happy, P., Loutan, C., & Young, R. J. D. (2018). Geographic Smoothing of Solar Photovoltaic Electric Power Production in the Western USA. *Journal of Renewable and Sustainable Energy*.

Klimko, J., Oráč, D., Miškufová, A., Vonderstein, C., Dertmann, C., Sommerfeld, M., Friedrich, B., & Havlík, T. (2020). A Combined Pyro- And Hydrometallurgical Approach to Recycle Pyrolyzed Lithium-Ion Battery Black Mass Part 2: Lithium Recovery From Li Enriched Slag—Thermodynamic Study, Kinetic Study, and Dry Digestion. *Metals*.

Koch, S., Birke, K. P., & Kuhn, R. (2018). Fast Thermal Runaway Detection for Lithium-Ion Cells in Large Scale Traction Batteries. *Batteries*.

Kong, Y. (2023). Intersectional Feminist Theory as a Non-Ideal Theory: Asian American Women Navigating Identity and Power. *Ergo an Open Access Journal of Philosophy*.

König, P., Müller, P., & Höschler, K. (2023). Assessment of (Hybrid)-Electric Drive-Train Architectures for Future Aircraft Applications. *Journal of Physics Conference Series*.

Koo, B., Sofen, L. E., Gisch, D. J., Kern, B., Rickard, M. A., & Francis, M. B. (2020). Lithium-Chelating Resins Functionalized With Oligoethylene Glycols Toward Lithium-Ion Battery Recycling. *Advanced Sustainable Systems*.

Korupp, S. E., & Szydlik, M. (2005). Causes and Trends of the Digital Divide. *European Sociological Review*.

Kousar, S., Afzal, M., Ahmed, F., & Bojnec, Š. (2022). Environmental Awareness and Air Quality: The Mediating Role of Environmental Protective Behaviors. *Sustainability*.

Krauss, F. T., Pantenburg, I., & Roling, B. (2022). Transport of Ions, Molecules, and Electrons Across the Solid Electrolyte Interphase: What Is Our Current Level of Understanding? *Advanced Materials Interfaces*.

Krebs, P., & Duncan, D. T. (2015). Health App Use Among US Mobile Phone Owners: A National Survey. *Jmir Mhealth and Uhealth*.

Kuek, T.-H., Puah, C. H., Arip, M. A., & Habibullah, M. S. (2020). Macroeconomic Perspective on Constructing Financial Vulnerability Indicator in China. *Journal of Business Economics and Management*.

Kühnel, R. S., Reber, D., & Battaglia, C. (2017). A High-Voltage Aqueous Electrolyte for Sodium-Ion Batteries. *ACS Energy Letters*.

Kühnel, R. S., Reber, D., Remhof, A., Figi, R., Bleiner, D., & Battaglia, C. (2016). "Water-in-Salt" Electrolytes Enable the Use of Cost-Effective Aluminum Current Collectors for Aqueous High-Voltage Batteries. *Chemical Communications*.

Kukah, A. S. K., Anafo, A., Kukah, R. M. K., Blay, A. V. K., Sinsa, D. B., Asamoah, E., & Korda, D. N. (2021). Exploring Innovative Energy Infrastructure Financing in Ghana: Benefits, Challenges and Strategies. *International Journal of Energy Sector Management*.

Kulkarni, S., Huang, T.-Y., Thapaliya, B. P., Luo, H., Dai, S., & Zhao, F. (2022). Prospective Life Cycle Assessment of Synthetic Graphite Manufactured via Electrochemical Graphitization. *Acs Sustainable Chemistry & Engineering*.

Kumar, A., Sujith, M., Valarmathi, K., Kumar, R. S., Al-Asbahi, B. A., & Ahmed, A. A. A. (2023). Double-Absorber $CZTS/Sb_2Se_3$ Architecture for High-Efficiency Solar-Cell Devices. *Physica Status Solidi (A)*.

Ladpli, P., Nardari, R., Kopsaftopoulos, F., & Chang, F. K. (2019). Multifunctional Energy Storage Composite Structures With Embedded Lithium-Ion Batteries. *Journal of Power Sources*.

Lai, I. K. W., & Liu, Y. (2020). The Effects of Content Likeability, Content Credibility, and Social Media Engagement on Users’ Acceptance of Product Placement in Mobile Social Networks. *Journal of Theoretical and Applied Electronic Commerce Research.*

Lai, I. K. W., Liu, Y., Sun, X., Zhang, H., & Xu, W. (2015). Factors Influencing the Behavioural Intention Towards Full Electric Vehicles: An Empirical Study in Macau. *Sustainability.*

Lakhani, N. (2023). Revealed: how US transition to electric cars threatens environmental havoc. *The Guardian.*

Landrigan, P. J., Böse-O'Reilly, S., Elbel, J., Nordberg, G. F., Lucchini, R., Bartrem, C., Grandjean, P., Mergler, D., Moyo, D., Nemery, B., Braun, M. v., & Nowak, D. (2022). Reducing Disease and Death From Artisanal and Small-Scale Mining (ASM) - The Urgent Need for Responsible Mining in the Context of Growing Global Demand for Minerals and Metals for Climate Change Mitigation. *Environmental Health.*

Larcher, D., & Jm, T. (2014). Towards Greener and More Sustainable Batteries for Electrical Energy Storage. *Nature Chemistry.*

Large. (2019). *Cylindrical Lithium Ion Battery.* Retrieved 23/12/2023 from

Larouche, F., Tedjar, F., Amouzegar, K., Houlachi, G., Bouchard, P., Demopoulos, G. P., & Zaghib, K. (2020). Progress and Status of Hydrometallurgical and Direct Recycling of Li-Ion Batteries and Beyond. *Materials.*

Lasley, S. (2023). Graphite demand outpaces EV sales. *North of 60 Mining News, September, 2023.*

Latosińska, J., & Czapik, P. (2020). The Ecological Risk Assessment and the Chemical Speciation of Heavy Metals in Ash After the Incineration of Municipal Sewage Sludge. *Sustainability.*

Laugesen, J., & Yuan, Y. (2010). What Factors Contributed to the Success of Apple's iPhone?

Laureto, J. J., & Pearce, J. M. (2016). Nuclear Insurance Subsidies Cost From Post-Fukushima Accounting Based on Media Sources. *Sustainability.*

Lawson, F. H., & Wallerstein, I. (1978). The Modern World System: Capitalist Agriculture and the Origins of the European World-Economy in the Sixteenth Century. *The Western Political Quarterly.*

Lee, B. S. (2020). A Review of Recent Advancements in Electrospun Anode Materials to Improve Rechargeable Lithium Battery Performance. *Polymers.*

Lee, C.-C., Chou, S. T.-H., & Huang, Y. (2014). A Study on Personality Traits and Social Media Fatigue-Example of Facebook Users. *Lecture Notes on Information Theory*.

Lee, C. Y., Lee, S.-J., Tang, M.-S., & Chen, P.-C. (2011). In Situ Monitoring of Temperature Inside Lithium-Ion Batteries by Flexible Micro Temperature Sensors. *Sensors*.

Lee, M.-s., & Oh, Y.-J. (2004). Estimation of Thermodynamic Properties and Ionic Equilibria of Cobalt Chloride Solution at 298 K. *Materials Transactions*.

Lee, M. C., & Wong, H. Y. (2013). Technical Solutions to Mitigate Reliability Challenges Due to Technology Scaling of Charge Storage NVM. *Journal of Nanomaterials*.

Lee, S.-h., Kim, H., Yun, Y.-S., Jeong, T.-H., Nam, S.-P., Kim, Y.-S. , Kim, J.-C., Lee, K.-T., & Im, I. (2014). Electrochemical Properties of $(Li_{0.5-x}Na_xLa_{0.5})Ti_{0.8}Zr_{0.2}O_3$ Ceramics as Improved Electrolyte Materials for Li-Ion Batteries. *Transactions on Electrical and Electronic Materials*.

Lee, S. M., Kang, D.-S., & Roh, J.-S. (2015). Bulk Graphite: Materials and Manufacturing Process. *Carbon Letters*.

Lee, T. (2020). From Nuclear Energy Developmental State to Energy Transition in South Korea: The Role of the Political Epistemic Community. *Environmental Policy and Governance*.

Lee, Y.-S., & Cheng, M.-W. (2005). Intelligent Control Battery Equalization for Series Connected Lithium-Ion Battery Strings. *Ieee Transactions on Industrial Electronics*.

Lei, Y., Yang, H., Liang, Y., Zhang, B., Wang, L., Lai, W. H., Wang, Y., Liu, H., & Dou, S. X. (2022). Progress and Prospects of Emerging Potassium–Sulfur Batteries. *Advanced Energy Materials*.

Lei, Z., Chen, X., Wang, J., Huang, Y., Du, F., & Yan, Z. (2022). Guite, the Spinel-Structured $Co^{2+}Co^{3+}_2O_4$, a New Mineral From the Sicomines Copper–cobalt Mine, Democratic Republic of Congo. *Mineralogical Magazine*.

Leonard, A., Wheeler, S., & McCulloch, M. (2023). Does Citizen Science Bring "Power to the People"? Evaluating a Remote Mapping Project to Identify Best Practices for Positive Impact on Volunteers. *Citizen Science Theory and Practice*.

Lettieri, R. (2005). The Ego Revisited. *Psychoanalytic Psychology*.

Levy, M. J., & Coser, L. A. (1957). The Functions of Social Conflict. *American Sociological Review*.

Li, B., Gao, X., & Li, J. (2014). Life Cycle Environmental Impact of High-Capacity Lithium Ion Battery With Silicon Nanowires Anode for Electric Vehicles. *Environmental Science & Technology*.

Li, C., Tang, F., & Li, S. (2016). Influences of Trace Water on Electrochemical Performances for Lithium Ion Batteries.

Li, D., Heimeriks, G., & Alkemade, F. (2020). The Emergence of Renewable Energy Technologies at Country Level: Relatedness, International Knowledge Spillovers and Domestic Energy Markets. *Industry and Innovation*.

Li, H., Zhang, W., Sun, K., Guo, J., Kuo, Y. A., Fu, J., Zhang, T., Zhang, X., Long, H., Zhang, Z., Lai, Y., & Sun, H. (2021). Manganese-Based Materials for Rechargeable Batteries Beyond Lithium-Ion. *Advanced Energy Materials*.

Li, H., & Zhou, Z. (2019). Numerical Simulation and Experimental Study of Fluid-Solid Coupling-Based Air-Coupled Ultrasonic Detection of Stomata Defect of Lithium-Ion Battery. *Sensors*.

Li, J. (2013). Addressing the Grand Challenges in Energy Storage. *Advanced Functional Materials*.

Li, J., Du, Z., Ruther, R. E., An, S. J., David, L., Hays, K. A., Wood, M., Phillip, N. D., Sheng, Y., Mao, C., Kalnaus, S., Daniel, C., & Wood, D. L. (2017). Toward Low-Cost, High-Energy Density, and High-Power Density Lithium-Ion Batteries. *Jom*.

Li, J., Gao, F., Yan, G., Zhang, T., & Li, J. (2018). Modeling and SOC Estimation of Lithium Iron Phosphate Battery Considering Capacity Loss. *Protection and Control of Modern Power Systems*.

Li, J., He, S.-C., He, H., Zou, W., & Cao, W. (2022). An Energy Management System for Second-Life Battery in Renewable Energy Systems Considering Battery Degradation Costs.

Li, J., Lu, Y., Yang, T., Ge, D., Wood, D. L., & Li, Z. (2020). Water-Based Electrode Manufacturing and Direct Recycling of Lithium-Ion Battery Electrodes—A Green and Sustainable Manufacturing System. *IScience*.

Li, J. J., & Chen, M. (2020). Simulation of the Electrochemical and Thermal Properties of Electric Vehicle Power Batteries. *Revista De Chimie*.

Li, M., Lü, J., Chen, Z., & Amine, K. (2018). 30 Years of Lithium-Ion Batteries. *Advanced Materials*.

Li, S., Gu, C., Xu, M., Li, J., Zhao, P., & Cheng, S. (2021). Optimal Power System Design and Energy Management for More Electric Aircrafts. *Journal of Power Sources*.

Li, S., Yan, J., Pei, Q., Sha, J., Mou, S., & Xiao, Y. (2019). Risk Identification and Evaluation of the Long-Term Supply of Manganese Mines in China Based on the VW-BGR Method. *Sustainability*.

Li, W., Ma, Z., & Jin, Y. (2023). Chemical Speciation Distribution and Thermal Stability of Heavy Metals Along Flue Gas Cleaning Systems in a Hazardous Waste Incinerator. *Energy & Fuels*.

Li, X., Deng, C., Wang, H., Si, J., Zhang, S., & Huang, B. (2021). Iron Nitride@C Nanocubes Inside Core–Shell Fibers to Realize High Air-Stability, Ultralong Life, and Superior Lithium/Sodium Storages. *Acs Applied Materials & Interfaces*.

Li, X., Huang, Z., Tian, Y., & Tian, J. (2020). Modeling and Comparative Analysis of a Lithium-ion Hybrid Capacitor Under Different Temperature Conditions. *International Journal of Energy Research*.

Li, X., Wu, H., Hu, Y., Liu, H., Yu, Y., Huang, K., Zhang, Z., Xue, B., & Gong, Y. (2022). Assessing the Environmental Benefits of Battery Packs From Multi-vehicle and Multi-region Perspective: Aiming for Lightweight and Carbon Neutrality. *Environmental Progress & Sustainable Energy*.

Li, Y., Jin, X., Wang, G., Ren, Y., Tan, H., & Li, N. (2022). Construction and Application of a Carbon Emission Model for China's Coal Production Enterprises and Result Analysis. *Frontiers in Energy Research*.

Li, Y., Kalnay, E., Motesharrei, S., Rivas, J., Kucharski, F., Kirk-Davidoff, D. B., Bach, E., & Zeng, N. (2018). Climate Model Shows Large-Scale Wind and Solar Farms in the Sahara Increase Rain and Vegetation. *Science*.

Li, Y., Liu, K., Foley, A., Zulke, A. A., Berecibar, M., Nanini-Maury, E., Mierlo, J. V., & Hoster, H. E. (2019). Data-Driven Health Estimation and Lifetime Prediction of Lithium-Ion Batteries: A Review. *Renewable and Sustainable Energy Reviews*.

Li, Z. (2020). Industrial Agglomeration and Regional Economic Growth<br/&Amp;gt;—Analysis of the Threshold Effect Based on Industrial Upgrading. *Open Journal of Business and Management*.

Lian, X., Piao, S., Chen, A., Huntingford, C., Fu, B., Li, L. Z., Huang, J., Sheffield, J., Berg, A. M., & Keenan, T. F. (2021). Multifaceted characteristics of dryland aridity changes in a warming world. *Nature Reviews Earth & Environment*, 2(4), 232-250.

Liang, Z., Cai, C., Peng, G., Hu, J., Hou, H., Liu, B., Liang, S., Xiao, K., Yuan, S., & Yang, J. (2021). Hydrometallurgical Recovery of Spent Lithium Ion Batteries: Environmental Strategies and Sustainability Evaluation. *Acs Sustainable Chemistry & Engineering*.

Liao-ji, Z., Chen, G., Liu, L., & Hu, Y. (2022). Tracing of Lithium Supply and Demand Bottleneck in China's New Energy Vehicle Industry—Based on the Chart of Lithium Flow. *Frontiers in Energy Research*.

Liao, Q., Zhang, Y., Tao, Y., Ye, J., & Li, C. (2019). Economic Analysis of an Industrial Photovoltaic System Coupling With Battery Storage. *International Journal of Energy Research*.

Liew, H. F., Chong, K. H., Fahmi, M. I., Ezanuddin, A. A. M., Junaidah, A. M. J., & Norhanisa, K. (2022). Improvement of Hybrid Energy Storage Wireless Charging System Performance and Efficiency. *Journal of Physics Conference Series*.

Lin, H., Yang, J., Feng, P., Liu, G., Liu, Y., Cui, J., Liu, X., & Xiao, Y. (2023). Structure Regulation and Application of Bagasse-Based Porous Carbon Material Based on H_2O_2-Assisted Hydrothermal Treatment. *Energy & Fuels*.

Lin, X., Fu, H., Pérez, H. E., Siege, J. B., Stefanopoulou, A. G., Ding, Y., & Castanier, M. P. (2013). Parameterization and Observability Analysis of Scalable Battery Clusters for Onboard Thermal Management. *Oil & Gas Science and Technology – Revue D'ifp Energies Nouvelles*.

Lin, Y. C., & Chung, K. J. (2019). Lifetime Prognosis of Lithium-Ion Batteries Through Novel Accelerated Degradation Measurements and a Combined Gamma Process and Monte Carlo Method. *Applied Sciences*.

Lin, Y. C., Yu, M.-Y., Xie, F., & Zhang, Y. L. (2022). Anthropogenic Emission Sources of Sulfate Aerosols in Hangzhou, East China: Insights From Isotope Techniques With Consideration of Fractionation Effects Between Gas-to-Particle Transformations. *Environmental Science & Technology*.

Liou, J. L., & Wu, P. I. (2021). Monetary Health Co-Benefits and GHG Emissions Reduction Benefits: Contribution From Private on-the-Road Transport. *International journal of environmental research and public health*.

Little, A. J., Sivarajah, B., Frendo, C., Sprague, D. D., Smol, J. P., & Vermaire, J. C. (2020). The Impacts of Century-Old, Arsenic-Rich Mine Tailings on Multi-Trophic Level Biological Assemblages in Lakes From Cobalt (Ontario, Canada). *The Science of the Total Environment*.

Liu, B., Zhang, J., Zhang, C., & Xu, J. (2018). Mechanical Integrity of 18650 Lithium-Ion Battery Module: Packing Density and Packing Mode. *Engineering Failure Analysis*.

Liu, C., & Guan-jun, X. (2018). Research on the Dynamic Interrelationship Among R&D Investment, Technological Innovation, and Economic Growth in China. *Sustainability*.

Liu, C., Wu, T., & He, C. (2020). State of Health Prediction of Medical Lithium Batteries Based on Multi-Scale Decomposition and Deep Learning. *Advances in Mechanical Engineering*.

Liu, D., Luo, Y., Le, G., & Yu, P. (2013). Uncertainty Quantification of Fusion Prognostics for Lithium-Ion Battery Remaining Useful Life Estimation.

Liu, D., Wang, H., Yu, P., Xie, W., & Liao, H. (2013). Satellite Lithium-Ion Battery Remaining Cycle Life Prediction With Novel Indirect Health Indicator Extraction. *Energies*.

Liu, H., Li, W., Shen, D., Zhao, D., & Wang, G. (2015). Graphitic Carbon Conformal Coating of Mesoporous TiO_2 Hollow Spheres for High-Performance Lithium Ion Battery Anodes. *Journal of the American Chemical Society*.

Liu, J., & Dai, W. (2019). Overview of Nuclear Waste Treatment and Management. *E3s Web of Conferences*.

Liu, J., Wang, Z., Gong, J., Liu, K., Wang, H., & Guo, L. (2017). Experimental Study of Thermal Runaway Process of 18650 Lithium-Ion Battery. *Materials*.

Liu, J., Yuan, Y., Guo, X., Li, B., Shahbazian-Yassar, R., Liu, D. H., Chen, Z., Amine, K., Yang, L., & Bai, Z. (2021). Mesocrystallizing Nanograins for Enhanced Li^+ Storage. *Advanced Energy Materials*.

Liu, J., Zhang, Y., Yao, M., Yang, Y., & Li, P. (2021). Alkaline Resins Enhancing Li^+/H^+ Ion Exchange for Lithium Recovery From Brines Using Granular Titanium-Type Lithium Ion-Sieves. *Industrial & Engineering Chemistry Research*.

Liu, W., Song, M. S., Kong, B., & Cui, Y. (2016). Flexible and Stretchable Energy Storage: Recent Advances and Future Perspectives. *Advanced Materials*.

Liu, Y., Dian, G., Xiang, H., Feng, X., & Yu, Y. (2021). Research Progress on Copper-Based Current Collector for Lithium Metal Batteries. *Energy & Fuels*.

Liu, Y., Liao, Y. G., & Lai, M. C. (2019). Transient Temperature Distributions on Lithium-Ion Polymer SLI Battery. *Vehicles*.

Liu, Y., Zhang, R., Wang, J., & Wang, Y. (2021). Current and future lithium-ion battery manufacturing. *IScience, 24*(4).

Liu, Z., Zhu, X., Song, M., Ali, R. N., & Tang, Y. (2023). Research Progress of Constructing Anode Materials for Potassium Ion Batteries Based on Electrospinning Technology. *Academic Journal of Science and Technology*.

Lü, C., & Chen, X. (2022). Learn From Nature: Bio-inspired Structure Design for Lithium-ion Batteries. *Ecomat*.

Lu, Q., Chen, L., Li, X., Chao, Y., Sun, J., Ji, H., & Zhu, W. (2021). Sustainable and Convenient Recovery of Valuable Metals From Spent Li-Ion Batteries by a One-Pot Extraction Process. *Acs Sustainable Chemistry & Engineering*.

Ludwig, B., Liu, J., Chen, I. M., Liu, Y., Shou, W., Wang, Y., & Pan, H. (2017). Understanding Interfacial-Energy-Driven Dry Powder Mixing for Solvent-Free Additive Manufacturing of Li-Ion Battery Electrodes. *Advanced Materials Interfaces*.

Lurie, C. Evaluation of Lithium Ion Cells for Space Applications.

Lutandula, M. S., Kime, M.-B., Mambwe, M. P., & Nyembo, T. K. (2019). A Review of the Beneficiation of Copper-Cobalt-Bearing Minerals in the Democratic Republic of Congo. *Journal of Sustainable Mining*.

Lv, S., Wang, X., Lu, W., Zhang, J., & Ni, H. (2021). The Influence of Temperature on the Capacity of Lithium Ion Batteries With Different Anodes. *Energies*.

Ly-Le, T.-M. (2022). Hiring for Gender Diversity in Tech. *The Journal of Management Development*.

Ma, X., Zhu, H., Xie, Z., Zheng, J., & Yuan, B. (2021). Advances in Simulation Research for Thermal Runaway of Lithium-Ion Batteries. *Destech Transactions on Materials Science and Engineering*.

Madani, S. S., Schaltz, E., & Kær, S. K. (2019). An Experimental Analysis of Entropic Coefficient of a Lithium Titanate Oxide Battery. *Energies*.

Madani, S. S., Schaltz, E., & Kær, S. K. (2020a). Thermal Analysis of Cold Plate With Different Configurations for Thermal Management of a Lithium-Ion Battery. *Batteries*.

Madani, S. S., Schaltz, E., & Kær, S. K. (2020b). Thermal Simulation of Phase Change Material for Cooling of a Lithium-Ion Battery Pack. *Electrochem*.

Madani, S. S., Schaltz, E., & Kær, S. K. (2021a). Applying Different Configurations for the Thermal Management of a Lithium Titanate Oxide Battery Pack. *Electrochem*.

Madani, S. S., Schaltz, E., & Kær, S. K. (2021b). Thermal Characterizations of a Lithium Titanate Oxide-Based Lithium-Ion Battery Focused on Random and Periodic Charge-Discharge Pulses. *Applied System Innovation*.

Madonna, V., Giangrande, P., & Galea, M. (2018). Electrical Power Generation in Aircraft: Review, Challenges, and Opportunities. *Ieee Transactions on Transportation Electrification*.

Mahmud, A., Huda, N., Farjana, S. H., & Lang, C. (2019). Comparative Life Cycle Environmental Impact Analysis of Lithium-Ion (LiIo) and Nickel-Metal Hydride (NiMH) Batteries. *Batteries*.

Majeau-Bettez, G., Hawkins, T. R., & Strømman, A. H. (2011). Life Cycle Environmental Assessment of Lithium-Ion and Nickel Metal Hydride Batteries for Plug-in Hybrid and Battery Electric Vehicles. *Environmental Science & Technology*.

Malik, R. (2017). Aqueous Li-Ion Batteries: Now in Striking Distance. *Joule*.

Mallard, K., Debusschere, V., & Garbuio, L. (2020). Multi-Criteria Method for Sustainable Design of Energy Conversion Systems. *Sustainability*.

Mamo, S. K., Nieman, C. L., & Lin, F. R. (2016). Prevalence of Untreated Hearing Loss by Income Among Older Adults in the United States. *Journal of Health Care for the Poor and Underserved*.

Man, H. (2023). *EV battery types explained: Lithium-ion vs LFP pros & cons*. Retrieved 27/12/2023 from

Mandić, A., & Praničević, D. G. (2019). Progress on the Role of ICTs in Establishing Destination Appeal. *Journal of Hospitality and Tourism Technology*.

Mankoff, J., Hayes, G. R., & Kasnitz, D. (2010). Disability Studies as a Source of Critical Inquiry for the Field of Assistive Technology.

Manthiram, A. (2011). Materials Challenges and Opportunities of Lithium Ion Batteries. *The Journal of Physical Chemistry Letters*.

Manthiram, A. (2017). An Outlook on Lithium Ion Battery Technology. *Acs Central Science*.

Manthiram, A. (2020). A Reflection on Lithium-Ion Battery Cathode Chemistry. *Nature Communications*.

Manthiram, A., Fu, Y., & Su, Y. S. (2012). Challenges and Prospects of Lithium–Sulfur Batteries. *Accounts of Chemical Research*.

Mari, M., Nadal, M., Ferré-Huguet, N., Schuhmacher, M., Borrajo, M. A., & Domingo, J. L. (2007). Monitoring PCDD/Fs in Soil and Herbage Samples Collected Near a Hazardous Waste Incinerator: Health Risks for the Population Living Nearby. *Human and Ecological Risk Assessment an International Journal*.

Marinaș, M.-C., Dinu, M., Socol, A.-G., & Socol, C. (2018). Renewable Energy Consumption and Economic Growth. Causality Relationship in Central and Eastern European Countries. *Plos One*.

Market Reports World. (2023, 27/12/2023). Li-ion Battery for Energy Storage Systems (ESS) Market Research Report2023-2030, Anticipated a CAGR of 16.3%.

Markopoulos, E., Bilibashi, A., & Vanharanta, H. (2021). Democratic Organizational Culture for SMEs Innovation Transformation and Corporate Entrepreneurship.

Martin, N. (2023). *Seven things you need to know about lithium-ion battery safety*. University of New South Wales. Retrieved 13/12/2023 from

Marwick, A. E. (2012). The Public Domain: Surveillance in Everyday Life. *Surveillance & Society*.

Marwick, A. E., & boyd, d. (2014). Networked Privacy: How Teenagers Negotiate Context in Social Media. *New Media & Society*.

Mashayekhi, M. (2021). Factors Influencing the Diffusion of Battery Electric Vehicles in Urban Areas.

Mashtalir, O., Nguyen, M. H., Bodoin, E., Swonger, L., & O'Brien, S. (2018). High-Purity Lithium Metal Films From Aqueous Mineral Solutions. *Acs Omega*.

Masias, A., Marcicki, J., & Paxton, W. A. (2021). Opportunities and Challenges of Lithium Ion Batteries in Automotive Applications. *ACS Energy Letters*, *6*(2), 621-630.

Matsuo, H., Ogihara, T., Aikiyo, H., & Yamanaka, S. (2009). Characterization of Large Lithium Ion Battery and Its Application to Railcar. *Journal of Asian Electric Vehicles*.

Matthews, R. (2022). *Nuclear Power Versus Renewable Energy*. Retrieved 5/1/2024 from

Mayer, C. H., & Louw, L. (2012). Managing Cross-Cultural Conflict in Organizations. *International Journal of Cross Cultural Management*.

Mazauric, A.-L., Sciora, P., Vincent, P., Droin, J.-B., Bésanger, Y., Hadjsaïd, N., & Tran, Q. T. (2022). Approach for the Adaptations of a Nuclear Reactor Model Towards More Flexibility in a Context of High Insertion of Renewable Energies. *Epj Nuclear Sciences & Technologies*.

McBeth, M. K., Wrobel, M. W., & Woerden, I. v. (2022). Political Ideology and Nuclear Energy: Perception, Proximity, and Trust. *Review of Policy Research*.

McCormack, A., & Fortnum, H. (2013). Why Do People Fitted With Hearing Aids Not Wear Them? *International Journal of Audiology*.

Medler, B., Fitzgerald, J., & Magerko, B. (2008). Using Conflict Theory to Model Complex Societal Interactions.

Meem, I. J., Osman, S., Bashar, K. M. H., Tushar, N. I., & Khan, R. (2022). Semi Wireless Underwater Rescue Drone With Robotic Arm. *Journal of Robotics and Control (Jrc)*.

Mehdi, A., & Moerenhout, T. (2023). *The IRA and the US Battery Supply Chain: Background and Key Drivers*. Center on Global Energy Policy. Retrieved 4/1/2024 from

Mehrasa, M., Pouresmaeil, E., Jôrgensen, B. N., & Catalão, J. P. S. (2015). A Control Plan for the Stable Operation of Microgrids During Grid-Connected and Islanded Modes. *Electric Power Systems Research*.

Meide, M. v. d., Harpprecht, C., Northey, S., Yang, Y., & Steubing, B. (2022). Effects of the Energy Transition on Environmental Impacts of Cobalt Supply: A Prospective Life Cycle Assessment Study on Future Supply of Cobalt. *Journal of Industrial Ecology*.

Meng, Y. (2021). Economic Analysis for Centralized Battery Energy Storage System With Reused Battery From EV in Australia. *E3s Web of Conferences*.

Mentus, S. (2021). The Role of Batteries in Near-Future Energetics. *Contemporary Materials*.

Mercer, K., Giangregorio, L., Schneider, E. C., Chilana, P. K., Li, M., & Grindrod, K. (2016). Acceptance of Commercially Available Wearable Activity Trackers Among Adults Aged Over 50 and With Chronic Illness: A Mixed-Methods Evaluation. *Jmir Mhealth and Uhealth*.

Merlo, A., Kaczan, W., Léonard, G., & Wirth, H. (2021). Assessing the Environmental Pertinence of Cobalt Exploitation From Slag in KGHM Mines.

Mervine, E. M., Valenta, R., Paterson, J., Mudd, G. M., Werner, T. T., & Sonter, L. J. (2023). Biomass Carbon Emissions From Nickel Mining Have Significant Implications for Climate Action.

Meshram, P., Pandey, B. D., & Mankhand, T. R. (2014). Extraction of Lithium From Primary and Secondary Sources by Pre-Treatment, Leaching and Separation: A Comprehensive Review. *Hydrometallurgy*.

Messagie, M., Boureima, F.-S., Matheys, J., Sergeant, N., Timmermans, J.-M., Macharis, C., & Mierlo, J. a. V. (2010). Environmental Performance of a Battery Electric Vehicle: A Descriptive Life Cycle Assessment Approach. *World Electric Vehicle Journal*.

Miles, M. H. (2001). Recent advances in lithium battery technology. GaAs IC Symposium. IEEE Gallium Arsenide Integrated Circuit Symposium. 23rd Annual Technical Digest 2001 (Cat. No. 01CH37191),

Miller, T. R., Baird, T. D., Littlefield, C. M., Kofinas, G. P., Chapin, F. S., & Redman, C. L. (2008). Epistemological Pluralism: Reorganizing Interdisciplinary Research. *Ecology and Society*.

Minami, H., Izumi, H., Hasegawa, T., Bai, F., Mori, D., Taminato, S., Takeda, Y., Yamamoto, O., & Imanishi, N. (2021). Aqueous Lithium--Air Batteries With High Power Density at Room Temperature Under Air Atmosphere. *Journal of Energy and Power Technology*.

Minami, K., Kobinata, K., & Yan, J. (2022). Generation of Si@C/SiC@C Core–shell Nanoparticles by Laser Irradiation of Silicon Grinding Waste. *Nano Select*.

Mo, F., Guo, B., Liu, Q., Ling, W., Liang, G., Chen, L., Yu, S., & Jun, W. (2022). Additive Manufacturing for Advanced Rechargeable Lithium Batteries: A Mini Review. *Frontiers in Energy Research*.

Mohammadian, S. K., Rassoulinejad-Mousavi, S. M., & Zhang, K. (2015). Thermal Management Improvement of an Air-Cooled High-Power Lithium-Ion Battery by Embedding Metal Foam. *Journal of Power Sources*.

Mond, H. G., & Freitag, G. (2014). The Cardiac Implantable Electronic Device Power Source: Evolution and Revolution. *Pacing and Clinical Electrophysiology*.

Montané, M., Cáceres, G., Villena, M. G., & O'Ryan, R. (2017). Techno-Economic Forecasts of Lithium Nitrates for Thermal Storage Systems. *Sustainability*.

Montenegro, L. M., & Bulgacov, S. (2014). Reflections on Actor-Network Theory, Governance Networks, and Strategic Outcomes. *Bar - Brazilian Administration Review*.

Moore Stephens. (2023). Retrieved 26/12/2023 from

Morse, J. C., Tittman, S. M., & Gelbard, A. (2018). Oropharyngeal Injury From Spontaneous Combustion of a Lithium-ion Battery: A Case Report. *The Laryngoscope*.

Mortenson, W. B., Demers, L., Fuhrer, M. J., Jutai, J. W., Lenker, J., & DeRuyter, F. (2012). How Assistive Technology Use by Individuals With Disabilities Impacts Their Caregivers. *American Journal of Physical Medicine & Rehabilitation*.

Mozaffarpour, F., Hassanzadeh, N., & Vahidi, E. (2022). Comparative Life Cycle Assessment of Synthesis Routes for Cathode Materials in Sodium-Ion Batteries. *Clean Technologies and Environmental Policy*.

Mudd, G. M., & Jowitt, S. M. (2014). A detailed assessment of global nickel resource trends and endowments. *Economic Geology*, *109*(7), 1813-1841.

Mudd, G. M., & Jowitt, S. M. (2022). The New Century for Nickel Resources, Reserves, and Mining: Reassessing the Sustainability of the Devil's Metal. *Economic Geology*.

Muimba-Kankolongo, A., Nkulu, C. B. L., Mwitwa, J., Kampemba, F. M., & Nabuyanda, M. M. (2022). Impacts of Trace Metals Pollution of Water, Food Crops, and Ambient Air on Population Health in Zambia and the DR Congo. *Journal of Environmental and Public Health*.

Muntaqin, A., Rahmasari, L., Wahyuni, N., & Sihombing, R. P. (2022). Effect of Sulfuric Acid Concentration on Nickel Recovery From Laterite Ore by Using Atmospheric Acid Leaching Method. *Jurnal Kimia Riset*.

Muralidharan, N., Self, E. C., Dixit, M., Du, Z., Essehli, R., Amin, R., Nanda, J., & Belharouak, I. (2022). Next-Generation Cobalt-Free Cathodes – A Prospective Solution to the Battery Industry's Cobalt Problem. *Advanced Energy Materials*.

Musolf, G. R. (1992). Structure, Institutions, Power, and Ideology: New Directions Within Symbolic Interactionism. *Sociological Quarterly*.

Musolf, G. R. (2003). Social Structure, Human Agency, and Social Policy. *International Journal of Sociology and Social Policy*.

Mutafela, R., Ye, F., Jani, Y., Dutta, J., & Hogland, W. (2020). Efficient and Low-Energy Mechanochemical Extraction of Lead From Dumped Crystal Glass Waste. *Environmental Chemistry Letters*.

Mutz, M., Perovic, M., Gümbel, P., Steinbauer, V., Taranovskyy, A., Li, Y., Beran, L., Käfer, T., Dröder, K., Knoblauch, V., Kwade, A., Presser, V., Werth, D., & Kraus, T. (2022). Toward a Li-Ion Battery Ontology Covering Production and Material Structure. *Energy Technology*.

Muzadi, H., Kamalia, N. Z., Lestariningsih, T., & Astuti, Y. (2023). Effect of LiTFSI Electrolyte Salt Composition on Characteristics of PVDF-PEO-LiTFSI-Based Solid Polymer Electrolyte (SPE) for Lithium-Ion Battery. *Molekul*.

Muzhikyan, A., & Farid, A. M. (2015). An Enterprise Control Assessment Method for Variable Energy Resource-Induced Power System Imbalances—Part II: Parametric Sensitivity Analysis. *Ieee Transactions on Industrial Electronics*.

Nakajima, K., Yokoyama, K., & Nagasaka, T. (2008). Substance Flow Analysis of Manganese Associated With Iron and Steel Flow in Japan. *Isij International*.

Nakaya, K., Nakata, A., Hirai, T., & Ogumi, Z. (2014). Oxidation of Nickel in $AlCl_3$-1-Butylpyridinium Chloride at Ambient Temperature. *Journal of the Electrochemical Society*.

Nalobile, P., Wachira, J. M., Thiong'o, J. K., & Marangu, J. M. (2020). A Review on Pyroprocessing Techniques for Selected Wastes Used for Blended Cement Production Applications. *Advances in Civil Engineering*.

Nandha, M. S., & Brooks, R. D. (2009). Oil Prices and Transport Sector Returns: An International Analysis. *Review of Quantitative Finance and Accounting*.

Narayanan, Y. (2019). "Cow Is a Mother, Mothers Can Do Anything for Their Children!" Gaushalas as Landscapes of Anthropatriarchy and Hindu Patriarchy. *Hypatia*.

Nature. (2021). Lithium-ion batteries need to be greener and more ethical. *Nature, 595, 7 (2021)*.

Nayak, S., Das, M., Pradhan, B., & Kundu, S. N. (2019). Environmental Impacts of Fossil Fuels.

Neamtiu, I. A., Al-Abed, S. R., McKernan, J., Baciu, C., Gurzău, E., Pogacean, A. O., & Bessler, S. (2017). Metal Contamination in Environmental Media in Residential Areas Around Romanian Mining Sites. *Reviews on Environmental Health*.

Nguyen, V. S., Van, H. N., Senzier, P. A., Pasquier, C., Wang, K., Rozenberg, M. J., Brun, N., March, K., Jomard, F., Giapintzakis, J., Mihailescu, I. N., Kyriakides, E., Nukala, P., Maroutian, T., Agnus, G., Lecoeur, P., Matzen, S., Aubert, P., Franger, S., . . . Schneegans, O. (2018). Direct Evidence of Lithium Ion Migration in Resistive Switching of Lithium Cobalt Oxide Nanobatteries. *Small*.

Nichols, E., Pavlidis, A., & Nowak, R. (2021). "It's Like Lifting the Power": Powerlifting, Digital Gendered Subjectivities, and the Politics of Multiplicity. *Leisure Sciences*.

Nicoll, K., Rose, A. M., Khan, M. A. A., Quaba, O., & Lowrie, A. G. (2016). Thigh Burns From Exploding E-Cigarette Lithium Ion Batteries: First Case Series. *Burns*.

Nie, M., Demeaux, J., Young, B., Heskett, D. R., Chen, Y., Bose, A., Woicik, J. C., & Lucht, B. L. (2015). Effect of Vinylene Carbonate and Fluoroethylene Carbonate on SEI Formation on Graphitic Anodes in Li-Ion Batteries. *Journal of the Electrochemical Society*.

Nishijima, M., Ootani, T., Kamimura, Y., Sueki, T., Esaki, S., Murai, S., Fujita, K., Tanaka, K., Ohira, K., Koyama, Y., & Tanaka, I. (2014). Accelerated Discovery of Cathode Materials With Prolonged Cycle Life for Lithium-Ion Battery. *Nature Communications*.

Nivetha, E. S., & Saravanathamizhan, R. (2019). Recovery of Nickel From Spent NiCd Batteries by Regular and Ultrasonic Leaching Followed by Electrodeposition. *Journal of Electrochemical Science and Engineering*.

Nizam, M., Maghfiroh, H., Ubaidilah, A., Inayati, I., & Adriyanto, F. (2022). Constant Current-Fuzzy Logic Algorithm for Lithium-Ion Battery Charging. *International Journal of Power Electronics and Drive Systems (Ijpeds)*.

Nkongolo, K. K., Spiers, G., Beckett, P., & Narendrula-Kotha, R. (2023). Levels of Metals and Microbial Biomass in Cobalt Coleman Mine Tailings (Canada) Three Decades After Land Reclamation. *American Journal of Environmental Sciences*.

Nkulu, C. B. L., Casas, L., Haufroid, V., Putter, T. D., Saenen, N. D., Kayembe-Kitenge, T., Obadia, P. M., Mukoma, D. K. w., Ilunga, J., Nawrot, T. S., Numbi, O. L., Smolders, E., & Nemery, B. (2018). Sustainability of Artisanal Mining of Cobalt in DR Congo. *Nature Sustainability*.

Noerochim, L., Suwarno, S., Idris, N. H., & Dipojono, H. K. (2021). Recent Development of Nickel-Rich and Cobalt-Free Cathode Materials for Lithium-Ion Batteries. *Batteries*.

Nøland, J. K., Leandro, M., Suul, J. A., Molinas, M., & Nilssen, R. (2019). Electrical Machines and Power Electronics for Starter-Generators in More Electric Aircrafts: A Technology Review.

Norgate, T., & Haque, N. (2010). Energy and Greenhouse Gas Impacts of Mining and Mineral Processing Operations. *Journal of Cleaner Production*.

Notter, D. A., Gauch, M., Widmer, R., Wäger, P., Stamp, A., Zah, R., & Althaus, H. J. (2010). Contribution of Li-Ion Batteries to the Environmental Impact of Electric Vehicles. *Environmental Science & Technology*.

Novas, N., Salvador, R. M. G., Camacho, J. M., & Alcayde, A. (2021). Advances in Solar Energy Towards Efficient and Sustainable Energy. *Sustainability*.

Nurqomariah, A., & Fajaryanto, R. (2018a). Leaching and kinetics process of cobalt from used lithium ion batteries with organic citric acid. E3s Web of Conferences,

Nurqomariah, A., & Fajaryanto, R. (2018b). Recovery of cobalt and nickel from spent lithium ion batteries with citric acid using leaching process: Kinetics study. E3s Web of Conferences,

O'Faircheallaigh, C., & Babidge, S. (2023). Negotiated Agreements, Indigenous Peoples and Extractive Industry in the Salar De Atacama, Chile: When Is an Agreement More Than a Contract? *Development and Change*.

Oh, H. J., Ozkaya, E., & LaRose, R. (2014). How Does Online Social Networking Enhance Life Satisfaction? The Relationships Among Online Supportive Interaction,

Affect, Perceived Social Support, Sense of Community, and Life Satisfaction. *Computers in Human Behavior*.

Ohba, N., Ogata, S., Tamura, T., Kobayashi, R., Yamakawa, S., & Asahi, R. (2012). Enhanced Thermal Diffusion of Li in Graphite by Alternating Vertical Electric Field: A Hybrid Quantum-Classical Simulation Study. *Journal of the Physical Society of Japan*.

Oliveira, R., Pei, X., Nilsson, E., Rouquette, J.-f., Rivenc, J., Ybanez, L., & Zeng, X. (2023). Performance Analysis of Resistive Superconducting Fault Current Limiter Using LN_2 and GHe Cooling. *Ieee Transactions on Applied Superconductivity*.

Olivetti, E., Ceder, G., Gaustad, G., & Fu, X. (2017). Lithium-Ion Battery Supply Chain Considerations: Analysis of Potential Bottlenecks in Critical Metals. *Joule*.

Ouyang, M., Feng, X., Han, X., Lu, L., Li, Z., & He, X. (2016). A Dynamic Capacity Degradation Model and Its Applications Considering Varying Load for a Large Format Li-Ion Battery. *Applied Energy*.

Overholt, W. H. (1977). An Organizational Conflict Theory of Revolution. *American Behavioral Scientist*.

Özcan, M. A., Yılmaz, C. H., Tuncay, R. N., & Kıvanç, Ö. C. (2023). Development of a Lithium-Ion – Supercapacitor Hybrid Battery for Electric Forklifts.

Özkan, D., Özekinci, M. C., Öztürk, Z. T., & Sulukan, E. (2020). Two Dimensional Materials for Military Applications. *Defence Science Journal*.

Palomares, V., Serras, P., Villaluenga, I., Hueso, K. B., Carretero-González, J., & Rojo, T. (2012). Na-Ion Batteries, Recent Advances and Present Challenges to Become Low Cost Energy Storage Systems. *Energy & Environmental Science*.

Panchal, S., Khasow, R., Dinçer, İ., Agelin-Chaab, M., Fraser, R., & Fowler, M. (2017). Thermal Design and Simulation of Mini-Channel Cold Plate for Water Cooled Large Sized Prismatic Lithium-Ion Battery. *Applied Thermal Engineering*.

Pandey, R. K., Chen, L., Teraji, S., Nakanishi, H., & Soh, S. (2019). Eco-Friendly, Direct Deposition of Metal Nanoparticles on Graphite for Electrochemical Energy Conversion and Storage. *Acs Applied Materials & Interfaces*.

Parimalam, B. S., & Lucht, B. L. (2018). Reduction Reactions of Electrolyte Salts for Lithium Ion Batteries: $LiPF_6$, $LiBF_4$, LiDFOB, LiBOB, and LiTFSI. *Journal of the Electrochemical Society*.

Parker, J. F., Chervin, C. N., Pala, I. R., Machler, M., Burz, M. F., Long, J. W., & Rolison, D. R. (2017). Rechargeable Nickel–3D Zinc Batteries: An Energy-Dense, Safer Alternative to Lithium-Ion. *Science*.

Passerini, S., Owens, B. B., & Coustier, F. (2000). Lithium-Ion Batteries for Hearing Aid Applications: I. Design and Performance. *Journal of Power Sources*.

Pata, U. K. (2020). Renewable and Non-Renewable Energy Consumption, Economic Complexity, CO2 Emissions, and Ecological Footprint in the USA: Testing the EKC Hypothesis With a Structural Break. *Environmental Science and Pollution Research*.

Patil, R., Phadatare, M. R., Blomquist, N., Örtegren, J., Hummelgård, M., Meshram, J., Dubal, D. P., & Olin, H. (2021). Highly Stable Cycling of Silicon-Nanographite Aerogel-Based Anode for Lithium-Ion Batteries. *Acs Omega*.

Patrício, P. d. R., Mesquita, M. C., Silva, L. H. M. d., & Silva, M. d. C. H. d. (2011). Application of Aqueous Two-Phase Systems for the Development of a New Method of Cobalt(II), Iron(III) and Nickel(II) Extraction: A Green Chemistry Approach. *Journal of Hazardous Materials*.

Pavlopoulos, C., Christoula, A., Patsidis, A. C., Semitekolos, D., Παπαδοπούλου, K., Psarras, G. C., Zoumpoulakis, L., & Lyberatos, G. (2023). Epoxy-Silicon Composite Materials From End-of-Life Photovoltaic Panels. *Waste and Biomass Valorization*.

Pearson, H. A., Montazami, A., & Dubé, A. K. (2022). Why This App: Can a Video-based Intervention Help Parents Identify Quality Educational Apps? *British journal of educational technology*.

Peled, E., & Menkin, S. (2017). Review—SEI: Past, Present and Future. *Journal of the Electrochemical Society*.

Peñalvo-López, E., Pérez-Navarro, A., Hurtado, E., & Carrasco, F. J. C. (2019). Comprehensive Methodology for Sustainable Power Supply in Emerging Countries. *Sustainability*.

Peng, Z., Freunberger, S., Chen, Y., & Bruce, P. G. (2012). A Reversible and Higher-Rate Li-O$_2$ Battery. *Science*.

Pete, C., McGowan, J. G., & Jaslanek, W. (2015). Evaluating the Underwater Compressed Air Energy Storage Potential in the Gulf of Maine. *Wind Engineering*.

Peters, D. R., Schnell, J. L., Kinney, P. L., Naik, V., & Horton, D. E. (2020). Public Health and Climate Benefits and Trade-Offs of U.S. Vehicle Electrification. *Geohealth*.

Peters, J. F., Baumann, M., Zimmermann, B., Braun, J., & Weil, M. (2017). The environmental impact of Li-Ion batteries and the role of key parameters–A review. *Renewable and Sustainable Energy Reviews, 67,* 491-506.

Philippot, M., Álvarez, G., Ayerbe, E., Mierlo, J. V., & Messagie, M. (2019). Eco-Efficiency of a Lithium-Ion Battery for Electric Vehicles: Influence of Manufacturing Country and Commodity Prices on GHG Emissions and Costs. *Batteries.*

Pholboon, S., Sumner, M., & Kounnos, P. (2016). Community Power Flow Control for Peak Demand Reduction and Energy Cost Savings.

Pietrosemoli, L., & Rodríguez-Monroy, C. (2013). The Impact of Sustainable Construction and Knowledge Management on Sustainability Goals. A Review of the Venezuelan Renewable Energy Sector. *Renewable and Sustainable Energy Reviews.*

Piippo, S., & Pongrácz, É. (2020). Sustainable Energy Solutions for Rural Communities.

Pikultong, P., Thongsan, S., & Jiajitsawat, S. (2022). The Study of Usable Capacity Efficiency and Lifespan of Hybrid Energy Storage (Lead-Acid With Lithium-Ion Battery) Under Office Building Load Pattern. *Journal of Advanced Research in Fluid Mechanics and Thermal Sciences.*

Ponce, V. H., Galvez-Aranda, D. E., & Seminario, J. M. (2017). Analysis of a Li-Ion Nanobattery With Graphite Anode Using Molecular Dynamics Simulations. *The Journal of Physical Chemistry C.*

Pop, R.-A., Dabija, D.-C., Pelău, C., & Dinu, V. (2022). Usage Intentions, Attitudes, and Behaviors Towards Energy-Efficient Applications During the Covid-19 Pandemic. *Journal of Business Economics and Management.*

Pouloudi, A., Gandecha, R., Atkinson, C., & Papazafeiropoulou, A. (2004). How Stakeholder Analysis Can Be Mobilized With Actor-Network Theory to Identify Actors.

Pouresmaeil, E., Mehrasa, M., Godina, R., Vechiu, I., Rodriguez, R. L., & Catalào, J. P. S. (2017). Double Synchronous Controller for Integration of Large-Scale Renewable Energy Sources Into a Low-Inertia Power Grid.

Prasetyo, E., Muryanta, W. A., Anggraini, A. G., Sudibyo, S., Amin, M., & Muttaqii, M. A. (2022). Tannic Acid as a Novel and Green Leaching Reagent for Cobalt and Lithium Recycling From Spent Lithium-Ion Batteries. *Journal of Material Cycles and Waste Management.*

Procentese, F., Gatti, F., & Napoli, I. D. (2019). Families and Social Media Use: The Role of Parents' Perceptions About Social Media Impact on Family Systems in the

Relationship Between Family Collective Efficacy and Open Communication. *International journal of environmental research and public health.*

Puente, D., Amelibia, J., Cumplido, I., Hernández, A., Ugarte, I., & Duo, A. (2023). Data-Driven Methodology for Optimal Lithium-Ion Battery RUL Prediction.

Puspitarini, H. D., François, B., Baratieri, M., Brown, C., Zaramella, M., & Borga, M. (2020). Complementarity Between Combined Heat and Power Systems, Solar PV and Hydropower at a District Level: Sensitivity to Climate Characteristics Along an Alpine Transect. *Energies.*

Qays, O., Buswig, Y. M., Basri, H. M., Hossain, L., Abu-Siada, A., Rahman, M. M., & Muyeen, S. M. (2020). An Intelligent Controlling Method for Battery Lifetime Increment Using State of Charge Estimation in PV-Battery Hybrid System. *Applied Sciences.*

Qian, J., & Ji, R. (2022). Impact of Energy-Biased Technological Progress on Inclusive Green Growth. *Sustainability.*

Qian, W., Yu, Z., & Zhang, T. (2019). Mechanical Properties of Carbon Fiber Reinforced Nanocrystalline Nickel Composite Electroforming Deposit. *Open Chemistry.*

Qian, X., Gu, X., Dresselhaus, M. S., & Yang, R. (2016). Anisotropic Tuning of Graphite Thermal Conductivity by Lithium Intercalation. *The Journal of Physical Chemistry Letters.*

Qiu, B., Liu, M., Qu, X., Zhang, B., Xie, H., Wang, D., Lee, Y. S., & Yin, H. (2023). Recycling Spent Lithium-Ion Batteries Using Waste Benzene-Containing Plastics: Synergetic Thermal Reduction and Benzene Decomposition. *Environmental Science & Technology.*

Qiu, X., Tang, C.-s., & Dong, W. (2022). Architecture Design of Battery Energy Storage Coordinated Control System Based on Multi-Agent Mechanism. *Journal of Physics Conference Series.*

Qu, J., Liu, F., Ma, Y., & Fan, J. (2019). A Neural-Network-Based Method for RUL Prediction and SOH Monitoring of Lithium-Ion Battery. *Ieee Access.*

Qu, Q., Fu, L., Zhan, X. Y., Samuelis, D., Maier, J., Li, L., Tian, S., Li, Z., & Wu, Y. (2011). Porous LiMn2O4 as Cathode Material With High Power and Excellent Cycling for Aqueous Rechargeable Lithium Batteries. *Energy & Environmental Science.*

Quann, C., & Bradley, T. H. (2017). Renewables Firming Using Grid Scale Battery Storage in a Real-Time Pricing Market.

Radhakrishnan, R. (2021). Experiments With Social Good: Feminist Critiques of Artificial Intelligence in Healthcare in India. *Catalyst Feminism Theory Technoscience*.

Radin, M. D., Vinckevičiūtė, J., Seshadri, R., & Ven, A. V. d. (2019). Manganese Oxidation as the Origin of the Anomalous Capacity of Mn-Containing Li-Excess Cathode Materials. *Nature Energy*.

Rahman, A., Afroz, R., & Safrin, M. (2017). Recycling and Disposal of Lithium Battery: Economic and Environmental Approach. *Iium Engineering Journal*.

Rahman, T., Mansur, A. A., Lipu, M. S. H., Rahman, M. S., Salam, Z., Houran, M. A., Elavarasan, R. M., & Hossain, E. (2023). Investigation of Degradation of Solar Photovoltaics: A Review of Aging Factors, Impacts, and Future Directions Toward Sustainable Energy Management. *Energies*.

Rahnamay-Naeini, M., & Hayat, M. M. (2013). On the Role of Power-Grid and Communication-System Interdependencies on Cascading Failures.

Raj, R., Das, S., & Ghangrekar, M. M. (2021). A Sustainable Approach for the Production of Green Energy With the Holistic Treatment of Wastewater Through Microbial Electrochemical Technologies: A Review. *Frontiers in Sustainability*.

Ramadesigan, V., Methekar, R., Latinwo, F., Braatz, R. D., & Subramanian, V. R. (2010). Optimal Porosity Distribution for Minimized Ohmic Drop Across a Porous Electrode. *Journal of the Electrochemical Society*.

Ramana, M. V., & Rao, D. B. (2010). The Environmental Impact Assessment Process for Nuclear Facilities: An Examination of the Indian Experience. *Environmental Impact Assessment Review*.

Ramelan, A., Adriyanto, F., Apribowo, C. H. B., Ibrahim, M. H., Iftadi, I., & Ajie, G. S. (2021). Simulation and Techno-Economic Analysis of on-Grid Battery Energy Storage Systems in Indonesia. *Journal of Electrical Electronic Information and Communication Technology*.

Ramkumar, M. S., Reddy, C. S. R., Ramakrishnan, A., Raja, K., Pushpa, S., Jose, S., & Jayakumar, M. (2022). Review on Li-Ion Battery With Battery Management System in Electrical Vehicle. *Advances in Materials Science and Engineering*.

Ramos, H. M., Vargas, B., & Saldanha, J. R. (2022). New Integrated Energy Solution Idealization: Hybrid for Renewable Energy Network (Hy4REN). *Energies*.

Rao, V., & Gould, C. (2022). Nuclear Physics, Nuclear Energy, and Nuclear Weapons.

Rasche, P., Wille, M., Bröhl, C., Theis, S., Schäfer, K., Knobe, M., & Mertens, A. (2018). Prevalence of Health App Use Among Older Adults in Germany: National Survey. *Jmir Mhealth and Uhealth*.

Rathore, R. S., Sangwan, S., Mazumdar, S., Kaiwartya, O., Adhikari, K., Kharel, R., & Song, H. (2020). W-Gun: Whale Optimization for Energy and Delay-Centric Green Underwater Networks. *Sensors*.

Ravichandran, R., Binukumar, J. P., Sreeram, R., & Arunkumar, L. S. (2011). An Overview of Radioactive Waste Disposal Procedures of a Nuclear Medicine Department. *Journal of Medical Physics*.

Reck, B. K., Müller, D. B., Rostkowski, K. H., & Graedel, T. E. (2008). Anthropogenic Nickel Cycle: Insights Into Use, Trade, and Recycling. *Environmental Science & Technology*.

Reddy, M. V., Mauger, A., Julien, C., Paolella, A., & Zaghib, K. (2020). Brief History of Early Lithium-Battery Development. *Materials*.

Reddy, M. V., Mauger, A., Julien, C. M., Paolella, A., & Zaghib, K. (2020). Brief History of Early Lithium-Battery Development. *Materials (Basel, Switzerland), 13(8)*(1884).

Renforth, P. (2019). The Negative Emission Potential of Alkaline Materials. *Nature Communications*.

Reportlinker. (2023). *Global Manganese Market to Reach 39.3 Million Metric Tons by 2030* Retrieved 26/12/2023 from

Rexhepi, J., & Torres, C. A. (2011). Reimagining Critical Theory. *British Journal of Sociology of Education*.

Reyes-Sosa, H., Martínez-Cueva, S., & Mondragón, N. I. (2022). Rape Culture, Revictimization, and Social Representations: Images and Discourses on Sexual and Violent Crimes in the Digital Sphere in Mexico. *Journal of Interpersonal Violence*.

Rezk, H., Kanagaraj, N., & Al-Dhaifallah, M. (2020). Design and Sensitivity Analysis of Hybrid Photovoltaic-Fuel-Cell-Battery System to Supply a Small Community at Saudi NEOM City. *Sustainability*.

Richter, V. (2023). Social Media Platforms as Multi-Sided Stakeholders: Youtube's Role in Shaping Imaginaries of Ai. *Aoir Selected Papers of Internet Research*.

Riofrancos, T. (2023). The Security–Sustainability Nexus: Lithium Onshoring in the Global North. *Global Environmental Politics*.

Risman, B. J. (2004). Gender as a Social Structure. *Gender & Society*.

Ritchie, H. (2021). *The price of batteries has declined by 97% in the last three decades*. Our World in Data. Retrieved 1/1/2024 from

Rivera, R. M., Xakalashe, B., Ounoughene, G., Binnemans, K., Friedrich, B., & Gerven, T. V. (2019). Selective Rare Earth Element Extraction Using High-Pressure Acid Leaching of Slags Arising From the Smelting of Bauxite Residue. *Hydrometallurgy*.

Rizky, M. A., Sukamto, U., & Setiawan, A. (2023). Literature Review: Comparison of Caron Process and RKEF on the Processing of Nickel Laterite Ore for Battery. *Jurnal Mineral Energi Dan Lingkungan*.

Robotham, D. (2005). Culture, Society and Economy: Bringing Production Back in. In. SAGE Publications Ltd.

Rofiuddin, M., Aisyah, S., Pratiwi, D. N., Annisa, A. A., Puspita, R. E., & Nabila, R. (2019). Does Economic Growth Reduce Pollution? Empirical Evidence From Low Income Countries. *E3s Web of Conferences*.

Rome, P. M. S. M. A. o., & Coser, L. A. (1956). The Functions of Social Conflicts. *The American Catholic Sociological Review*.

Rothermel, S., Evertz, M., Kasnatscheew, J., Qi, X., Grützke, M., Winter, M., & Nowak, S. (2016). Graphite Recycling From Spent Lithium-Ion Batteries. *Chemsuschem*.

Roy, J. J., Rarotra, S., Krikstolaityte, V., Wu, Z., Cindy, Y. D.-I., Tan, X. Y., Carboni, M., Meyer, D., Yan, Q., & Srinivasan, M. (2021). Green Recycling Methods to Treat Lithium-Ion Batteries E-Waste: A Circular Approach to Sustainability. *Advanced Materials*.

Rubanovich, C. K., Mohr, D. C., & Schueller, S. M. (2017). Health App Use Among In-dividuals With Symptoms of Depression and Anxiety: A Survey Study With Thematic Coding. *Jmir Mental Health*.

Ruismäki, R., Dańczak, A., Klemettinen, L., Taskinen, P., Lindberg, D., & Jokilaakso, A. (2020). Integrated Battery Scrap Recycling and Nickel Slag Cleaning With Methane Reduction. *Minerals*.

Saez-de-Ibarra, A., Martinez-Laserna, E., Stroe, D. I., Świerczyński, M., & Rodriguez, P. (2016). Sizing Study of Second Life Li-Ion Batteries for Enhancing Renewable Energy Grid Integration. *Ieee Transactions on Industry Applications*.

Şahin, S., & Ün, M. B. (2021). Counter-Hegemonic Struggle and the Framing Practices of the Anti-Nuclear Platform in Turkey (2002–2018). *Environment and Planning C Politics and Space*.

Sahoo, B. P., Sahu, H. B., & Pradhan, D. K. (2021). Hydrogeochemistry and Surface Water Quality Assessment of IB Valley Coalfield Area, India. *Applied Water Science*.

Sahoo, L. K., Bandyopadhyay, S., & Banerjee, R. (2014). Benchmarking Energy Consumption for Dump Trucks in Mines. *Applied Energy*.

Sahoo, M., & Sahoo, J. (2020). Effects of Renewable and <scp>non-renewable</Scp> Energy Consumption on <scp>CO₂</Scp> Emissions in India: Empirical Evidence From Disaggregated Data Analysis. *Journal of Public Affairs*.

Sakuda, A., Takeuchi, T., Shikano, M., Sakaebe, H., & Kobayashi, H. (2016). High Reversibility of "Soft" Electrode Materials in All-Solid-State Batteries. *Frontiers in Energy Research*.

Salazar, C. C., Barahona, C., Viruel, A. R., Zuñiga, J. D. V., Palma, B., Meza, K. J., & Moreno, R. (2022). Research Collectives With, For, and by Undocumented Scholars: Creating Counterspaces for Revelation, Validation, Resistance, Empowerment, and Liberation in Higher Education. *Journal of Hispanic Higher Education*.

Sandler, S., Williams, E., Hittinger, E., & Elenes, A. G. N. (2020). The Nonlinear Shift to Renewable Microgrids: Phase Transitions in Electricity Systems. *International Journal of Energy Research*.

Santanilla, A. J. M., Tenório, J. A. S., & Espinosa, D. C. R. (2014). Synergistic Effect on Extraction of Nickel and Cobalt From Synthetic Sulfate Solution Using DEHPA and CYANEX 272 as Extractants.

Santos, M. C., Kesler, O., & Reddy, A. L. M. (2012). Nanomaterials for Energy Conversion and Storage. *Journal of Nanomaterials*.

Santoso, B., Ammarullah, M. I., Haryati, S., Sofijan, A., & Bustan, M. D. (2023). Power and Energy Optimization of Carbon Based Lithium-Ion Battery From Water Spinach (≪i>Ipomoea Aquatica</I>). *Journal of Ecological Engineering*.

Santoso, N. A., Alwiyah, & Nabila, E. A. (2021). Social Media Factors and Teen Gadget Addiction Factors in Indonesia. *Adi Journal on Recent Innovation (Ajri)*.

Saraswati, A. R., & Giantari, I. G. A. K. (2022). Brand Image Mediation of Product Quality and Electronic Word of Mouth on Purchase Decision. *International Research Journal of Management It and Social Sciences*.

Sari, N., Rofianto, W., & Dhewi, T. S. (2022). The Influence of Product Quality, Brand Image on Purchase Decisions and Brand Trust as Mediation Variables (Study on iPhone Users in Malang City). *International Journal of Humanities Education and Social Sciences (Ijhess)*.

Sato, M., Lü, L., & Nagai, H. (2020). *Lithium-Ion Batteries - Thin Film for Energy Materials and Devices*.

Saxena, S., Roman, D., Robu, V., Flynn, D., & Pecht, M. (2021). Battery Stress Factor Ranking for Accelerated Degradation Test Planning Using Machine Learning. *Energies*.

Schäfer, A., Barrett, S. R. H., Doyme, K., Dray, L., Gnadt, A. R., Self, R. H., O'Sullivan, A., Synodinos, A. P., & Torija, A. J. (2018). Technological, Economic and Environmental Prospects of All-Electric Aircraft. *Nature Energy*.

Schauser, N. S., Lininger, C. N., Leland, E. S., & Sholklapper, T. Z. (2022). An Open Access Tool for Exploring Machine Learning Model Choice for Battery Life Cycle Prediction. *Frontiers in Energy Research*.

Schleussner, C., Rogelj, J., Schaeffer, M., Lissner, T., Licker, R., Fischer, E. M., Knutti, R., Levermann, A., Frieler, K., & Hare, B. (2016). Science and Policy Characteristics of the Paris Agreement Temperature Goal. *Nature Climate Change*.

Schlosberg, D., & Carruthers, D. (2010). Indigenous Struggles, Environmental Justice, and Community Capabilities. *Global Environmental Politics*.

Schneider-Kamp, A., & Askegaard, S. (2019). Putting Patients Into the Centre: Patient Empowerment in Everyday Health Practices. *Health an Interdisciplinary Journal for the Social Study of Health Illness and Medicine*.

Schomberg, A., Bringezu, S., & Flörke, M. (2021). Extended Life Cycle Assessment Reveals the Spatially-Explicit Water Scarcity Footprint of a Lithium-Ion Battery Storage. *Communications Earth & Environment*.

Schulz, A., Eder, A., Tiberius, V., Solorio, S. C., Fabro, M., & Brehmer, N. (2021). The Digitalization of Motion Picture Production and Its Value Chain Implications. *Journalism and Media*.

Schurr, C., Marquardt, N., & Militz, E. (2023). Intimate Technologies: Towards a Feminist Perspective on Geographies of Technoscience. *Progress in Human Geography*.

Searle, S., & Malins, C. (2014). A Reassessment of Global Bioenergy Potential in 2050. *GCB Bioenergy*.

Secrist, E., & Fehring, T. K. (2022). Cobalt Mining in the Democratic Republic of the Congo for Orthopaedic Implants. *The Journal of Bone and Joint Surgery (American)*.

Sehil, K., Alamri, B., Alqarni, M., Sallama, A., & Darwish, M. (2021). Empirical Analysis of High Voltage Battery Pack Cells for Electric Racing Vehicles. *Energies*.

Sehol, M., Hentihu, I., Yusuf, N., & Zakariah, M. I. (2022). Public Perception of the Socio-Economic and Environmental Impact of Gold Mining in Buru Regency. *South Asian Journal of Social Studies and Economics*.

Selwyn, N. (2004). Reconsidering Political and Popular Understandings of the Digital Divide. *New Media & Society*.

Sen, R., & Bhattacharyya, S. C. (2014). Off-Grid Electricity Generation With Renewable Energy Technologies In India: An Application of HOMER. *Renewable Energy*.

Seo, M., Park, M., Song, Y., & Kim, S. W. (2020). Online Detection of Soft Internal Short Circuit in Lithium-Ion Batteries at Various Standard Charging Ranges. *Ieee Access*.

Sessa, S. D., Tortella, A., Andriollo, M., & Benato, R. (2018). Li-Ion Battery-Flywheel Hybrid Storage System: Countering Battery Aging During a Grid Frequency Regulation Service. *Applied Sciences*.

Shafiei, S., & Salim, R. (2014). Non-Renewable and Renewable Energy Consumption and CO2 Emissions in OECD Countries: A Comparative Analysis. *Energy Policy*.

Shahjalal, M., Roy, P. K., Shams, T., Fly, A., Chowdhury, J. I., Ahmed, M., & Liu, K. (2022). A Review on Second-Life of Li-Ion Batteries: Prospects, Challenges, and Issues. *Energy*.

Shaju, K. M., & Bruce, P. G. (2006). Macroporous Li(Ni1/3Co1/3Mn1/3)o2: A High-Power and High-Energy Cathode for Rechargeable Lithium Batteries. *Advanced Materials*.

Shan, M.-x. (2017). Analysis on Thermal Runaway of Lithium Ion Battery. *Destech Transactions on Materials Science and Engineering*.

Shang, C., & Cheng, K. (2022). The Present Applications and Prospects of Lithium-Ion Battery.

Shao, M., Dou, L., & Mu, M. (2022). Design of Thermal Management System for Lithium Battery at Low Temperature. *Itm Web of Conferences*.

Sharif, A., Mishra, S., Sinha, A., Jiao, Z., Shahbaz, M., & Afshan, S. (2020). The Renewable Energy Consumption-Environmental Degradation Nexus in Top-10 Polluted Countries: Fresh Insights From Quantile-on-Quantile Regression Approach. *Renewable Energy*.

Sharifi-Asl, S., Lü, J., Amine, K., & Shahbazian-Yassar, R. (2019). Oxygen Release Degradation in Li-Ion Battery Cathode Materials: Mechanisms and Mitigating Approaches. *Advanced Energy Materials*.

Sharp, B. E., & Miller, S. E. (2016). Potential for Integrating Diffusion of Innovation Principles Into Life Cycle Assessment of Emerging Technologies. *Environmental Science & Technology*.

Shaw, L., & Gant, L. M. (2002). Users Divided? Exploring the Gender Gap in Internet Use. *Cyberpsychology & Behavior*.

Shen, N., Wang, Y., Peng, H., & Hou, Z. (2020). Renewable Energy Green Innovation, Fossil Energy Consumption, and Air Pollution—Spatial Empirical Analysis Based on China. *Sustainability*.

Shi, H., Wang, S., Fernandez, C., Yu, C., Fan, Y., & Cao, W. (2021). Improved Splice-electrochemical Circuit Polarization Modeling and Optimized Dynamic Functional Multi-innovation Least Square Parameter Identification for Lithium-ion Batteries. *International Journal of Energy Research*.

Shi, W., Shen, J., Shen, L., Hu, W., Xu, P., Baucom, J., Ma, S., Yang, S., Chen, X., & Lu, Y. (2020). Electrolyte Membranes With Biomimetic Lithium-Ion Channels. *Nano Letters*.

Shinoda, Y., Tsuchida, S., & Kimurâ, H. (2021). Periodical Public Opinion Survey on Nuclear Energy (Inhabitants Living in the Tokyo Metropolitan Area).

Shreenivasa, L., R.T, Y., Viswanatha, R., Yogesh, K., & Ashoka, S. (2020). Sucrose-Assisted Rapid Synthesis of Multifunctional CrVO4 Nanoparticles: A New High-Performance Cathode Material for Lithium Ion Batteries. *Ionics*.

Shudo, T., & Suzuki, K. (2008). Performance Improvement in Direct Methanol Fuel Cells Using a Highly Porous Corrosion-Resisting Stainless Steel Flow Field. *International Journal of Hydrogen Energy*.

Silver, D. (2022). Learning and Organising for Social Transformation: Lessons From WEB Du Bois and Jane Addams. *Methodological Innovations*.

Šimić, Z., Topić, D., Knežević, G., & Pelin, D. (2021). Battery Energy Storage Technologies Overview. *International Journal of Electrical and Computer Engineering Systems*.

Simon, P., & Gogotsi, Y. (2008). Materials for Electrochemical Capacitors. *Nature Materials*.

Singh, A. K., Mahato, M. K., Neogi, B., & Singh, K. K. (2010). Quality Assessment of Mine Water in the Raniganj Coalfield Area, India. *Mine Water and the Environment*.

Sivarajah, B., Vermaire, J. C., & Smol, J. P. (2021). Assessing the Potential Environmental Factors Affecting Cladoceran Assemblage Composition in Arsenic-Contaminated Lakes Near Abandoned Silver Mines. *Journal of Limnology*.

Skousen, J., Gorman, J., Pena-Yewtukhiw, E. M., King, J., Stewart, J., Emerson, P., & DeLong, C. (2009). Hardwood Tree Survival in Heavy Ground Cover on Reclaimed Land in West Virginia: Mowing and Ripping Effects. *Journal of Environmental Quality*.

Slack, J. F., Kimball, B. A., & Shedd, K. B. (2017). Cobalt.

Smolders, E., Roels, L., Kuhangana, T. C., Coorevits, K., Vassilieva, E., Nemery, B., & Nkulu, C. B. L. (2019). Unprecedentedly High Dust Ingestion Estimates for the General Population in a Mining District of DR Congo. *Environmental Science & Technology*.

Şoavă, G., Mehedinţu, A., Sterpu, M., & Raduteanu, M. (2018). Impact of Renewable Energy Consumption on Economic Growth: Evidence From European Union Countries. *Technological and Economic Development of Economy*.

Sobianowska-Turek, A., Urbańska, W., Janicka, A., Zawiślak, M., & Matla, J. (2021). The Necessity of Recycling of Waste Li-Ion Batteries Used in Electric Vehicles as Objects Posing a Threat to Human Health and the Environment. *Recycling*.

Sofyan, N., Alfaruq, S., Zulfia, A., & Subhan, A. (2018). Characteristics of Vanadium Doped and Bamboo Activated Carbon Coated LiFePO4 and Its Performance for Lithium Ion Battery Cathode. *Jurnal Kimia Dan Kemasan*.

Sone, Y., Hoshino, T., & Kawaguchi, J. i. (2010). Long-Term Operation of the Energy Storage System for Lunar and Planetary Missions. *Transactions of the Japan Society for Aeronautical and Space Sciences Aerospace Technology Japan*.

Song, W., Liu, X., Li, T., Wang, Y., Zhang, X., Lou, C. W., & Lin, J. H. (2021). The Strategy of Achieving Flexibility in Materials and Configuration of Flexible Lithium-Ion Batteries. *Energy Technology*.

Spira, C., Kirkby, A., Kujirakwinja, D., & Plumptre, A. J. (2017). The Socio-Economics of Artisanal Mining and Bushmeat Hunting Around Protected Areas: Kahuzi–Biega National Park and Itombwe Nature Reserve, Eastern Democratic Republic of Congo. *Oryx*.

Sposato, R. G., & Hampl, N. (2020). 3 Social Acceptance of Renewable Energy Technologies.

Sprague, D. D. (2017). *Lake Sediment Geochemistry in Northeastern Ontario: the geologic controls on natural background variation and investigating lakes contaminated with arsenic-rich mine tailings in Cobalt, ON* Carleton University].

Srivastava, A., & Willoughby, J. (2022). Capital, Caste, and Patriarchy: Theory of Marriage Formation in India. *Review of Radical Political Economics*.

Statista. (2023a). *Lithium-ion battery price worldwide from 2013 to 2023* Statista Research Department. Retrieved 1/1/2024 from

Statista. (2023b). *Smartphones - statistics & facts*. Retrieved 27/12/2023 from

Statista. (2023c). *Worldwide primary aluminum production from 2000 to 2022*. Retrieved 26/12/2023 from

Stevens, R. J. A. M., Hobbs, B. F., Ramos, A., & Meneveau, C. (2016). Combining Economic and Fluid Dynamic Models to Determine the Optimal Spacing in Very Large Wind Farms. *Wind Energy*.

Stock, N. (2020). The Postmodern Condition: A Report on Knowledge. *Educational Review*.

Stoppato, A., Benato, A., & Vanna, F. D. (2021). Environmental Impact of Energy Systems Integrated With Electrochemical Accumulators and Powered by Renewable Energy Sources in a Life-Cycle Perspective. *Applied Sciences*.

Stringfellow, W. T., & Dobson, P. (2021). Technology for the Recovery of Lithium From Geothermal Brines. *Energies*.

Stroe, D. I., Świerczyński, M., Stroe, A.-I., Teodorescu, R., Lærke, R., & Kjær, P. C. (2015). Degradation Behaviour of Lithium-Ion Batteries Based on Field Measured Frequency Regulation Mission Profile.

Stumpf, L., Fernández, M., Miguel, P., Pinto, L. F. S., Schubert, R. N., Filho, L. C. I. d. O., Montiel, T. H., Barbosa, L. d. S., Leidemer, J. D., & Duarte, T. B. (2023). Impact of Revegetation on Ecological Restoration of a Constructed Soil in a Coal Mining in Southern Brazil.

Subramanian, A. (2016). Teaching-Learning Approaches and Strategies in Peace Education. *Ira International Journal of Education and Multidisciplinary Studies*.

Suchet, D., Jeantet, A., Elghozi, T., & Jehl, Z. (2020). Defining and Quantifying Intermittency in the Power Sector. *Energies*.

Sun, H., Wang, J., Ren, J., Zhang, W., Tang, W., Wu, X., & Gu, A. (2021). Current Situation of Global Manganese Resources and Suggestions for Sustainable Development in China. *Natural Resources Conservation and Research*.

Sun, X., Luo, X., Zhang, Z., Meng, F., & Yang, J. (2020). Life Cycle Assessment of Lithium Nickel Cobalt Manganese Oxide (NCM) Batteries for Electric Passenger Vehicles. *Journal of Cleaner Production*.

Suresh, R., None, N., Jasim, A. A., Nagdev, O. R. A., Sabin, B. B., Nagaraj, V., Jeyasimman, D., None, N., None, N., None, N., None, N., & None, N. (2020). Design and Analysis of Remotely Amphibious Drone. *International Journal of Innovative Technology and Exploring Engineering*.

Surovtseva, D., Crossin, E., Pell, R., & Stamford, L. (2022). Toward a Life Cycle Inventory for Graphite Production. *Journal of Industrial Ecology*.

Suryoatmojo, H., & Pratama, I. A. (2021). Non-Inverting Cascaded Bidirectional Buck-Boost DC-DC Converter with Average Current Mode Control for Lithium-Ion Battery Charger. *JAREE (Journal on Advanced Research in Electrical Engineering)*, 5(2).

Suzuki, K., Kanai, R., Tsuji, N., Yamashige, H., Orikasa, Y., Uchimoto, Y., Sakurai, Y., & Sakurai, H. (2018). Dependency of the Charge–Discharge Rate on Lithium Reaction Distributions for a Commercial Lithium Coin Cell Visualized by Compton Scattering Imaging. *Condensed Matter*.

Tadaros, M., Migdalas, A., Samuelsson, B., & Segerstedt, A. (2020). Location of Facilities and Network Design for Reverse Logistics of Lithium-Ion Batteries in Sweden. *Operational Research*.

Takahashi, M., Watanabe, T., Yamamoto, K., Ohara, K., Kimura, T., Yang, S. D., Nakanishi, K., Uchiyama, T., Kimura, M., Sakuda, A., Tatsumisago, M., & Uchimoto, Y. (2021). Investigation of the Suppression of Dendritic Lithium Growth With a Lithium-Iodide-Containing Solid Electrolyte. *Chemistry of Materials*.

Takeda, Y., & Yamamoto, O. (2016). Lithium Dendrite Formation on a Lithium Metal Anode From Liquid, Polymer and Solid Electrolytes. *Electrochemistry*.

Takeuchi, E. S., & Leising, R. A. (2002). Lithium Batteries for Biomedical Applications. *Mrs Bulletin*.

Talens Peiró, L., & Villalba Méndez, G. (2013). Material and energy requirement for rare earth production. *Jom*, 65, 1327-1340.

Tan, V., Dias, P., Chang, N. L., & Deng, R. (2022). Estimating the Lifetime of Solar Photovoltaic Modules in Australia. *Sustainability*.

Tang, J., & Hai, L. (2021). Construction and Exploration of an Intelligent Evaluation System for Educational APP Through Artificial Intelligence Technology. *International Journal of Emerging Technologies in Learning (Ijet)*.

Tang, S., Chen, Y., Wang, X., Guo, X., & Si, X. (2014). Remaining Useful Life Prediction of Lithium-Ion Batteries Based on the Wiener Process With Measurement Error. *Energies*.

Tanoto, Y. (2011). Optimum Configuration of Stand-Alone Hybrid Distributed Generation: A Case of Eastern Indonesia.

Tao, Y., Xue, J., Xia, M., Tao, J., Zhang, Q., Li, X., Liao, Q., Li, C., & Tang, H. (2020). Economic Feasibility of Echelon Utilization Battery in Photovoltaic Energy Storage. *E3s Web of Conferences*.

Tarascon, J.-M., & Armand, M. (2001). Issues and challenges facing rechargeable lithium batteries. *Nature*, *414*(6861), 359-367.

Tarascon, J. M., & Armand, M. (2001). Issues and Challenges Facing Rechargeable Lithium Batteries. *Nature*.

Tariq, F., & Mahmood, T. (2021). Modeling and Analysis of Voltage Source Based Battery Energy Storage System in Microgrid to Improve Power Quality. *Pakistan Journal of Engineering and Technology*.

Tastanova, A., Abdykirova, G., Temirova, S., & Biryukova, A. A. (2021). Processing and Production of Pellets From Poor-Grade Manganese-Containing Raw Materials.

Tedesco, S., Barton, J., & O'Flynn, B. (2017). A Review of Activity Trackers for Senior Citizens: Research Perspectives, Commercial Landscape and the Role of the Insurance Industry. *Sensors*.

Theobald, E., Hosken, D. J., Foster, P., & Moyes, K. (2020). Mines and Bats: The Impact of Open-Pit Mining on Bat Activity. *Acta Chiropterologica*.

Thomas, J. W., Qidwai, M. A., Pogue, W. M., & Pham, G. T. (2012). Multifunctional Structure-Battery Composites for Marine Systems. *Journal of Composite Materials*.

Tinambunan, A., Ahmad, F., Sakti, A. W., Putro, P. A., Syafri, & Alatas, H. (2022). Effects of Nickel/Manganese Variation on Na2Mn3–z Ni z O7 for Sodium-Ion Battery Cathodes. *The Journal of Physical Chemistry C*, *126*(49), 20754-20761.

Tischner, U., & Charter, M. (2017). Sustainable Product Design.

Tisserant, A., & Pauliuk, S. (2016). Matching Global Cobalt Demand Under Different Scenarios for Co-Production and Mining Attractiveness. *Journal of Economic Structures*.

Toh, L. (2021). *Let's Come Clean: The Renewable Energy Transition Will Be Expensive*. Columbia Climate School. Retrieved 5/1/2024 from

Tong, S. Y., Chang, A., Lan, S.-H., & Li, S.-C. (2020). Analysis of the Dividend Policy Decision-Making Mechanism of Chinese and Taiwanese Lithium Battery Industries. *Mathematics*.

Tool Cobler. (2023). *The Best Cordless Power Tool Batteries Australia Has To Offer, Information, Buyers Guides, Comparisons and Reviews*. Retrieved 27/12/2023 from

Trieu, T., Jadun, P., Logan, J., McMillan, C., Muratori, M., Steinberg, D., Vimmerstedt, L., Haley, B., Jones, R. M., & Nelson, B. (2018). Electrification Futures Study: Scenarios of Electric Technology Adoption and Power Consumption for the United States.

Trinca, D. (2022). Near Future Submarine: Development of a Combined Air Independent and Lithium Battery Propulsion System (AI-LiB Propulsion System).

Tsai, T. L., Chiou, Y.-F., & Tsai, S.-C. (2020). Overview of the Nuclear Fuel Cycle Strategies and the Spent Nuclear Fuel Management Technologies in Taiwan. *Energies*.

Turedi, S., & Turedi, N. (2021). The Effects of Renewable and Non-Renewable Energy Consumption and Economic Growth on CO2 Emissions: Empirical Evidence From Developing Countries. *Business and Economics Research Journal*.

Türk, B. E., Sarul, M. H., Cengelci, E., Karadağ, Ç. İ., San, F. G. B., Kılıç, M., Okumuş, E., & Yazici, M. S. (2021). Integrated Process Control-Power Management System Design and Flight Performance Tests for Fuel Cell Powered Mini-Unmanned Aerial Vehicle. *Energy Technology*.

Udianto, P., Perdana, F. A., Viyus, V., & Alia, N. (2022). Experimental Research and Energy Consumption Analysis of a 350 Watt Brushless Electric Motor With LiFePO4 Battery.

Um, J., & Han, N.-h. (2020). Understanding the Relationships Between Global Supply Chain Risk and Supply Chain Resilience: The Role of Mitigating Strategies. *Supply Chain Management an International Journal*.

Ünal, B., Sel, O., & Demir-Cakan, R. (2023). Current Collectors Corrosion Behaviours and Rechargeability of TiO2 in Aqueous Electrolyte Aluminium-Ion Batteries.

Utgikar, V., Lattin, W., & Jacobsen, R. T. (2006). Nanometallic Fuels for Transportation: A Well-to-Wheels Analysis. *International Journal of Energy Research*.

Vallati, A., Grignaffini, S., & Romagna, M. (2015). A New Method to Energy Saving in a Micro Grid. *Sustainability*.

Vandecasteele, I., Rivero, I. M. i., Sala, S., Claudia, B., Barranco, R., Batelaan, O., & Lavalle, C. (2015). Impact of Shale Gas Development on Water Resources: A Case Study in Northern Poland. *Environmental Management*.

Vavoulioti, A. R., Stylos, G., & Kotsis, K. T. (2023). Acceptance of Nuclear Energy by Pre-Service Teachers in Greece. *Aquademia.*

Vera, M. L., Torres, W., Galli, C. I., Chagnes, A., & Flexer, V. (2023). Environmental Impact of Direct Lithium Extraction From Brines. *Nature Reviews Earth & Environment.*

Verma, K., & Agarwal, D. (2022). Sustainable Energy Solutions.

Verma, Y. P., & Kumar, A. (2013). Potential Impacts of Emission Concerned Policies on Power System Operation With Renewable Energy Sources. *International Journal of Electrical Power & Energy Systems.*

Vidhi, R., & Shrivastava, P. (2018). A Review of Electric Vehicle Lifecycle Emissions and Policy Recommendations to Increase EV Penetration in India. *Energies.*

Vikström, H., Davidsson, S., & Höök, M. (2013). Lithium Availability and Future Production Outlooks. *Applied Energy.*

Viola, D. M. (2021). Negative Health Review of Cell Phones and Social Media. *Journal of Mental Health and Clinical Psychology.*

Vissers, D. R., Chen, Z., Shao, Y., Engelhard, M. H., Das, U., Redfern, P. C., Curtiss, L. A., Pan, B., Liu, J., & Amine, K. (2016). Role of Manganese Deposition on Graphite in the Capacity Fading of Lithium Ion Batteries. *Acs Applied Materials & Interfaces.*

Voronin, V., Nepsha, F., & Ermakov, A. Y. (2021). Analysis of Electric Consumption: A Study in the Excavation Area of the Modern Coal Mine. *E3s Web of Conferences.*

Wagemaker, M., Huijben, M., & Tromp, M. (2021). Where Are Those Promising Solid-State Batteries? *Europhysics News.*

Wagh, P., Islam, S. Z., Deshmane, V. G., Gangavarapu, P. R. Y., Poplawsky, J. D., Yang, G., Sacci, R. L., Evans, S. F., Mahajan, S., Paranthaman, M. P., Moyer, B. A., Harrison, S. C., & Bhave, R. R. (2020). Fabrication and Characterization of Composite Membranes for the Concentration of Lithium Containing Solutions Using Forward Osmosis. *Advanced Sustainable Systems.*

Walsham, G. (1997). Actor-Network Theory and IS Research: Current Status and Future Prospects.

Wang, A., Zhang, Y., & Zuo, H. (2019). Assessing the Performance Degradation of Lithium-Ion Batteries Using an Approach Based on Fusion of Multiple Feature Parameters. *Mathematical Problems in Engineering.*

Wang, G., Zhao, Y., Yang, B., & Song, Y. (2018). A Thermodynamic and Kinetic Study of Trace Iron Removal From Aqueous Cobalt Sulfate Solutions Using Monophos Resin. *Journal of Applied Biomaterials & Functional Materials*.

Wang, H., Liu, J., Gu, Z., Xiao, H., Zhu, Y., & Zhang, Q. (2021). Microanalysis of the Substances in Positive and Negative Electrodes of a $LiMn_2O_4$ Battery for Road Vehicles Under Over-Charge and Over-Discharge. *E3s Web of Conferences*.

Wang, H., Sun, Y., Li, M., Li, G., Xue, K., Chen, Z., & Yu, Y. (2020). Engineering Solvation Complex–Membrane Interaction to Suppress Cation Crossover in 3 v Cu-Al Battery. *Small*.

Wang, J., Huang, J., Yan, R., Wang, F., Cheng, W., Guo, Q., & Wang, J. (2015). Graphene Microsheets From Natural Microcrystalline Graphite Minerals: Scalable Synthesis and Unusual Energy Storage. *Journal of Materials Chemistry A*.

Wang, L., Zhao, J., Sun, J., & Dong, Z. (2019). The Impact of Biased Technology on Employment Distribution and Labor Status in Income Distribution. *Chinese Management Studies*.

Wang, Q., Sun, J., & Chu, G. (2005). Lithium Ion Battery Fire and Explosion. *Fire Safety Science*.

Wang, Q., Zhao, C., Yao, Z., Wang, J., Wu, F., Kumar, S. G. H., Ganapathy, S., Eustace, S., Bai, X., Li, B., & Wagemaker, M. (2023). Entropy-Driven Liquid Electrolytes for Lithium Batteries. *Advanced Materials*.

Wang, S. (2022). Multi-Angle Analysis of Electric Vehicles Battery Recycling and Utilization. *Iop Conference Series Earth and Environmental Science*.

Wang, S., Jin, Y., Li, Z., Chen, S., Hu, Y., Cui, Y., Svanberg, R., Tang, C., Jiang, J., Yang, W., Jönsson, P. G., & Han, T. (2023). Establishment of Green Graphite Industry: Graphite From Biomass and Its Various Applications. *Susmat*.

Wang, S., Li, K., Gao, M., & Wang, J. (2020). Experimental Exploration of Finned Cooling Structure for the Thermal Management of Lithium Batteries With Different Discharge Rate and Materials. *Thermal Science*.

Wang, S., Zhou, Z., & Tian, K. (2022). Environmental Awareness and Environmental Information Disclosure: An Empirical Study Based on Energy Industry. *Frontiers in psychology*.

Wang, T., Ren, K., Miao, H., Dong, W., Xiao, W., Yang, J., Yang, Y., Liu, P., Cao, Z., Ma, X., & Wang, H. (2020). Synthesis and Manipulation of Single-Crystalline Lithium

Nickel Manganese Cobalt Oxide Cathodes: A Review of Growth Mechanism. *Frontiers in Chemistry*.

Wang, W., Hu, L., Ge, J., Hu, Z., Sun, H., Sun, H., Zhang, H., Zhu, H., & Jiao, S. (2014). In Situ Self-Assembled $FeWO_4$/Graphene Mesoporous Composites for Li-Ion and Na-Ion Batteries. *Chemistry of Materials*.

Wang, X., Chen, J., Dong, C., Wang, D., & Mao, Z. (2022). Hard Carbon Derived From Graphite Anode by Mechanochemistry and the Enhanced Lithium-Ion Storage Performance. *Chemelectrochem*.

Wang, X., Hou, Y., Zhu, Y., Wu, Y., & Holze, R. (2013). An Aqueous Rechargeable Lithium Battery Using Coated Li Metal as Anode. *Scientific Reports*.

Wang, X., Sun, T., Chen, C., & Hu, T. (2017). Current Studies of Treating Processes for Nickel Laterite Ores.

Wang, Y., Huang, Y., Huang, L., & Zhang, D. (2017). A Systematic Review of Application and Effectiveness of mHealth Interventions for Obesity and Diabetes Treatment and Self-Management. *Advances in Nutrition*.

Wang, Y. H., Walter, R., White, C., Kehrli, M., Hamilton, S. F., Soper, P. H., & Ruttenberg, B. I. (2019). Spatial and Temporal Variation of Offshore Wind Power and Its Value Along the Central California Coast. *Environmental Research Communications*.

Wang, Y. H., Wu, J., Hu, G., & Ma, W. (2023). Recovery of Li, Mn, and Fe From LiFePO4/LiMn2O4 Mixed Waste Lithium-Ion Battery Cathode Materials. *Journal of Mining and Metallurgy Section B Metallurgy*.

Wang, Z., Zhang, H., Li, N., Shi, Z., Gu, Z., & Cao, G. (2010). Laterally Confined Graphene Nanosheets and Graphene/SnO2 Composites as High-Rate Anode Materials for Lithium-Ion Batteries. *Nano Research*.

Wanger, T. C. (2011). The Lithium Future-Resources, Recycling, and the Environment. *Conservation Letters*.

Wantzen, K. M., & Mol, J. H. (2013). Soil Erosion From Agriculture and Mining: A Threat to Tropical Stream Ecosystems. *Agriculture*.

Warren, P. (2021). Techno-Economic Analysis of Lithium Extraction From Geothermal Brines.

Warsame, Z. A. (2023). The Significance of FDI Inflow and Renewable Energy Consumption in Mitigating Environmental Degradation in Somalia. *International Journal of Energy Economics and Policy*.

Wei, W., He, L. L., Li, X., Cui, Q., & Chen, H. (2022). The Effectiveness and Trade-Offs of Renewable Energy Policies in Achieving the Dual Decarbonization Goals in China: A Dynamic Computable General Equilibrium Analysis. *International journal of environmental research and public health.*

Wei, X. Y., Xu, Y., Zhang, C., Leonard, D. P., Markir, A., Lü, J., & Ji, X. (2019). Reverse Dual-Ion Battery via a $ZnCl_2$ Water-in-Salt Electrolyte. *Journal of the American Chemical Society.*

Wessells, C. D., Peddada, S. V., Huggins, R. A., & Cui, Y. (2011). Nickel Hexacyanoferrate Nanoparticle Electrodes for Aqueous Sodium and Potassium Ion Batteries. *Nano Letters.*

Whiteside, N., Aleti, T., Pallant, J., & Zeleznikow, J. (2018). Helpful or Harmful? Exploring the Impact of Social Media Usage on Intimate Relationships. *Australasian Journal of Information Systems.*

Whittingham, M. S. (2021). Solid-State Ionics: The Key to the Discovery and Domination of Lithium Batteries: Some Learnings From B-Alumina and Titanium Disulfide. *Mrs Bulletin.*

Wickramasinghe, P. M., Palandagama, P. G. K. D., & Dharmagunawardhane, H. A. (2018). A Combined Electromagnetic and Resistivity Survey for Exploration for Vein Graphite: A Case Study Over a Potential Graphite Field in the Sabargamuwa Province, Sri Lanka. *Journal of the Geological Society of Sri Lanka.*

Wijesinghe, H. D. W. M. A. M., Manathunga, C. H., & Perera, V. P. S. (2019). Development of Sodium-Ion Rechargeable Battery Using Sodium Cobalt Phosphate Cathode. *International Journal of Multidisciplinary Studies.*

Williams, J. (2010). Doing Feminist-demography. *International journal of social research methodology.*

Wilner, K. B., Wiber, M. G., Charles, A., Kearney, J. F., Landry, M., & Wilson, L. (2012). Transformative Learning for Better Resource Management: The Role of Critical Reflection. *Journal of Environmental Planning and Management.*

Wilson, B. M. (2021). When Numbers Eclipse Narratives: A Cultural-Political Critique of the 'Ethical' Impacts of Short-Term Experiences in Global Health in Dominican Republic Bateyes. *Medical Humanities.*

Winarti, Y., Sarkum, S., & Halim, A. (2021). Product Innovation on Customer Satisfaction and Brand Loyalty of Smartphone Users. *Journal of Applied Business Administration.*

Wirth, J. (2023, 28/12/2023). Deafness And Hearing Loss Statistics. *Hearing Aids*.

Wolde-Rufael, Y., & Menyah, K. (2010). Nuclear Energy Consumption and Economic Growth in Nine Developed Countries. *Energy Economics*.

Woo, J., Lim, S., Lee, Y.-G., & Huh, S.-Y. (2018). Financial Feasibility and Social Acceptance for Reducing Nuclear Power Plants: A Contingent Valuation Study. *Sustainability*.

Woodall, C. M., Lu, X., Dipple, G. M., & Wilcox, J. (2021). Carbon Mineralization With North American PGM Mine Tailings—Characterization and Reactivity Analysis. *Minerals*.

World Nuclear Association. (2022). *Electricity and Energy Storage*. Retrieved 5/1/2024 from

World Population Review. (2023). *Energy Consumption by Country 2023*. Retrieved 26/12/2023 from

Wu, L., Fu, X., & Guan, Y. (2016). Review of the Remaining Useful Life Prognostics of Vehicle Lithium-Ion Batteries Using Data-Driven Methodologies. *Applied Sciences*.

Wu, W., Hasegawa, T., Ohashi, H., Hanasaki, N., Liu, J., Matsui, T., Fujimori, S., Masui, T., & Toko, K. (2019). Global Advanced Bioenergy Potential Under Environmental Protection Policies and Societal Transformation Measures. *GCB Bioenergy*.

Wu, Y., Chen, Z., Wang, Z., Chen, S., Ge, D., Chen, C., Jia, J., Li, Y., Jin, M., Zhou, T., Wang, F., & Hu, L. (2019). Nuclear Safety in the Unexpected Second Nuclear Era. *Proceedings of the National Academy of Sciences*.

Wu, Z., Huang, A. C., Tang, Y., Yang, Y., Liu, Y.-C., Zhi-ping, L. I., Zhou, H.-L., Huang, C.-F., Zhang, X., Shu, C. M., & Jiang, J. (2021). Thermal Effect and Mechanism Analysis of Flame-Retardant Modified Polymer Electrolyte for Lithium-Ion Battery. *Polymers*.

Wu, Z., & Pagell, M. (2010). Balancing Priorities: Decision-making in Sustainable Supply Chain Management. *Journal of Operations Management*.

Xia, L., Chen, C., Li, B., Ren, H., & Yao, Y. (2023). Calculation Modeling and Analysis of Carbon Emission Reduction of Battery Electric Passenger Vehicles in Use Phase.

Xia, Q., & Xin, Z. (2022). Surface Engineering of Anode Materials for Improving Sodium-Ion Storage Performance. *Journal of Materials Chemistry A*.

Xiao, J., Gao, R., Zhan, L., & Xu, Z. (2021). Unveiling the Control Mechanism of the Carbothermal Reduction Reaction for Waste Li-Ion Battery Recovery: Providing Instructions for Its Practical Applications. *Acs Sustainable Chemistry & Engineering*.

Xie, H., Yu, H., Jalbout, A. F., Yang, G., Pan, X., & Wang, R. (2006). Li$_x$C$_n$ as Anode Material for Lithium Ion Batteries. *Mrs Proceedings*.

Xie, X. (2022). Insurance Industry and China's Regional Economic Development. *Finance Theory and Practice*.

Xing, X., Chen, Y., Tang, S., Sun, X., Si, X., & Wu, L. (2019). State-of-Health Estimation for Lithium-Ion Batteries Based on Wiener Process With Modeling the Relaxation Effect. *Ieee Access*.

Xiong, Y., Wang, Y., Ma, N., Zhang, Y., Luo, S., & Fan, J. (2023). First Principles Study of B$_7$N$_5$ as a High Capacity Electrode Material for K-Ion Batteries. *Physical Chemistry Chemical Physics*.

Xu, C., Chen, Y., Shi, S., Li, J., Kang, F., & Su, D. S. (2015). Secondary Batteries With Multivalent Ions for Energy Storage. *Scientific Reports*.

Xu, C., Dai, Q., Gaines, L., Hu, M., Tukker, A., & Steubing, B. (2020). Future Material Demand for Automotive Lithium-Based Batteries. *Communications Materials*.

Xu, K. (2014). Electrolytes and Interphases in Li-Ion Batteries and Beyond. *Chemical Reviews*.

Xu, N., Xu, C., Jin, Y., & Yu, Z. (2022). Research on the Operating Mechanism of E-Commerce Poverty Alleviation in Agricultural Cooperatives: An Actor Network Theory Perspective. *Frontiers in psychology*.

Xu, P., Hong, J., Qian, X., Xu, Z., Xia, H., Tao, X., Xu, Z., & Ni, Q. Q. (2020). Materials for Lithium Recovery From Salt Lake Brine. *Journal of Materials Science*.

Xu, W., Wang, J., Ding, F., Chen, X., Nasybulin, E., Zhang, Y., & Zhang, J. G. (2014). Lithium Metal Anodes for Rechargeable Batteries. *Energy & Environmental Science*.

Xu, X., Lv, X., & Han, L. (2019). Carbon Asset of Electrification: Valuing the Transition From Fossil Fuel-Powered Buses to Battery Electric Buses in Beijing. *Sustainability*.

Xu, Y., Liu, F., Guo, J., Li, M., & Han, B. (2021). Mechanical Properties and Thermal Runaway Study of Automotive Lithium-Ion Power Batteries. *Ionics*.

Yang, J., Xu, W., Xu, S., & Ma, S. (2022). Study on Graphitization of Anthracite and Petroleum Coke Catalyzed by Boric Acid. *Journal of Physics Conference Series*.

Yang, S., Zhang, S., Dong, W., & Xia, Y. (2022). Purification Mechanism of Microcrystalline Graphite and Lithium Storage Properties of Purified Graphite. *Materials Research Express*.

Yang, X., & Rogach, A. L. (2020). Anodes and Sodium-Free Cathodes in Sodium Ion Batteries. *Advanced Energy Materials*.

Yang, X., Zhang, G., Ge, S., & Wang, C. Y. (2018). Fast Charging of Lithium-Ion Batteries at All Temperatures. *Proceedings of the National Academy of Sciences*.

Yang, Y., Hu, N., Jin, Y., Ma, J., & Cui, G. (2023). Research Advance of Lithium-Rich Cathode Materials in All-Solid-State Lithium Batteries. *Acta Physica Sinica*.

Yang, Z., Qiao, Q., Kang, X., & Yang, W. (2012). Facile Synthesis of Nanostructured LiFePO4/C Cathode Material for Lithium-Ion Batteries. *Chinese Science Bulletin*.

Ye, Z. (2022). Performance of Lithium Ion Battery With Graphene Microstructure in Cathode. *Journal of Physics Conference Series*.

Yi, L., Huang, Z., Jiang, T., Zhao, P., Zhong, R., & Liang, Z. (2017). Carbothermic Reduction of Ferruginous Manganese Ore for Mn/Fe Beneficiation: Morphology Evolution and Separation Characteristic. *Minerals*.

Yi, M. S., Li, W., & Manthiram, A. (2022). Delineating the Roles of Mn, Al, and Co by Comparing Three Layered Oxide Cathodes With the Same Nickel Content of 70% for Lithium-Ion Batteries. *Chemistry of Materials*.

Yıldız, A., & Arı, E. (2019). An Investigation on the Social Acceptance of Nuclear Energy: A Case Study on University Students. *Dokuz Eylul Universitesi Iktisadi Ve Idari Bilimler Dergisi*.

Ying-hao, X., Yu, H., & Li, C. (2014). Present Situation and Prospect of Lithium-Ion Traction Batteries for Electric Vehicles Domestic and Overseas Standards.

Yoho, R., Foster, T., Urban-Lurain, M., Merrill, J., & Haudek, K. C. (2019). Interdisciplinary Insights From Instructor Interviews Reconciling "Structure and Function" in Biology, Biochemistry, and Chemistry Through the Context of Enzyme Binding. *Disciplinary and Interdisciplinary Science Education Research*.

Yokokawa, H. (2011). Current Status of NEDO Project on Durability/Reliability of Solid Oxide Fuel Cell Stacks/Systems. *Ecs Transactions*.

Yoo, S. H., & Ku, S.-J. (2009). Causal Relationship Between Nuclear Energy Consumption and Economic Growth: A Multi-Country Analysis. *Energy Policy*.

Yoon, M., & Yun, H. (2021). Relationships Between Adolescent Smartphone Usage Patterns, Achievement Goals, and Academic Achievement. *Asia Pacific Education Review*.

Yu, X., Li, W., Gupta, V., Gao, H., Tran, D., Sarwar, S., & Chen, Z. (2022). Current Challenges in Efficient Lithium-Ion Batteries' Recycling: A Perspective. *Global Challenges*, *6*(12), 2200099.

Yu, Y., Chen, B., Huang, K., Wu, X., & Wang, D. (2014). Environmental Impact Assessment and End-of-Life Treatment Policy Analysis for Li-Ion Batteries and Ni-Mh Batteries. *International journal of environmental research and public health*.

Yu, Y., Chen, C., Shui, J., & Xie, S. (2005). Nickel-Foam-Supported Reticular CoO–Li$_2$O Composite Anode Materials for Lithium Ion Batteries. *Angewandte Chemie*.

Yuan, H., Dai, H., Wei, X., & Dai, Y. (2019). Sensitivity of Impedance Parameters of Li-Ion Batteries Under Different State of Health. *Destech Transactions on Environment Energy and Earth Science*.

Yun, K. (2015). A Critical Review of the Premises Underlying Korea's Nuclear Energy Policy. *Energy & Environment*.

Yusof, A. A., Nor, M. K. M., Aras, M. S. M., & Razali, W. Z. B. W. (2021). Installation and Testing for Wireless Control and Communication Capability for DugongBot 2.0.

Zackrisson, M., Avellán, L., & Orlenius, J. (2010). Life Cycle Assessment of Lithium-Ion Batteries for Plug-in Hybrid Electric Vehicles – Critical Issues. *Journal of Cleaner Production*.

Zaidi, S. A. H., Khan, D., Hou, F., & Mirza, F. M. (2018). The Role of Renewable and Non-Renewable Energy Consumption in CO2 Emissions: A Disaggregate Analysis of Pakistan. *Environmental Science and Pollution Research*.

Zainurin, N. A., Anas, S. A., & Singh, R. (2021). A Review of Battery Charging - Discharging Management Controller: A Proposed Conceptual Battery Storage Charging – Discharging Centralized Controller. *Engineering Technology & Applied Science Research*.

Zarei, M., Lee, G., Lee, S. G., & Cho, K. (2022). Advances in Biodegradable Electronic Skin: Material Progress and Recent Applications in Sensing, Robotics, and Human–Machine Interfaces. *Advanced Materials*.

Zeng, L., Huang, L., Zhu, J., Li, P., Chu, P. K., Wang, J., & Yu, X. F. (2022). Phosphorus-Based Materials for High-Performance Alkaline Metal Ion Batteries: Progress and Prospect. *Small*.

Zeng, L., Qiu, L., & Cheng, H. M. (2019). Towards the Practical Use of Flexible Lithium Ion Batteries. *Energy Storage Materials*.

Zhang, B., He, L., Zhang, R., Yuan, W., Wang, J., Hu, Y., Zhao, Z., Zhou, L., Wang, J., & Wang, Z. L. (2023). Achieving Material and Energy Dual Circulations of Spent Lithium-Ion Batteries via Triboelectric Nanogenerator. *Advanced Energy Materials*.

Zhang, C., Li, F., Zhu, X., & Yu, J. (2022). Triallyl Isocyanurate as an Efficient Electrolyte Additive for Layered Oxide Cathode Material-Based Lithium-Ion Batteries With Improved Stability Under High-Voltage. *Molecules*.

Zhang, C., Xia, Z., Gao, H., Wen, J., Chen, S.-R., Dang, M., Gu, S., & Zhang, J. (2020). A Coolant Circulation Cooling System Combining Aluminum Plates and Copper Rods for Li-Ion Battery Pack. *Energies*.

Zhang, C., Zhao, S., Zhong, Y., & Chen, Y. (2022). A Reliable Data-Driven State-of-Health Estimation Model for Lithium-Ion Batteries in Electric Vehicles. *Frontiers in Energy Research*.

Zhang, F., Zhu, Y., & Ge, Z. (2022). Thermal Performance of Reverse-Layered Air-Cooled Cylindrical Lithium Battery Pack Integrated With Staggered Battery Arrangement and Spoiler. *Energy Technology*.

Zhang, G., Du, Z., He, Y., Wang, H., Xie, W., & Zhang, T. (2019). A Sustainable Process for the Recovery of Anode and Cathode Materials Derived From Spent Lithium-Ion Batteries. *Sustainability*.

Zhang, G., Wei, X., Chen, S., Han, G., Zhu, J., & Dai, H. (2022). Investigation the Degradation Mechanisms of Lithium-Ion Batteries Under Low-Temperature High-Rate Cycling. *Acs Applied Energy Materials*.

Zhang, H., Fu, Y., & Cheng, X. (2017). Effect of Heating Temperature on Thermal Stability of Lithium-Ion Battery. *Destech Transactions on Environment Energy and Earth Science*.

Zhang, J., Li, Z., Su, X., Tan, Y., Li, S., Su, Y., Li, J., Deng, P., Xu, L., & Ziqiang, P. (2019). Development Strategy of Nuclear Energy Mineral Resources. *Chinese Journal of Engineering Science*.

Zhang, K., Yin, J., & He, Y. (2021). Acoustic Emission Detection and Analysis Method for Health Status of Lithium Ion Batteries. *Sensors*.

Zhang, L., Mu, Z., & Sun, C. (2018). Remaining Useful Life Prediction for Lithium-Ion Batteries Based on Exponential Model and Particle Filter. *Ieee Access*.

Zhang, L., Peng, H., Ning, Z., Mu, Z., & Sun, C. (2017). Comparative Research on RC Equivalent Circuit Models for Lithium-Ion Batteries of Electric Vehicles. *Applied Sciences*.

Zhang, R., Li, N., Cheng, X. B., Yin, Y. X., Zhang, Q., & Guo, Y. G. (2017). Advanced Micro/Nanostructures for Lithium Metal Anodes. *Advanced Science*.

Zhang, W., Wu, L., Du, J., Tian, J., Li, Y., Zhao, Y., Wu, H., Zhong, Y., Cao, Y. C., & Cheng, S. (2021). Fabrication of a Microcapsule Extinguishing Agent With a Core–shell Structure for Lithium-Ion Battery Fire Safety. *Materials Advances*.

Zhang, X., Fang, S., Zhang, Z., & Yang, L. (2011). Li/LiFePO4 Battery Performance With a Guanidinium-Based Ionic Liquid as the Electrolyte. *Chinese Science Bulletin*.

Zhang, Y., Liu, P., Zhang, Q., & Wen, C. (2010). Separation of Cadmium(II) From Spent Nickel/Cadmium Battery by Emulsion Liquid Membrane. *The Canadian Journal of Chemical Engineering*.

Zhang, Y., Zhu, X., Chen, J., Zhu, X., & Xu, Q. (2023). Experimental Study on Electrical Properties of Power Lithium-Ion Battery. *Journal of Physics Conference Series*.

Zhao, C., Guo, Z., Zhang, G., Zhang, T., Liu, Z. Q., & Líu, H. (2021). Effect and Mechanism of C_2HF_5 on Premixed Flame of Runaway Gas (Syngas) in Lithium-Ion Batteries. *E3s Web of Conferences*.

Zhao, K., Pharr, M., Vlassak, J. J., & Suo, Z. (2010). Fracture of Electrodes in Lithium-Ion Batteries Caused by Fast Charging. *Journal of Applied Physics*.

Zhao, R., Liu, J., Gu, J., Zhai, L., & Ma, F. (2020). Experimental Study of a Direct Evaporative Cooling Approach for Li-ion Battery Thermal Management. *International Journal of Energy Research*.

Zhao, Y., Yang, J., Ma, J., Wu, Q., Qian, W., Wang, Z., Zhang, H., & He, D. (2021). Highly Reduced Graphene Assembly Film as Current Collector for Lithium Ion Batteries. *Acs Sustainable Chemistry & Engineering*.

Zhdanov, V., Sokolova, M., Smirnov, P. S., Andrzejewski, L., Bondareva, J., & Evlashin, S. A. (2021). A Comparative Analysis of Energy and Water Consumption of Mined Versus Synthetic Diamonds. *Energies*.

Zheng, C., Ge, Y., Chen, Z., Huang, D., Liu, J., & Zhou, S. (2017). Diagnosis Method for Li-Ion Battery Fault Based on an Adaptive Unscented Kalman Filter. *Energies*.

Zheng, D., Qiu, D., Ding, T., & Qu, D. (2022). Examining the Chemical Stability of Battery Components With Polysulfide Species by High-Performance Liquid Chromatography and X-Ray Photoelectron Spectroscopy. *Industrial & Engineering Chemistry Research*.

Zheng, P., Young, D. L., Yang, T., Xiao, Y., & Li, Z. (2023). Powering Battery Sustainability: A Review of the Recent Progress and Evolving Challenges in Recycling Lithium-Ion Batteries. *Frontiers in Sustainable Resource Management*.

Zheng, Y., Hirayama, M., Taminato, S., Lee, S. Y., Oshima, Y., Takayanagi, K., Suzuki, K., & Kanno, R. (2015). Reversible Lithium Intercalation in a Lithium-Rich Layered Rocksalt Li2RuO3 Cathode Through a Li3PO4 Solid Electrolyte. *Journal of Power Sources*.

Zheng, Y., Sun, G., Wei, Z., Zhao, F., & Sun, Y. (2013). A Novel Power System Reliability Predicting Model Based on PCA and RVM. *Mathematical Problems in Engineering*.

Zheng, Z., Wang, Z., Song, X., Xun, S., Battaglia, V., & Liu, G. (2014). Biomimetic Nanostructuring of Copper Thin Films Enhances Adhesion to the Negative Electrode Laminate in Lithium-Ion Batteries. *Chemsuschem*.

Zhou, L., Tian, Y., Roy, S. B., Thorncroft, C. D., Bosart, L. F., & Hu, Y. (2012). Impacts of Wind Farms on Land Surface Temperature. *Nature Climate Change*.

Zhu, L., Xi, X., Zhao, L., & Yuan, Q. (2021). Comparative Analysis of Thermal Runaway Characteristics of Lithium-ion Battery Under Oven Test and Local High Temperature. *Fire and Materials*.

Zhu, P., Jiang, Z., Sun, W., Yang, Y., Silvester, D. S., Hou, H., Banks, C. E., Hu, J., & Ji, X. (2023). Built-in Anionic Equilibrium for Atom-Economic Recycling of Spent Lithium-Ion Batteries. *Energy & Environmental Science*.

Zhu, W., Zhou, P., Ren, D., Yang, M., Rui, X., Jin, C., Shen, T., Han, X., Zheng, Y., Lu, L., & Ouyang, M. (2022). A Mechanistic Calendar Aging Model of Lithium-ion Battery Considering Solid Electrolyte Interface Growth. *International Journal of Energy Research*.

Zhu, Y., Xiao, S. Y., Shi, Y., Yang, Y., & Wu, Y. (2013). A Trilayer Poly(vinylidene Fluoride)/Polyborate/Poly(vinylidene Fluoride) Gel Polymer Electrolyte With Good Performance for Lithium Ion Batteries. *Journal of Materials Chemistry A*.

Ziadat, A. H. (2009). Major Factors Contributing to Environmental Awareness Among People in a Third World Country/Jordan. *Environment Development and Sustainability*.

Zou, G., Yan, Z., Zhang, C., & Song, L. (2022). Transfer Learning With CNN-LSTM Model for Capacity Prediction of Lithium-Ion Batteries Under Small Sample. *Journal of Physics Conference Series*.

Zubi, G., Dufo-López, R., Carvalho, M., & Pasaoglu, G. (2018). The lithium-ion battery: State of the art and future perspectives. *Renewable and Sustainable Energy Reviews, 89*, 292-308.

Zubi, G., Dufo-López, R., Carvalho, M., & Paşaoğlu, G. (2018). The Lithium-Ion Battery: State of the Art and Future Perspectives. *Renewable and Sustainable Energy Reviews*.

Zuo, W., Li, R., Zhou, C., Li, Y., Xia, J., & Liu, J. (2017). Battery-Supercapacitor Hybrid Devices: Recent Progress and Future Prospects. *Advanced Science*.

Also By

Other books by Richard Skiba

Redefining Work Health and Safety: Systems, Strategies, and Progressive Approaches

Shadows of Catastrophe: Navigating Modern Suffering Risks in a Vulnerable Society

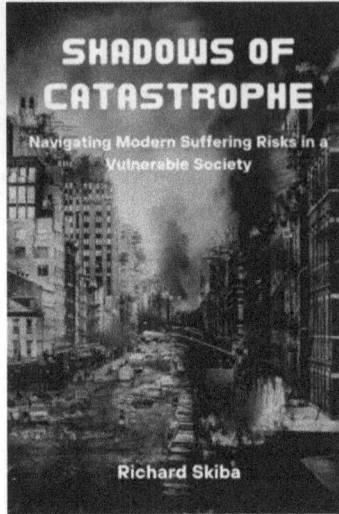

Gun Control: International Views, Perspectives and Comparisons

Sports Shooting and Hunting Australia: Introductory Guide to Safe, Responsible and Legal Use of Firearms

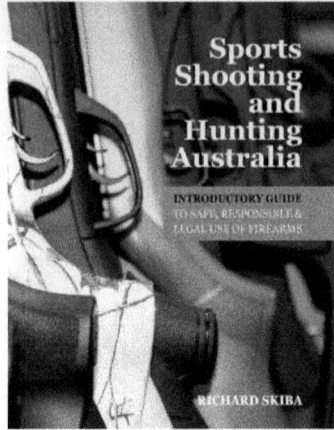

Index